QA3 .P8 v. 2
D1072408

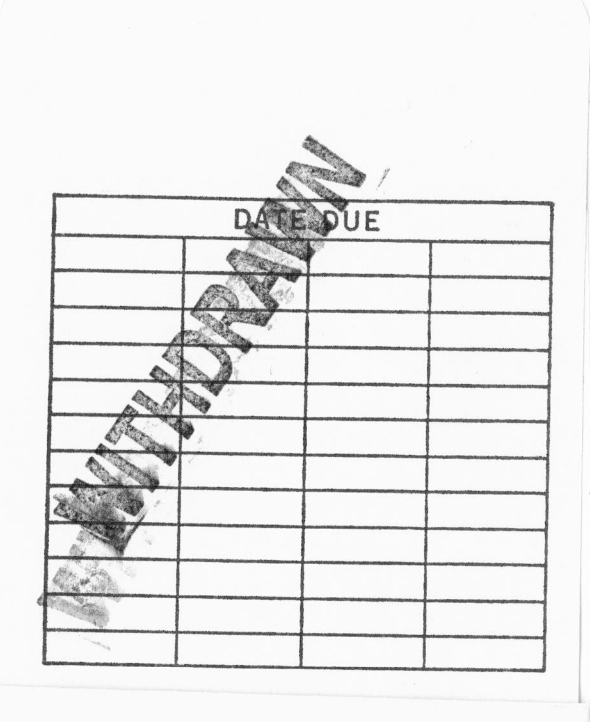
DATE DUE
WITHDRAWN

LINEAR ALGEBRA
and
PROJECTIVE GEOMETRY

Pure and Applied Mathematics

A Series of Monographs and Textbooks

EDITED BY

PAUL A. SMITH AND SAMUEL EILENBERG

Columbia University, New York, N.Y.

VOLUME II

LINEAR ALGEBRA AND PROJECTIVE GEOMETRY

BY REINHOLD BAER

1952

ACADEMIC PRESS INC., PUBLISHERS

NEW YORK, N.Y.

Linear Algebra and Projective Geometry

BY REINHOLD BAER

University of Illinois, Urbana, Illinois

1952

ACADEMIC PRESS INC., PUBLISHERS

NEW YORK, N.Y.

ACADEMIC PRESS INC.
125 EAST 23RD STREET, NEW YORK 10, N.Y.

Library of Congress Catalog Card Number: 52-7480

PREFACE

In this book we intend to establish the essential structural identity of projective geometry and linear algebra. It has, of course, long been realized that these two disciplines are identical. The evidence substantiating this statement is contained in a number of theorems showing that certain geometrical concepts may be represented in algebraic fashion. However, it is rather difficult to locate these fundamental existence theorems in the literature in spite of their importance and great usefulness. The core of our discussion will consequently be formed by theorems of just this type. These are concerned with the representation of projective geometries by linear manifolds, of projectivities by semi-linear transformations, of collineations by linear transformations and of dualities by semi-bilinear forms. These theorems will lead us to a reconstruction of the geometry which was the starting point of our discourse within such (apparently) purely algebraic structures as the endomorphism ring of the underlying linear manifold or the full linear group.

Dimensional restrictions will be imposed only where they are necessary for the validity of the theorem under consideration. It is, for instance, well known that most of these existence theorems cease to be true if the dimension is too low. Thus we will have to exclude the low dimensions quite often. But finiteness of dimension will have to be assumed only in exceptional cases; and this will lead us to a group of finiteness criteria. Similarly we will obtain quite a collection of criteria for the commutativity of the field of scalars; the Index lists all of them. Only the characteristic two will be treated in rather a cavalier fashion, being excluded from our discussion whenever it threatens to be inconvenient.

From the remarks in the last paragraph it is apparent that certain topics ordinarily connected with linear algebra cannot appear in our presentation. Determinants are ruled out, since the existence of determinants enjoying all the desirable properties implies commutativity of the field of scalars. Matrices will make only fleeting appearances, mainly to show that they really have no place in our discussion. The invariant concept is after all that of linear transformation or bilinear form and any choice of a representative matrix would mean an inconvenient and unjustifiable fixing of a not-at-all distinguished system of coordinates.

All considerations of continuity have been excluded from our discussion in spite of the rather fascinating possibilities arising from the interplay of algebraic and topological concepts. But the founders of projective geometry conceived it as the theory of intersection and joining, purely algebraic concepts. Thus we felt justified in restricting our discussion to topics of an algebraic nature and to show how far one may go by purely algebraic methods.

Some sections have been labeled "Appendix" since the topics treated in them are not needed for the main body of our discussion. In these appendices either we discuss applications to special problems of particular interest or we investigate special situations of the general theory in which deeper results may be obtained. No subsequent use will be made of these so the reader may omit them at his discretion.

Little actual knowledge is presupposed. We expect the reader to be familiar with the basic concepts and terms of algebra like group, field, or homomorphism, but the facts needed will usually be derived in the form in which we are going to use them. Ample use will be made of the methods of transfinite set theory—no metaphysical prejudice could deter the author from following the only way to a complete understanding of the situation. For the convenience of the reader not familiar with this theory we have collected the concepts and principles that we need in a special appendix at the end of the book. No proofs are given in this appendix; for these the reader is referred to the literature.

No formal exercises are suggested anywhere in the book. But many facts are stated without proof. To supply the missing arguments will give the reader sufficient opportunity to test his skill.

The references are designed almost exclusively to supply "supplementary reading," rounding off what has been said or supplying what has been left unsaid. We have not tried to trace every concept and result to its origin. What we present here is essentially the combined achievement of a generation of algebraists who derived their inspiration from Dedekind, Hilbert, and Emmy Noether; what little the author may have added to the work of his predecessors will presumably be clear to the expert.

We turn finally to the pleasant task of thanking those who helped us: the editors of this series, in particular Professor S. Eilenberg who read a draft of the manuscript and gave freely of his advice; Professors Eckmann and Nakayama and Dr. Wolfson who helped with reading the proofs; and last but not least my wife who drew all the figures and read all the proofs. The publishers and their staff helped us greatly and we are extremely obliged to them for the way they treated our wishes concerning the book's make-up.

REINHOLD BAER

February 1952

TABLE OF CONTENTS

PREFACE		v
CHAPTER I.	MOTIVATION	1
I. 1.	The Three-Dimensional Affine Space as Prototype of Linear Manifolds	2
I. 2.	The Real Projective Plane as Prototype of the Lattice of Subspaces of a Linear Manifold	5
CHAPTER II.	THE BASIC PROPERTIES OF A LINEAR MANIFOLD	7
II. 1.	Dedekind's Law and the Principle of Complementation	7
II. 2.	Linear Dependence and Independence; Rank	13
II. 3.	The Adjoint Space	25
Appendix I.	Application to Systems of Linear Homogeneous Equations	33
Appendix II.	Paired Spaces	34
II. 4.	The Adjunct Space	36
Appendix III.	Fano's Postulate	37
CHAPTER III.	PROJECTIVITIES	39
III. 1.	Representation of Projectivities by Semi-linear Transformations	40
Appendix I.	Projective Construction of the Homothetic Group	52
III. 2.	The Group of Collineations	62
III. 3.	The Second Fundamental Theorem of Projective Geometry	66
Appendix II.	The Theorem of Pappus	69
III. 4.	The Projective Geometry of a Line in Space; Cross Ratios	71
Appendix III.	Projective Ordering of a Space	93
CHAPTER IV.	DUALITIES	95
IV. 1.	Existence of Dualities; Semi-bilinear Forms	95
IV. 2.	Null Systems	106
IV. 3.	Representation of Polarities	109
IV. 4.	Isotropic and Non-isotropic Subspaces of a Polarity; Index and Nullity	112
Appendix I.	Sylvester's Theorem of Inertia	127
Appendix II.	Projective Relations between Lines Induced by Polarities	131
Appendix III.	The Theorem of Pascal	138
IV. 5.	The Group of a Polarity	144
Appendix IV.	The Polarities with Transitive Group	153
IV. 6.	The Non-isotropic Subspaces of a Polarity	158

CHAPTER V.	THE RING OF A LINEAR MANIFOLD	167
V. 1.	Definition of the Endomorphism Ring	168
V. 2.	The Three Cornered Galois Theory	172
V. 3.	The Finitely Generated Ideals	177
V. 4.	The Isomorphisms of the Endomorphism Ring	182
V. 5.	The Anti-isomorphisms of the Endomorphism Ring	188
Appendix I.	The Two-sided Ideals of the Endomorphism Ring	197
CHAPTER VI.	THE GROUPS OF A LINEAR MANIFOLD	200
VI. 1.	The Center of the Full Linear Group	201
VI. 2.	First and Second Centralizer of an Involution	203
VI. 3.	Transformations of Class 2	207
VI. 4.	Cosets of Involutions	219
VI. 5.	The Isomorphisms of the Full Linear Group	228
Appendix I.	Groups of Involutions	237
VI. 6.	Characterization of the Full Linear Group within the Group of Semi-linear Transformations	244
VI. 7.	The Isomorphisms of the Group of Semi-linear Transformations	247
CHAPTER VII.	INTERNAL CHARACTERIZATION OF THE SYSTEM OF SUBSPACES	257
	A Short Bibliography of the Principles of Geometry	257
VII. 1.	Basic Concepts, Postulates and Elementary Properties	258
VII. 2.	Dependent and Independent Points	263
VII. 3.	The Theorem of Desargues	266
VII. 4.	The Imbedding Theorem	268
VII. 5.	The Group of a Hyperplane	292
VII. 6.	The Representation Theorem	302
VII. 7.	The Principles of Affine Geometry	303
APPENDIX S.	A SURVEY OF THE BASIC CONCEPTS AND PRINCIPLES OF THE THEORY OF SETS	308
	A Selection of Suitable Introductions into the Theory of Sets	308
	Sets and Subsets	308
	Mappings	309
	Partially Ordered Sets	309
	Well Ordering	310
	Ordinal Numbers	310
	Cardinal Numbers	311
Bibliography		313
Index		315

CHAPTER I

Motivation

The objective of this introductory chapter is to put well-known geometrical facts and concepts into a form more suitable to the ways of present day algebraical thinking. In this way we shall obtain some basic connections between geometrical and algebraical structures and concepts that may serve as justification and motivation for the fundamental concepts: linear manifold and its lattice of subspaces which we are going to introduce in the next chapter. All the other concepts will be derived from these; and when introducing these derived concepts we shall motivate them by considerations based on the discussion of this introductory chapter.

Since what we are going to do in this chapter is done only for the purposes of illustration and connection of less familiar concepts with such parts of mathematics as are part of everybody's experience, we shall choose for discussion geometrical structures which are as special as is compatible with our purposes. Reading of this chapter might be omitted by all those who are already familiar with the essential identity of linear algebra and affine and projective geometry. We add a list of works which elaborate this point.

A SHORT BIBLIOGRAPHY OF INTRODUCTORY WORKS EMPHASIZING THE MUTUAL INTERDEPENDENCE OF LINEAR ALGEBRA AND GEOMETRY

L. Bieberbach: Analytische Geometrie. Leipzig, Berlin, 1930.

L. Bieberbach: Projektive Geometrie. Leipzig, Berlin, 1930.

G. Birkhoff and S. MacLane: A Survey of Modern Algebra. New York, 1948.

W. Blaschke: Projektive Geometrie. Wolfenbüttel, 1948.

P. Halmos: Finite Dimensional Vector Spaces. *Ann. Math. Studies* 7. Princeton, N. J., 1940.

L. Heffter and C. Köhler: Lehrbuch der analytischen Geometrie, Bd. 1, 2. Karlsruhe, Leipzig, 1929.

W. W. D. Hodge and D. Pedoe: Methods of Algebraic Geometry. Cambridge, 1947.

C. C. MacDuffee: Vectors and Matrices. Carus Mathematical Monographs 7. Ithaca, N. Y., 1943.

O. Schreier and E. Sperner: Einführung in die analytische Geometrie. Bd. 1, 2. Leipzig, Berlin, 1931-1935.

O. Schreier and E. Sperner: Introduction to Modern Algebra and Matrix Theory. Translated by M. Davis and M. Hausner. New York, 1951.

B. Segre: Lezioni di geometria moderna. Vol. I: Fondamenti di geometria sopra un corpo qualsiasi. Bologna, 1948.

I.1. The Three-Dimensional Affine Space as Prototype of Linear Manifolds

The three-dimensional real affine space may be defined as the totality E [$= E_3$] of triplets (x,y,z) of real numbers x,y,z. This definition is certainly short, but it has the grave disadvantage of giving preference to a definite system of coordinates, a defect that will be removed in due course of time.

The triplets (x,y,z) are usually called the points of this space. Apart from these points we shall have to consider lines and planes, but we shall not discuss such concepts as distance or angular measurement as we want to adhere to the affine point of view. It is customary to define a plane as the totality of points (x,y,z) satisfying a linear equation

$$xa + yb + zc + d = 0$$

where a,b,c,d are real numbers and where at least one of the numbers a,b,c is different from 0; and a line may then be defined as the intersection of two different but intersecting planes. It is known that the points on a line as those on a plane may be represented in the soc. parametric form; and we find it more convenient to make these parametric representations the starting point of our discussion.

The points of a line L may be represented in the form:

$$L: \begin{cases} x = tu + a \\ y = tv + b \\ z = tw + c \end{cases}$$

where (a,b,c) is some point on the line L, where (u,v,w) is a triplet of real numbers, not all 0, and where the parameter t ranges over all the real numbers. As t ranges over all the real numbers, $(tu + a, tv + b, tw + c)$ ranges over all the points of the line L. To obtain a concise notation for this we let $P = (a,b,c)$ and $D = (u,v,w)$, and then we put

$$(tu + a,\ tv + b,\ tw + c) = tD + P.$$

Algebraically we have used, and introduced, two operations: the addition of triplets according to the rule

$$(x,y,z) + (x',y',z') = (x + x', y + y', z + z');$$

and the (scalar) multiplication of a triplet by a real number according to the rule

$$t(x,y,z) = (tx,ty,tz).$$

Indicating by $R(x,y,z)$ the totality of triplets of the form $t(x,y,z)$, we may denote the totality of points on the line L by

$$L = RD + P$$

where we have identified the line L with the set of its points.

Using the operations already introduced we may now treat planes in a similar fashion. Consider three triplets $P = (a,b,c)$, $D' = (u',v',w')$ and $D'' = (u'',v'',w'')$. Then the totality N of points of the form:

$$t'D' + t''D'' + P$$

where the parameters t' and t'' may range independently of each other over all the real numbers may be designated by

$$N = RD' + RD'' + P.$$

If both D' and D'' are the 0-triplet $[D' = D'' = (0,0,0) = 0]$, then N degenerates into the point P; if D' or D'' is 0 whereas not both D' and D'' are 0, then N degenerates into a line. More generally N will be a line whenever D' is a multiple of D'' or D'' is a multiple of D' [and not both are 0]. But if N is neither a point nor a line, then N is actually the totality of points on a plane; or as we shall say more shortly: N is a plane.

In this treatment of lines and planes we have considered the line $L = RD + P$ as the line through the two points P and $P + D$ and the plane $N = RD' + RD'' + P$ as the plane spanned by the three not collinear points P, $P + D'$, $P + D''$. The question arises under which circumstances two pairs of points determine the same line, or two triplets of points span the same plane, and more generally how to characterize by internal properties of the set those sets of points which form a line or a plane.

With this in mind we introduce the following

DEFINITION: The not vacuous set S of points in E is a *flock* of points, if $sU - sV + W$ belongs to S whenever s is a real number and U, V, W are in S.

Note that

$$s(u,u',u'') - s(v,v',v'') + (w,w',w'') = (su - sv + w, su' - sv' + w', su'' - sv'' + w'').$$

A set consisting of one point only certainly has this property; and the reader will find it easy to verify that lines and planes too are flocks of points. Trivially the totality of points in E is a flock. Consider now conversely some flock S of points. This flock contains at least one point P. If P is the only point in S, then we have finished our argument. Assume therefore that S contains a second point Q. It follows from the flock property that S contains the whole line

$$L = P + R(P - Q)$$

—note that $P - Q \neq (0,0,0)$. If this line exhausts S, then we have again reached our goal; and thus we may assume that S contains a further point K, not on L. It follows from the flock property that S contains the totality

$$N = P + R(Q - P) + R(K - P);$$

and N is a plane, since K is not on L. If N exhausts S, then again we have achieved our end. If, however, there exists a point M in S, but not in N, then one may prove that $S = E$ [by realizing that the four points P,Q,K,M are "linearly independent," and that therefore every further point "depends" on them]; we leave the details to the reader.

Now we may exhibit those features of the space E which are "coordinate-free." The space E consists of elements, called points. These points may be added and subtracted [$P \pm Q$] and they form an additive abelian group with respect to addition. There exists furthermore a scalar multiplication rP of real numbers r by points P with the properties:

$$(r + s)P = rP + sP, r(P + Q) = rP + rQ, (rs)P = r(sP), 1P = P.$$

There exist furthermore distinguished sets of points, called flocks in the preceding discussion; they are characterized by the closure property:

If U, V, W are in the flock F, and if r is a real number, then $rU - rV + W$ belongs to F.

Affine geometry may then be defined (in a somewhat preliminary fashion) as the study of the flocks in the space E.

Among the flocks those are of special interest which contain the origin (the null element with respect to the addition of points). It is easy to see that a set S of points is a flock containing the origin if, and only if,

(*a*) S contains $P + Q$ and $P - Q$ whenever S contains P and Q, and

(*b*) S contains rP whenever r is a real number and P is a point in S.

In other words the flocks through the origin are exactly the subsets of E which are closed under addition, subtraction and multiplication or, as we shall always say, the flocks through the origin are exactly the subspaces of E.

If T is a subspace of E (closed under addition, subtraction and multiplication) and if P is a point, then $T + P$ is a flock. If S is a flock, then the totality T of points of the form $P - Q$ for P and Q in S is a subspace, and S has the form $S = T + P$ for some P in S. This shows that we know all the flocks once we know the subspaces; and so in a way it may suffice to investigate the subspaces of E. But the observation has been made that the totality of lines and planes through the origin of a three-dimensional affine space has essentially the same structure as the real projective plane; and this remark we want to substantiate in the next section.

I.2. The Real Projective Plane as Prototype of the Lattice of Subspaces of a Linear Manifold

We begin by stating the following definition of the real projective plane which has the advantage of being short and in accordance with customary terminology, but has the disadvantage of giving preference to a particular system of coordinates.

Every triplet (x_0,x_1,x_2) of real numbers, not all 0, represents a point, and every point may be represented in this fashion.

The triplets (x_0,x_1,x_2) and (y_0,y_1,y_2) represent the same point if, and only if, there exists a number $c \neq 0$ such that $x_i = cy_i$ for $i = 0,1,2$.

Every triplet (u_0,u_1,u_2) of real numbers, not all 0, represents a line, and every line may be represented in this fashion.

The triplets $(u_0,u_1,u_2,)$ and (v_0,v_1,v_2) represent the same line if, and only if, there exists a number $d \neq 0$ such that $u_i = v_i d$ for $i = 0,1,2$.

The point represented by (x_0,x_1,x_2) is on the line represented by (u_0,u_1,u_2) if, and only if,

$$x_0u_0 + x_1u_1 + x_2u_2 = 0.$$

If we use notations similar to those used in I.1, then we may say that the triplets x and y represent the same point if, and only if, $x = cy$, and that the triplets u and v represent the same line if, and only if, $u = vd$. The principal reason for writing the scalar factor d on the right will become apparent much later [II.3]; at present we can only say that some of our formulas will look a little better. If we define the scalar product of the triplets x and u by the formula

$$xu = \sum_{i=0}^{2} x_iu_i,$$

then the incidence relation "point x on line u" is defined by $xu = 0$. We note that $xu = 0$ implies $(cx)u = 0$ and $x(ud) = 0$.

If x is a triplet, not 0, then the totality of triplets cx with $c \neq 0$ represents the same point, and thus we may say without any danger of confusion that Rx is a point. Likewise uR may be termed a line whenever u is a triplet not 0.

If Rx is a point, then this is a set of triplets closed under addition and multiplication by real numbers. If uR is a line, then we may consider the totality S of triplets x such that $xu = 0$. It is clear that S too is closed under addition and multiplication by real numbers; and that S is composed of all the points Rx on the line uR. Thus we might identify the line uR with the totality S.

But now we ought to remember that the totality of triplets $x = (x_0,x_1,x_2)$ is exactly the three-dimensional affine space discussed in I.1, and that the points Rx and the lines S, discussed in the preceding paragraph, are just what we called in I.1 subspaces of the three-dimensional affine space. That all subspaces—apart from 0 and E—are just points and lines in this projective sense, the reader will be able to verify without too much trouble. Once he has done this, he will realize the validity of the contention we made at the end of I.1:

The real projective plane is essentially the same as the system of subspaces (= flocks through the origin) of the three-dimensional real affine space.

Consequently all our algebraical discussion of linear manifolds admits of two essentially different geometrical interpretations: the affine interpretation where the elements (often called vectors) are the basic atoms of discussion and the projective interpretation where the subspaces are the elementary particles. We shall make use of both interpretations feeling free to use whichever is the more suitable one in a special situation, but in general we shall give preference to projective ways of thinking.

The real affine space and the real projective plane are just two particularly interesting members in a family of structures which may be obtained from these special structures by generalization in two directions: first, all limitations as to the number of dimensions will be dropped so that the dimension of the spaces under consideration will be permitted to take any finite and infinite value (though sometimes we will have to exclude the very low dimensions from our discussion); secondly we will substitute for the reals as field of coordinates any field whatsoever whether finite or infinite, whether commutative or not. But in all these generalizations the reader will be wise to keep in mind the geometrical picture which we tried to indicate in this introductory chapter.

CHAPTER II

The Basic Properties of a Linear Manifold

In this chapter the foundations will be laid for all the following investigations. The concepts introduced here and the theorems derived from them will be used almost continuously. Thus we prove the principle of complementation and the existence of a basis which contains a basis of a given subspace; we show that any two bases contain the same number of elements which number (finite or infinite) is the rank of the space. It is then trivial to derive the fundamental rank identities, which contain as a special case the theory of systems of homogeneous linear equations, as we show in Appendix I, and to relate the rank of a space with the rank of its adjoint space (= space of hyperplanes).

II.1. Dedekind's Law and the Principle of Complementation

A *linear manifold* is a pair (F,A) consisting of a (not necessarily commutative) field F and an additive abelian group A such that the elements in F operate on the elements in A in a way subject to the following rules:

(*a*) If f is an element in F and a an element in A, then their product fa is a uniquely determined element in A.

(*b*) $(f' + f'')a = f'a + f''a$, $f(a' + a'') = fa' + fa''$ for f,f',f'' in F and a,a',a'' in A.

(*c*) $1a = a$ for every a in A [where 1 designates the identity element in F].

(*d*) $(f'f'')a = f'(f''a)$ for f',f'' in F and a in A.

From these rules one deduces readily such further rules as

(*e*) $0a = f0 = 0$ for f in F and a in A [where the first 0 is the null element in F whereas the second and third 0 stand for the null element in A];

(*f*) $(-f)a = f(-a) = -(fa)$ for f in F and a in A.

REMARK ON TERMINOLOGY: It should be noted that we use the word "field" here in exactly the same fashion as other authors use terms like division ring, skew field, and sfield. Thus a field is a system of at least two elements with two compositions, addition and multiplication. With respect to addition the field is a commutative group; the elements, not 0, in the field form a group, which need not be commutative, with respect to multiplication; and addition and multiplication are connected by the distributive laws. A good example of a field which is not commutative is provided by the real quaternions; see, for instance, Birkhoff-MacLane [1], p. 211 for a discussion.

In a linear manifold we have two basic classes of elements: those in the additive group A (the vectors) and those in the field F (the scalars). To keep these two classes of elements apart it will sometimes prove convenient to refer to the elements in the field F as to "numbers in F," a terminology that seems to be justified by the fact that numbers in F may be added, subtracted, multiplied and divided.

Instead of linear manifold we shall use expressions like F-space A, etc; and we shall often say that F is the field of coordinates of the space A. Note that in the literature also terms like F-group A, F-modulus A, vector space A over F are used.

A *linear submanifold* or *subspace* of (F,A) is a non-vacuous subset S of A meeting the following requirements:

(g) $s'-s''$ belongs to S whenever s',s'' are in S; and fs belongs to S whenever s is in S and f in F.

If we indicate as usual by $X+Y$ and $X-Y$ respectively the sets of all the sums $x+y$ and $x-y$ with x in X and y in Y, and by GX the totality of products gx for g in G and x in X, then one sees easily the equivalence of (g) with the following conditions:

$$S = S + S = S - S = FS.$$

We note a few simple examples of such linear submanifolds: 0; the points Fp with $p \neq 0$; the lines $Fp + Fq$ [where Fp and Fq are distinct points]; the planes $L + Fp$ where L is a line and Fp is a point, not part of L. A justification for these terms, apart from the reasons already given in Chapter I, will be given in the next section.

Instead of linear submanifold we may also use terms like subspace, F-subgroup and admissible subspace.

Our principal objective is the study of the totality of subspaces of a given linear manifold. This totality has a certain structure, since subspaces are connected by a number of relations.

Containedness or Inclusion: If S and T are subspaces, and if every element in S belongs to T, then we write $S \leqslant T$ and say that S is part of T or S is on T or S is contained in T. If $S \leqslant T$, but $S \neq T$, then we write $S < T$. If H contains K and K contains L, then K may be said to be "between" H and L.

INTERSECTION: If S and T are subspaces, then $S \cap T$ is the set of all the elements which belong to both S and T. It is readily seen that $S \cap T$ is a subspace too.

If Φ is a set of subspaces, then we define as the intersection of the subspaces in Φ the set of all the elements which belong to each of the subspaces in Φ. This intersection is again a subspace, and will be indicated in a variety of ways. For instance, if Φ consists only of a finite number of subspaces $S_1, \cdots, S_n$, then their intersection will be written as $S_1 \cap \cdots \cap S_n$; if the subspaces in Φ are indicated by subscripts: $\Phi = [\cdots S_\sigma \cdots]$, then we write the intersection as $\bigcap_\sigma S_\sigma$ and so on.

Instead of intersection the term cross cut is used too.

SUM. If S and T are subspaces, then their sum $S + T$ consists of all the elements $s + t$ with s in S and t in T. One verifies easily that $S + T$ too is a subspace, the subspace "spanned" by S and T.

If $S_1, \cdots, S_n$ are a finite number of subspaces, then their sum

$$S_1 + \cdots + S_n = \sum_{i=1}^{n} S_i$$

consists of all the sums $s_1 + \cdots + s_n = \sum_{i=1}^{n} s_i$ with s_i in S_i. Again it is clear that the sum of the S_i is a subspace, the subspace spanned by the S_i.

If finally Φ is any set of subspaces, then their sum consists of all the sums $s_1 + \cdots + s_k$ where each s_i belongs to some subspace S in Φ. This again is a subspace which may be indicated in a variety of ways, like $\sum_\nu S_\nu$. Note that only finite sums of elements in A may be formed, though we may form the sum of an infinity of subspaces. Furthermore it should be verified that the definition of the sum of a finite number of subspaces is a special case of our definition of the sum of the subspaces in Φ.

Intersection and sum of subspaces are connected by the following rule which is easily verified:

The sum of the subspaces in Φ is the intersection of all the subspaces which contain every subspace in Φ; the intersection of the subspaces in Φ is the sum of all the subspaces which are contained in every subspace in Φ.

We turn now to the derivation of more fundamental relations.

Dedekind's Law: *If R, S, T are subspaces, and if $R \leqslant S$, then*

$$S \cap (R + T) = R + (S \cap T).$$

PROOF: From $S \cap T \leqslant S$ and $R \leqslant R + T$ and $R \leqslant S$ we deduce

$$R + (S \cap T) \leqslant S \cap (R + T).$$

If conversely the element s belongs to $S \cap (R + T)$, then $s = r + t$ with r in R and t in T. From $R \leqslant S$ we infer now that $s - r$ belongs to S. Hence $t = s - r$ belongs to $S \cap T$ so that $s = r + t$ belongs to $R + (S \cap T)$. Consequently $S \cap (R + T) \leqslant R + (S \cap T)$; and this proves the desired equation.

The reader should construct examples which show that the above equation fails to hold without the hypothesis $R \leqslant S$.

Two more concepts are needed for the enunciation of the next law.

Quotient Spaces: If M is a subspace, then we define congruence modulo M by the following rule:

The elements x and y in A are congruent modulo M, in symbols: $x \equiv y$ modulo M, if their difference $x - y$ belongs to M.

One verifies that congruence modulo M is reflexive, symmetric, and transitive; and thus we may divide A into mutually exclusive classes of congruent elements. Congruences may be added and subtracted, since $x \equiv y$ modulo M and $x' \equiv y'$ modulo M imply $x + x' \equiv y + y'$ modulo M and $x - x' \equiv y - y'$ modulo M. Since for every f in F we may deduce from $x \equiv y$ modulo M the congruence $fx \equiv fy$ modulo M, we may multiply congruences by elements in F.

Complete classes of congruent elements modulo M are often called cosets modulo M. The totality of these cosets modulo M we designate by A/M. Addition and subtraction of cosets modulo M is defined by the corresponding operations with the elements in the cosets; and the product fX of f in F by X in A/M is just the totality of all fx for x in X, unless $f = 0$ in which case we let $fX = 0X = M = 0$. Then it is clear from the preceding discussion that $(F,A/M)$ is likewise a linear manifold. It may be said that the F-space A/M arises from the F-space A by substituting congruence modulo M for the original equality.

If the subspace S of A contains M, then the elements in S form complete classes of [modulo M] congruent elements. We may form S/M; and one sees easily that S/M is a subspace of A/M.

Conversely let T be some subspace of A/M. Every element in T is a class of congruent elements in A; and thus we may form the set T^* of all the elements in A which belong to some class of congruent elements in T. One verifies that T^* is a subspace of A which contains M and which satisfies $T^*/M = T$.

The reader ought to discuss the example where A is the real projective plane and M some point in it. Then the subspaces of A/M correspond essentially to the lines of A which pass through the point M; and their totality has "the structure of a line."

An Isomorphism of the F-Space A upon the F-Space B is a one-to-one correspondence σ mapping the elements in A upon the elements in B in such a way that

$$A\sigma = B,\ (a + b)\sigma = a\sigma + b\sigma,\ (fa)\sigma = f(a\sigma)$$

for a,b in A and f in F.—It is clear that the inverse σ^{-1} may be formed, and that σ^{-1} is an isomorphism of B upon A.

This concept of isomorphism may be applied in particular upon subspaces and their quotient spaces.

The existence of an isomorphism between the F-spaces A and B we indicate by saying that A and B are isomorphic and by writing $A \sim B$. Instead of isomorphism we are going to say usually "*linear transformation*." Thus linear transformation signifies what in classical terminology is called a non-singular linear transformation. The concept of isomorphism is going to be extended later when we introduce the more comprehensive concept of semi-linear transformation [III.1].

Isomorphism Law: *If S and T are subspaces of the F-space A, then $(S + T)/S \sim T/(S \cap T)$.*

PROOF: Every element x in $S + T$ has the form $x = s + t$ with s in S and t in T. Clearly $x \equiv t$ modulo S. Thus every element X in $(S+T)/S$ contains elements in T; and we may form the non-vacuous intersection $X \cap T$ of the sets X and T. If x' and x'' belong both to $X \cap T$, then $x' \equiv x''$ modulo S so that $x' - x''$ belongs to $S \cap T$; and now one verifies that $X \cap T$ is an element in $T/(S \cap T)$.

If Y is an element in $T/(S \cap T)$, then $S + Y$ is easily seen to be an element in $(S + T)/S$. Since

$$X = S + (X \cap T) \quad \text{for } X \text{ in } (S + T)/S,$$
$$Y = T \cap (S + Y) \quad \text{for } Y \text{ in } T/(S \cap T),$$

we see that the mappings: $X \to X \cap T$ and $Y \to S + Y$ are reciprocal mappings between $T/(S \cap T)$ and $(S + T)/S$; and thus they are in particular one-to-one correspondences between the two quotient spaces. That they are actually isomorphisms is now quite easily verified (so that we may leave the verification to the reader). This completes the proof of the Isomorphism Law.

Lemma: *The join of an ordered set of subspaces is a subspace.*

PROOF: If Φ is an ordered set of subspaces of A, and if $S \neq T$ are distinct subspaces in Φ, then one and only one of the relations $S < T$ and $T < S$ is valid. Denote by J the join of all the subspaces in Φ so that an element belongs to J if, and only if, it belongs to at least one subspace in Φ. If x and y are elements in J, then there exist subspaces X and Y in Φ such that x is in X and y in Y. It follows from our hypothesis that one of these subspaces contains the other one, say $X \leqslant Y$. Then x and y and consequently $x - y$ are in Y so that $x - y$ belongs to J. Since x is in X, so is fx for f in F; and this shows that fx belongs to J too. Thus J is a subspace.

Complementation Theorem: *To every subspace S of A there exists a subspace T of A such that $S \cap T = 0$ and $S + T = A$.*

PROOF: Form the set Θ of all the subspaces W of A with the property $S \cap W = 0$. This set Θ contains for instance 0 and is consequently not vacuous. If Φ is an ordered subset of Θ then we may form [according to the preceding lemma] the join J of the subspaces in Φ. It is clear from its construction that $S \cap J = 0$; and it follows from the lemma that J is a subspace of A too. Thus J belongs to Θ. We have shown therefore that the Maximum Principle of Set Theory (see Appendix S) may be applied on Θ. Hence there exists a subspace M of A with the following two properties:

$$S \cap M = 0; \ M < Q \text{ implies } S \cap Q \neq 0.$$

Form now the subspace $S + M$ of A. Consider an element a in A which does not belong to M. Then $M < M + Fa$; and it follows that $S \cap (M + Fa) \neq 0$. Hence there exists an element $s \neq 0$ in this intersection so that $s = m + fa$ with m in M and f in F. If f were 0, then $s \neq 0$ would belong to $S \cap M = 0$ which is impossible. Thus $f \neq 0$. Hence $a = f^{-1}(s - m)$ belongs to $S + M$; and this proves $S + M = A$. Thus M is a desired subspace.

If S and T are subspaces of A such that $0 = S \cap T$ and $A = S + T$, then we say that *A is the direct sum of S and T*, and we indicate this by writing $A = S \oplus T$. If A is the direct sum of S and T, then T is termed a *complement* of S in A; and it follows from the Isomorphism Law that $A/S = (S + T)/S \sim T/(S \cap T) = T$; and thus the complement of S is essentially [= up to isomorphisms] uniquely determined by S and A.

Suppose that T is a complement of S and that $S \leqslant V$ where V is a subspace of A. Then it follows from Dedekind's Law that

$$V = V \cap (S + T) = S + (V \cap T)$$

whereas $(V \cap T) \cap S = 0$ is obvious. Hence $V = S \oplus (V \cap T)$; and we have shown

Corollary: *If the subspace S is part of the subspace V, and if T is a complement of S in A, then $V \cap T$ is a complement of S in V.*

The principal result of this section may be stated as follows: The totality of subspaces of a linear manifold (F,A) is a complete, complemented, modular lattice. Here we use the language customary in lattice theory; see, for instance, Birkhoff [1].

II.2. Linear Dependence and Independence

It is customary to term the finitely many elements $b_1,$

F-space A linearly independent, if $\sum_{i=1}^{n} f_i b_i = 0$ [with f_i in

$f_1 = \cdots = f_n = 0$. This implies in particular that these n elements are actually distinct elements. Likewise we say that *the subset B of A is linearly independent*, if all its finite subsets are linearly independent in the sense just defined.

Conversely we say that the subset B of A is *linearly dependent*, if B is not linearly independent; and this is equivalent to the following property: There exist finitely many distinct elements $b_1, \cdots, b_n$ in B and numbers $f_1, \cdots, f_n$ not all 0 in F such that $0 = \sum_{i=1}^{n} f_i b_i$.

If D is any subset of A, then we may form the sum $\{D\} = \sum Fd$ [where the summation ranges over all the d in D]; and we say that every element in this subspace $\{D\}$ *is linearly dependent of the subset D.*

Lemma 1: *The subset L of A is linearly independent if, and only if, none of the elements in L is linearly dependent on the remaining elements in L.*

Proof: If L is linearly dependent, then there exist distinct elements $b_1, \cdots, b_n$ in L and numbers $f_1, \cdots, f_n$ not all 0 in F such that $0 = \sum_{i=1}^{n} f_i b_i$. If, as we may assume without loss in generality, $f_1 \neq 0$, then

$$b_1 = \sum_{i=2}^{n} - f_1^{-1} f_i b_i$$

so that b_1 depends on $b_2, \cdots, b_n$ and hence on the elements, not b_1, in L.—The converse of our statement is obvious.

Similar, though still simpler, is the proof of the fact that the element x depends on the independent set B if, and only if, either x is in B or else the set composed of B and x is dependent.

We define now *a basis of the F-space A* as a linearly independent subset of A on which every element in A depends.

Lemma 2: *If B is a basis of A, then every element in A may be represented in one and only one way in the form*

$$a = \sum_{b \in B} a(b) b$$

with $a(b)$ in F where only a finite number of the $a(b)$'s is different from 0.

The proof is an almost immediate consequence of our definitions. For its execution one has just to remember two facts: $\sum_{b \in B} Fb$ consists of all the elements $\sum_{b \in B} f(b)b$ where only a finite number of the elements $f(b)$ in F are different from 0; from $\sum_{b \in B} f'(b)b = \sum_{b \in B} f''(b)b$ one deduces the linear relation $0 = \sum_{b \in B} [f'(b) - f''(b)]b$.

Existence Theorem: *Every linear manifold possesses a basis.*

PROOF: Suppose that the F-space A is not 0 [otherwise the empty set would be a basis]. Then every element, not 0, in A forms an independent set. Thus the set Θ of all the independent subsets of A is not vacuous. If Φ is an ordered [by inclusion] subset of Θ, then we form the join J of all the sets in Φ. Suppose that $b_1, \cdots, b_n$ are finitely many elements in J and that $f_1, \cdots, f_n$ are elements in F, satisfying $0 = \sum_{i=1}^{n} f_i b_i$. Every b_i belongs to some set B_i in Φ. Since Φ is ordered [by inclusion], there exists an index m such that $B_i \leqslant B_m$ for every i. But then $b_1, \cdots, b_n$ belong to the linearly independent set B_m; and this implies $f_1 = \cdots = f_n = 0$. Hence J is linearly independent. Now we may apply the Maximum Principle of Set Theory (Appendix S) on Θ to prove the existence of a set B in Θ such that B is not a proper subset of an independent set. Assume now that the element a in A does not belong to B. Then the set $[B,a]$ is dependent; and it follows from a previous remark that the element a depends on B. Hence B is a basis of A, as we intended to show.

Uniqueness Theorem: *Any two bases of a linear manifold possess the same number of elements.*

As is customary in set theory we say that two sets contain the same number of elements, if there exists a one-to-one correspondence between them; for further information see Appendix S.

It will be convenient to precede the proof of our theorem by the proof of the following special case which is well suited for inductive handling.

($U.n$) *If a linear manifold possesses an n-element basis, then all its bases contain exactly n elements.*

Here n indicates a positive integer.

If the F-space A possesses a one element basis, then $A = Fb$ for some $b \neq 0$; and the validity of ($U.1$) is immediately clear. Hence we may assume now that $1 < n$ and that ($U.i$) is true for every $i < n$. Suppose now that the F-space possesses an n-element basis $b_1, \cdots, b_n$; and assume that B is some basis of A. Then

$$\sum_{i=1}^{n-1} Fb_i < \sum_{i=1}^{n} Fb_i = A = \sum_b Fb$$

where b ranges over all the elements b in B. It is consequently impossible that every b in B belongs to

$$\sum_{i=1}^{n-1} Fb_i = S.$$

Denote by w some element in B, not in S. Then it is easily seen that the elements $b_1, \cdots, b_{n-1}, w$ form an independent set, since otherwise w would belong to S. Now $A/S = (S + Fb_n)/S \sim Fb_n/(Fb_n \cap S) = Fb_n$ (by the Isomorphism Law). But Fb_n does not possess any subspaces except 0 and Fb_n. Hence A and S are the only subspaces between S and A. Since $S < S + Fw \leqslant A$, it follows that $A = S + Fw$ and that $b_1, \cdots, b_{n-1}, w$ is an n-element basis W of A.

Now we consider the quotient space A/Fw; and denote, for x in A, by x^* the class of elements congruent to x modulo Fw. Then one verifies that $\sum_{i=1}^{n-1} f_i b_i^* = 0$ implies $\sum_{i=1}^{n-1} f_i b_i = fw$; and we infer $f_1 = \cdots = f_{n-1} = 0$ from the independence of W. Hence a basis of A/Fw is formed by the $n-1$ elements $b_1^*, \cdots, b_{n-1}^*$. Denote now by B^* the set of all elements t^* with $t \neq w$ in B. Then one verifies likewise that B^* is a basis of A/Fw. Applying, as we may, $(U.n-1)$ on A/Fw it follows that B^* is an $(n-1)$-element set. Hence B is, by the construction of B^*, an n-element set. Thus $(U.n)$ is a consequence of $(U.n-1)$; and this completes the inductive proof of $(U.n)$.

Now we turn to the general *proof of the Uniqueness Theorem*. Suppose that B' and B'' are bases of the F-space A. Suppose first that one of these bases contains a finite number of elements. If, for instance, B' contains n elements, then it follows from $(U.n)$ that B'' contains n elements too. Consequently it suffices to consider the case where neither of the sets B' and B'' is finite. (See Appendix S for the set theoretical facts employed in the remainder of the proof.)

Suppose that T is a finite subset of B'. We form

$$T^{**} = B'' \cap \sum_{t \in T} Ft.$$

It follows from the Complementation Theorem that there exists a complement C of $\sum_{t \in T^{**}} Fx$ in $\sum_{t \in T} Ft$. Combining T^{**}, which is independent, with a basis of C, we obtain a basis of $\sum_{t \in T} Ft$. Since the latter F-space possesses the finite basis T, all its bases are finite; and it follows that T^{**} contains at most as many elements as T. Thus we have obtained a single valued mapping $T \to T^{**}$ of the set Φ' of all the finite subsets of B' upon a subset Θ' of the set Φ'' of all the finite subsets of B''. If w is some element in B'', then there exist finitely many elements $b_1, \cdots, b_k$ in B' such that w belongs to $\sum_{j=1}^{k} Fb_j$. Letting W be the set consisting of $b_1, \cdots, b_k$, it follows that w belongs to W^{**}. The join of the sets in Θ' is consequently exactly the set B''.

Since B' is infinite, B' and Φ' contain the same number of elements. Since $T \to T^{**}$ is a single-valued mapping of Φ' upon Θ', the set Θ' contains at most as many elements as Φ', and hence as B'. Since Θ' is part of Φ'', and since Φ'' and B'' contain the same number of elements, Θ' contains at most as many elements as B''. Since Θ' is a set of finite sets covering the infinite set B'', Θ' contains at least as many elements as B''. Thus Θ' and B'' contain the same number of elements. Hence we have shown that B'' contains at most as many elements as B'. For reasons of symmetry it follows now that B' contains at most as many elements as B''; and thus we have shown that B' and B'' contain the same number of elements. This completes the proof of the Uniqueness Theorem.

Historical note. The first theorem of this type has been proved by E. Steinitz (the invariance of the degree of transcendency); the present version of the proof is due to S. Bochner.

The uniquely determined number of elements in all the bases of the F-space A is called *the rank* $r(A)$ *of* A. If $r(A)$ happens to be finite, then $r(A) - 1 = \dim A$ is the *dimension* of A.

Structure Theorem: *The F-spaces A and B are isomorphic if, and only if, they have the same rank* $r(A) = r(B)$. *To every field F and to every cardinal number c there exists one and essentially only one F-space of rank c.*

Proof: It is obvious that isomorphic F-spaces have the same rank, since isomorphisms map bases onto bases. Suppose next that the F-spaces A and B have the same rank c. Then A possesses a c-element basis A' and B possesses a c-element basis B'. There exists consequently a one-to-one correspondence τ mapping A' upon B'. If x is an element in A, then there exists a uniquely determined representation [Lemma 2]

$$x = \sum_{t \in A'} x(t)t$$

with $x(t)$ in F where only a finite number of the $x(t)$ may be different from 0. Then we let

$$x^{\sigma} = \sum_{t \in A'} x(t)t^{\tau};$$

and it is easy to verify that σ is an isomorphism of A upon B, which induces τ in A'.

To construct an F-space of rank c one produces first a set C of marks which contains exactly c elements. The elements of our space A are then all the functions $f(v)$ from C to F, subject to the following restriction:

(+) There exists at most a finite number of elements in C which are not mapped upon 0 by f.

If we define addition of the functions f' and f'' by the formula

$$(f' + f'')(v) = f'(v) + f''(v) \text{ for } v \text{ in } C,$$

and multiplication by the rule

$$(fg)(v) = f[g(v)] \text{ for every } v \text{ in } C \text{ and } f \text{ in } F,$$

then it is clear that A has been made into an F-space.

If w is some definite element in C, then we define a function f_w by the rule:

$$f_w(v) = \begin{cases} 1 \text{ for } v = w \\ 0 \text{ for } v \neq w. \end{cases}$$

It follows from condition (+) that these functions f_w in A form a basis of the F-space A. Since this set of functions contains as many elements as the set C, it follows that $r(A) = c$, as desired.

When contemplating the preceding construction, the reader should remember such examples of function spaces as the system of all real valued functions of a real variable.

In case the rank c is supposed to be a finite integer n, then we may show easily that an F-space of rank n may be given in the following form: Its elements are the n-tuplets $(x_1, \cdots, x_n)$ with coordinates x_i in F; and addition, subtraction, and multiplication are defined "coordinatewise."

General Rank Formula: *If S and T are subspaces of the linear manifold (F,A), then*

$$r(S + T) = r(S \cap T) + r(S/S \cap T) + r(T/S \cap T).$$

PROOF: From the Complementation Theorem we deduce the existence of subspaces U and V such that $S = (S \cap T) \oplus U$ and $T = (S \cap T) \oplus V$. It follows from the Isomorphism Laws (II.1) that $U \sim S/S \cap T$; and this implies $r(U) = r(S/S \cap T)$. Likewise $r(V) = r(T/S \cap T)$. Next we note that $S + T = (S \cap T) \oplus U \oplus V$, and that consequently a basis of $S + T$

may be formed by joining a basis of $S \cap T$, a basis of U and a basis of V. Hence

$$r(S+T) = r(S \cap T) + r(U) + r(V)$$
$$= r(S \cap T) + r(S/S \cap T) + r(T/S \cap T),$$

as we desired to show.

Special Rank Formula: *If S and T are subspaces such that $S \leqslant T$, then*

$$r(T) = r(S) + r(T/S).$$

This is an immediate special case of the General Rank Formula.

Rank Formulas for Subspaces of Finite Rank: *If S and T are subspaces of finite rank of A, then*

(*a*) $S \leqslant T$ *and* $r(S) = r(T)$ *imply* $S = T$, *and*

(*b*) $r(S) + r(T) = r(S \cap T) + r(S + T)$.

PROOF: If $S \leqslant T$ and $r(S) = r(T)$, then we deduce from the finiteness of all the ranks and from the Special Rank Formula that $0 = r(T/S)$. But this implies $T/S = 0$ or $S = T$, proving (*a*). To prove (*b*) we apply the Special Rank Formula on the subspace $S \cap T$ of S and find that, because of the finiteness of rank, $r(S/S \cap T) = r(S) - r(S \cap T)$. Likewise we see that $r(T/S \cap T) = r(T) - r(S \cap T)$; and now it follows from the General Rank Formula that

$$r(S+T) = r(S \cap T) + r(S/S \cap T) + r(T/S \cap T)$$
$$= r(S) + r(T) - r(S \cap T);$$

and this proves (*b*).

Rank Formula for Subspaces of Infinite Rank: *If at least one of the subspaces S and T has infinite rank, then*

$$r(S+T) = \max\,[r(S), r(T)].$$

PROOF: Suppose without loss in generality that $r(T) \leqslant r(S)$. Then it follows from our hypothesis that $r(S)$ is infinite. It follows from the Special Rank Formula that $r(T/S \cap T) \leqslant r(T)$; and now we deduce from a combined application of General and Special Rank Formula and the "absorption law" for the addition of infinite cardinal numbers [see Appendix S] that

$$r(S+T) = r(S \cap T) + r(S/S \cap T) + r(T/S \cap T)$$
$$= r(S) + r(T/S \cap T) = r(S) = \max\,[r(S), r(T)].$$

REMARK: It is easy to deduce the general validity of the above formula (*b*) from the Rank Formula for Subspaces of Infinite Rank together with the absorption law for infinite cardinal numbers. The reader is advised to consider the problem of determining the rank of a sum of infinitely many subspaces.

Now that we have assigned to every linear manifold a definite rank, we may also introduce the customary geometrical names for spaces and subspaces with special ranks. Thus *points*, *lines*, and *planes* are spaces of ranks 1, 2, and 3 respectively. Since the rank of the space under consideration may be infinite, it is impossible to define a hyperplane as a subspace H of A such that $r(A) = r(H) + 1$ (after all in case of infinite rank A itself has this property); but we may define a *hyperplane* as a subspace H of A such that A/H has rank 1 [is a point]. The Rank Formulas assure us that distinct lines in a plane meet in a point; and that the space A is spanned by each of its hyperplanes H together with any subspace S, not part of H. Note furthermore that $A/(H' \cap H'')$ is a line whenever H' and H'' are distinct hyperplanes.

The following result is easily deduced from the preceding theorems.

The following properties of the linear manifold A are equivalent.

(I) *A has finite rank.*

(II) *Every descending sequence of subspaces*

$$\cdots \leqslant S_{i+1} \leqslant S_i \leqslant \cdots \leqslant S_2 \leqslant S_1$$

terminates after a finite number of steps.

(III) *Every ascending sequence of subspaces*

$$S_1 \leqslant S_2 \leqslant \cdots \leqslant S_i \leqslant S_{i+1} \leqslant \cdots$$

terminates after a finite number of steps.

In the course of the proof of the Structure Theorem we considered a space of functions which was subject to a finiteness condition (+). For future applications we need also the space of "unrestricted" functions which we define as follows:

Suppose that the set C of "marks" contains exactly c elements where c is a finite or infinite cardinal number. Suppose that F is a field. Then we denote by $[F,C]$ the set of *all* single-valued mappings from C into F; and we define addition and multiplication by the rules:

If f' and f'' are elements in $[F,C]$, then the function $f' + f''$ is defined by the formula:

$$(f' + f'')(x) = f'(x) + f''(x) \text{ for every } x \text{ in } C.$$

If f is an element in $[F,C]$ and y an element in F, then the function yf is defined by the formula:

$$(yf)(x) = y[f(x)] \text{ for every } x \text{ in } C.$$

It is not difficult to verify that in this fashion $[F,C]$ has been made into an F-space. It is fairly obvious that the structure of this F-space depends only on the number c and not on the special nature of the set C of marks. This becomes apparent, if we consider the next theorem in conjunction with the Structure Theorem.

Proposition: (*a*) *If c is finite, then* $c = r([F,C])$.

(*b*) *If c is infinite, and if d is the number of elements in the field F, then* $d^c = r([F,C])$.

The proof of (*a*) is effected by proving that the following set of functions f_x is a basis of $[F,C]$, in case c is finite.

$$f_x(y) = \begin{cases} 1 \text{ for } x = y \\ 0 \text{ for } x \neq y \text{ in } C. \end{cases}$$

The verification of this fact we leave to the reader.

The proof of (*b*) will be preceded by the proofs of various lemmas some of which are of independent interest. (The arguments used in the proof of (*b*) are mainly of a set theoretical nature. Readers not familiar with set theoretical argumentations are therefore advised to omit this proof in a first reading.)

Lemma 3: *If the F-space A has infinite rank, and if F contains d elements, then A contains dr(A) elements.*

PROOF: Consider some basis B of A. Then B contains $r(A)$ elements and every element in A may be represented in one and only one way in the form

(*) $w = \sum_{b \in B} w(b)b$ with $w(b)$ in F and only a finite number of the $w(b)$ different from 0.

Denote by d^* the number of elements, not 0, in F; then $d = d^* + 1$. If $b_1, \cdots, b_k$ are finitely many distinct elements in B, then A contains exactly d^{*k} elements of the form $\sum_{i=1}^{k} w_i b_i$ with $w_i \neq 0$ in F. Since the infinite set B contains $r(A)$ elements, there exist for every positive [finite] k exactly $r(A)$ different k-element subsets. Consider now the elements in A with the property that exactly k of their coefficients [in the representation (*)] are different from 0. Their number is by the preceding considerations exactly $d^{*k}r(A)$. Consequently the number of elements in A is exactly $\sum_k d^{*k}r(A)$.

Case 1: If d and d^* are finite, then d^{*k} is finite too so that $r(A) = d^{*k}r(A)$ because of the infinity of $r(A)$. Hence

$$\sum_k d^{*k}r(A) = \sum_k r(A) = r(A)\aleph_0 = r(A) = r(A)d.$$

Case 2: If d and d^* are infinite, then $d = d^* = d^{*k}$; and we find that

$$\sum_k d^{*k}r(A) = \sum_k dr(A) = dr(A)\aleph_0 = dr(A);$$

and this completes the proof of our lemma.

Lemma 4: *If c is infinite, and if d is the number of elements in F, then the number of elements in $[F,C]$ is $dr([F,C]) = d^c$.*

PROOF: If c is infinite, then one verifies easily that the rank of $[F,C]$ is infinite too. The functions f_x, for instance, which are defined by

$$f_x(y) = \begin{cases} 1 \text{ for } x = y \\ 0 \text{ for } x \neq y \end{cases} \text{ in } C$$

form a set of independent elements in $[F,C]$ which contains just c elements so that $c \leqslant r([F,C])$. Hence we may apply Lemma 3 and see that the number of elements in $[F,C]$ is just $dr([F,C])$.

On the other hand we remember that $[F,C]$ consists exactly of all the single-valued mappings of the c-element set C into the d-element set F; and the number of such mappings is exactly d^c; this fact is practically identical with the definition of the cardinal number d^c. Consequently $dr([F,C]) = d^c$, as we wanted to show.

PROOF OF PROPOSITION (*b*) IN CASE $d \neq d^c$.

In this case we have $d < d^c$ so that $r([F,C]) < d^c$ would imply $dr([F,C]) < d^c$. The equation $r([F,C]) = d^c$ is now an immediate consequence of Lemma 4.

In the future we need only consider the case where $d = d^c$. Since c is infinite and $1 < d$, this implies that d is infinite. We show next the

Lemma 5: *If the field F contains d elements where d is infinite, and if A is an F-space of finite positive rank n, then there exists a subset W of A which contains d elements such that any n elements in W are independent.*

PROOF: Consider the set Θ of all the subsets T of A with the property

($T.n$) Any n elements in T are independent.

If, for instance, $b_1, \cdots, b_n$ is a basis of A, then they form a set in Θ; and we note furthermore that the $(n + 1)$-element set $b_1, \cdots, b_n, \sum_{i=1}^{n} b_i$ belongs to Θ too. If Φ is a subset of Θ which is ordered by inclusion, then we form the join J of all the sets in Φ. If $j_1, \cdots, j_n$ are n distinct elements in J, then each of them belongs to a set in Φ. If j_i belongs to the set S_i in Φ, then there exists a subscript m such that $0 < m \leqslant n$ and such that $S_i \leqslant S_m$ for every i, since Φ is ordered by inclusion. Hence all the j_i are in S_m. But S_m meets requirement ($T.n$); and so the elements $j_1, \cdots, j_n$ are independent. Thus J itself meets requirement ($T.n$). Hence the Maximum Principle of Set Theory (Appendix S) may be applied on Θ. Consequently there exists a subset W of A with the following properties:

(*a*) W contains at least $n + 1$ elements.

(*b*) Any n elements in W are independent.

(*c*) If W is a subset of the set W', and if W' has Property ($T.n$), then $W = W'$.

(That we may require W to satisfy condition (a) follows from the fact, pointed out before, that Θ contains at least one $(n+1)$-element set.)

Assume now that W contains less than d elements; and denote by W' the set of all $(n-1)$-tuplets of elements in W. Denote by t' the number of $(n-1)$-tuplets in W'. Then we show that

(1) $$1 < n < t' < d.$$

This is certainly true in the [impossible] case that t' is finite; and if t' is infinite, then t' is exactly the number of elements in W [since the number of finite subsets of an infinite set is exactly the number of elements in this infinite set] (see Appendix S).

(2) t' is the number of subspaces spanned by $(n-1)$-tuplets in W'.

We note first that every $(n-1)$-tuplet in W' is independent [by (b)] so that every $(n-1)$-tuplet in W' spans a subspace of rank $n-1$ of A. If $h_1, \cdots, h_{n-1}$ and $k_1, \cdots, k_{n-1}$ are $(n-1)$-tuplets in W' which span the same subspace, then each of the k_i depends on the h's. If one of them, say k_j, were distinct from all the h_j, then the set $h_1, \cdots, h_{n-1}, k_j$ of n elements in W were dependent which contradicts (b). Thus different $(n-1)$-tuplets in W' span different subspaces in A; and this proves (2).

(3) If S is a subspace of rank i, and if $1 < i$, then S contains at least d subspaces of rank $i-1$.

Denote by $s_1, \cdots, s_i$ a basis of S. If x is an element in F, then let

$$S_x = \sum_{j=1}^{i-2} Fs_j + F(s_{i-1} + xs_i).$$

It is easy to see that $r(S_x) = i-1$, and that $S_x = S_y$ if, and only if, $x = y$ (since the elements s_j are independent). Thus we have constructed exactly d distinct subspaces of rank $i-1$. (It would not be difficult to prove now that d is the exact number of such subspaces, since d is infinite.)

Denote by W'' the set of all the subspaces of A which are spanned by $(n-1)$-tuplets in W'. We have shown in (2) that W'' contains exactly t' subspaces.

(4) If $0 < i < n$, then there exists a subspace V_i of A whose rank is $n-i$ and which is not part of any subspace in W''.

It follows from (3) that A contains at least d subspaces of rank $n-1$, and it follows from (1) and (2) that W'' contains less than d subspaces. Thus there exists at least one subspace V_1 of A, not in W'', whose rank is $n-1$; and this proves (4) for $i=1$. Assume now that we have verified (4) already for some i such that $0 < i < n-1$. Then there exists a subspace V_i of rank $n-i$ which is not part of any subspace in W''. If H is some subspace in W'', then we have consequently $H < H + V_i \leqslant A$; and it follows therefore from the Rank Formulas that

$$n - 1 = r(H) < r(H + V_i) \leqslant r(A) = n \quad \text{or} \quad r(H + V_i) = n = r(A)$$

and hence $H + V_i = A$. We may deduce therefore from the Rank Formula (*b*) that

$$n - 1 + n - i = r(H) + r(V_i) = r(H + V_i) + r(H \cap V_i) = n + r(H \cap V_i)$$

or

$$r(H \cap V_i) = n - i - 1.$$

Since the number of intersections $H \cap V_i$ with subspaces H in W'' is less than d, and since it follows from (3) that V_i contains at least d subspaces of rank $n - i - 1$, there exists at least one subspace V_{i+1} of rank $n - (i + 1)$ of V_i which is distinct of all the intersections $H \cap V_i$ with H in W''. It is clear that V_{i+1} is not part of any subspace H in W'', since otherwise it would be one of those intersections $H \cap V_i$. This completes the inductive proof of (4).

(5) There exists an element v in A which does not belong to any subspace H in W''.

This is the same as the special case $i = n - 1$ of (4).

Form now the set M composed of W and the element v [of (5)]. It follows from (*b*) and (5) that M has Property $(T.n)$ (since otherwise v would be on one of the subspaces H in W''). But $W < M$; and this contradicts (*c*). Thus we have been led to a contradiction by assuming that W contains less than d elements. Since A itself contains d elements (as d is infinite), it follows that W contains just d elements; and this completes the proof of our lemma.

Lemma 6: *If C is the set of positive integers, and if the number d of elements in F is infinite, then $d \leqslant r([F,C])$.*

PROOF: Denote by A_n an F-space of rank n; and denote by

$$b[\tfrac{1}{2}n(n-1) + 1], \cdots, b[\tfrac{1}{2}n(n+1)]$$

a basis of A_n. It follows from Lemma 5 that there exists a subset V_n of A_n which contains exactly d elements and which has the further property that any n elements in V_n are linearly independent. There exists a one-to-one correspondence σ_n of F upon V_n. If x is an element in F, then the element corresponding to x is

$$x^{\sigma_n} = x[\tfrac{1}{2}n(n-1) + 1]\, b[\tfrac{1}{2}n(n-1) + 1] + \cdots + x[\tfrac{1}{2}n(n+1)]\, b[\tfrac{1}{2}n(n+1)]$$

where the coefficients $x[i]$ are uniquely determined elements in F. Thus $x[i]$ is for every x in F and for every positive integer i an element in F; and we may consider each of the $x[i]$ as a single-valued mapping of the set

C of positive integers into F. Thus each of the $x[i]$ is an element in $[F,C]$. Suppose now that $f_1, \cdots, f_n$ are elements in F and that $x_1, \cdots, x_n$ are distinct elements in F such that $0 = \sum_{j=1}^{n} f_j x_j[i]$ holds in $[F,C]$. This is equivalent to saying that

$$0 = \sum_{j=1}^{n} f_j x_j[i] \qquad \text{for } i = 1, \cdots$$

holds in F. But then we find that

$$\sum_{j=1}^{n} f_j x_j^{\sigma_n} = \sum_{j=1}^{n} f_j \sum_i x[i] b[i] = \sum_i \left[\sum_{j=1}^{n} f_j x_j[i] \right] b[i] = 0$$

where the i-summation ranges from $i = \frac{1}{2}n(n-1) + 1$ to $i = \frac{1}{2}n(n+1)$. But the elements $x_j^{\sigma_n}$ are n distinct elements in V_n and therefore are independent. Hence $f_1 = \cdots = f_n = 0$. Thus we have shown that the elements $x[i]$ in $[F,C]$ are independent. But the number of these elements is just d, the number of elements in F; and hence we have shown that the rank of $[F,C]$ is at least d, as we intended to prove.

PROOF OF PROPOSITION (*b*) IN CASE $d = d^c$.

Since c is infinite, C contains a countably infinite part C^*. Taking all the functions f in $[F,C]$ which satisfy $f(x) = 0$ for x in C, but not in C^*, we obtain a subspace D of $[F,C]$ which is clearly isomorphic to $[F,C^*]$. From $d = d^c$ and the infinity of c we infer the infinity of d; and hence it follows from Lemma 6 that the rank of $[F,C^*]$ is at least d. Hence we have a fortiori that

$$d \leqslant r([F,C]).$$

It follows from Lemma 4 that $[F,C]$ contains $d^c = d$ elements; and so the rank of $[F,C]$ is at most d. Consequently $r([F,C]) = d = d^c$; and this completes the proof.

Corollary: $c = r([F,C])$ *if, and only if,* c *is finite.*

This is an immediate consequence of the preceding proposition, since for infinite c we have the well-known set theoretical inequalities

$$c < 2^c \leqslant d^c,$$

as every field contains at least two elements.

II.3. The Adjoint Space

A linear form over the F-space A is a single-valued mapping f which maps the element a in A upon the element af in F and satisfies

$$(a' + a'')f = a'f + a''f \qquad \textit{for } a', a'' \textit{ in } A$$
$$(xa)f = x(af) \qquad \textit{for } x \textit{ in } F \textit{ and } a \textit{ in } A.$$

The present way of writing linear forms should remind the reader of the formation of scalar products [of a by f]. The formerly customary functional way of writing these linear forms [linear functional $f(a)$ instead of product af] has a number of formal inconveniences arising in connection with the definition of multiplication (see below). Note, for instance, that the "associative law" $(xa)f = x(af) = xaf$ would read in the functional notation: $f(xa) = xf(a)$.

The totality of linear forms over A we denote by $L(A)$. In $L(A)$ we define *addition* by the following rule:

If f' and f'' are linear forms over A, then $f' + f''$ is the mapping defined by

$$a(f' + f'') = af' + af'' \qquad \textit{for every } a \textit{ in } A.$$

It is readily verified that the sum (and also the difference) of linear forms is again a linear form; and that $L(A)$ *is an abelian group* with respect to the addition just defined.

Next we define *multiplication* of linear forms by elements in F by the following rule:

If f is a linear form over A, and if x is an element in F, then fx is the mapping defined by

$$a(fx) = (af)x \qquad \textit{for every } a \textit{ in } A.$$

Since both af and x are elements in F, it is easily seen that fx is likewise a linear form over A. It should be noted that elements in A are multiplied from the left by elements in F, whereas linear forms are multiplied from the right by elements in F.

Addition and multiplication satisfy the following formal properties.

(a') If f is an element in $L(A)$ and x an element in F, then their product fx is a uniquely determined element in $L(A)$.

(b') $(f' + f'')x = f'x + f''x$, $f(x' + x'') = fx' + fx''$ for f, f', f'' in $L(A)$ and x, x', x'' in F.

(c') $f1 = f$ for f in $L(A)$ [where 1 designates the identity element in the field F].

(d') $f(x'x'') = (fx')x''$ for f in $L(A)$ and x', x'' in F.

The verification of these rules is simple enough to be left to the reader.

(On the basis of these formulas it becomes apparent why we denoted the product of a form f by a number x in F by the symbol fx instead of xf. If we had chosen the second notation, then formula (d') would read

$$(x'x'')f = x''(x'f);$$

and the reader might find other examples of this kind.) When comparing the properties (a') to (d') with the rules defining a linear manifold (F,A) in II.1, then we notice that they are practically identical apart from the interchange of right and left which is necessitated by the fact that multiplication by elements in F is this time right-multiplication. We may consequently consider $L(A)$ as an F-space too; and whenever we want to emphasize the fact that the elements in F are right-multipliers, we shall speak of the linear manifold $(L(A),F)$. We shall call $(L(A),F)$ *the adjoint space of* A; and it is clear that all the results deduced for F-spaces A may be applied mutatis mutandis on the adjoint space $L(A)$ of A. Thus in particular $L(A)$ has a rank which we are going to relate to the rank of A.

Theorem 1: (a) *If $r(A)$ is finite, then $r(A) = r[L(A)]$.*

(b) *If $r(A)$ is infinite, then $r[L(A)] = c^{r(A)}$ where c is the number of elements in F.*

PROOF: Denote by B some basis of A. If f' and f'' are elements in $L(A)$ such that $bf' = bf''$ for every b in B, then we have $f' = f''$ because of their linearity. If furthermore we assign to every element b in B in some fashion an element $v(b)$ in F, then we may define a linear form f such that $bf = v(b)$ for every b in B as follows:

$$\left[\sum_{b \in B} w(b)b\right]f = \sum_{b \in B} w(b)v(b)$$

where we have to remember that only a finite number of the $w(b)$ is different from 0 [see II.2, Lemma 2]. Thus we have seen that $L(A)$ is essentially the same as the F-space of all single-valued mappings of B into F [defining multiplication by elements in F as right multiplication]; and now our theorem is an immediate consequence of II.2, Proposition.

Corollary 1: *A linear space and its adjoint have equal rank if, and only if, their rank is finite.*

This is a fairly immediate consequence of the preceding theorem [see in this context also II.2, Corollary].

Corollary 2: *F-spaces of finite rank are isomorphic if, and only if, their adjoint spaces are isomorphic.*

This is an immediate consequence of the fact that F-spaces of finite rank have the same rank as their adjoint spaces and that F-spaces of equal rank are isomorphic [II.2, Structure Theorem].—In the case of infinite rank we cannot make a similar statement, since $c^{r(A)} = c^{r(B)}$ for infinite

$r(A)$ and $r(B)$ does not imply $r(A) = r(B)$. If, for instance, $c \doteq 2^{2^{\aleph_0}}$, $r(A) = \aleph_0$ and $r(B) = 2^{\aleph_0}$, then $c = c^{r(A)} = c^{r(B)}$.

We note furthermore that the equality of rank of A and $L(A)$ does not imply their isomorphy, since the elements in F are left-multipliers for A and right-multipliers for $L(A)$. If A has finite rank n, then A may be represented as the F-space of all n-tuplets $(a_1, \cdots, a_n)$ and $L(A)$ may be represented in the same fashion. Thus we might be tempted to map the n-tuplet $(a_1, \cdots, a_n)$ in A upon the same n-tuplet in $L(A)$. This mapping will certainly be an isomorphism of the additive group of A upon the additive group of $L(A)$; but the reader should realize that in general it will be impossible to prove a rule for the multiplicative behavior of this mapping.

In this context the following remark will be in order which makes it clear that the distinction between right- and left-spaces is more than a notational device. For suppose that the space A admits F as a field of left-multipliers and that the space B admits F as a field of right-multipliers. Then we might be tempted to define an "isomorphism" σ as an isomorphism of the additive group of A upon the additive group of B which satisfies the rule $(fa)^\sigma = a^\sigma f$ for a in A and f in F. Then we have for any pair of elements x,y in F and every a in A:

$$a^\sigma xy = (a^\sigma x)y = (xa)^\sigma y = [y(xa)]^\sigma = [(yx)a]^\sigma = a^\sigma yx;$$

and this implies that either $A = 0$ or F is commutative. For a further analysis of this phenomenon compare Chapter IV below.

A substitute for the "missing isomorphism" between A and $L(A)$ will be obtained by the consideration of the adjoint of the adjoint. The linear forms over an F-space where multiplication by elements in F is right-multiplication may be defined exactly as the linear forms over linear manifolds with left-multiplication apart from some obvious changes necessitated by the order of multiplication; and these linear forms may then again be combined into an F-space. No harm is done if we use the same symbols throughout. If we begin with an F-space A (multiplication by elements in F is left-multiplication) we obtain first its adjoint space $L(A)$ (multiplication by elements in F is right-multiplication), and we form next the adjoint of the adjoint space to be denoted by $L[L(A)]$ or $L^2(A)$ (multiplication by elements in F is left-multiplication). Thus A and $L^2(A)$ are F-spaces of the same type. There is a fundamental relation between A and $L^2(A)$ which we are going to establish now.

If t is an element in A, then we may define a linear form t^* over $L(A)$ by the formula

$$t^*f = tf \qquad \textit{for every } f \textit{ in } L(A).$$

It is quite easy to see that every t^* is really a linear form over $L(A)$; and that the mapping of t upon t^* is an isomorphism of A upon a subspace A^* of $L^2(A)$. Thus it is justified to term this mapping *the natural isomorphism of A into $L^2(A)$.*

Theorem 2: *The F-space A is mapped upon $L^2(A)$ by the natural isomorphism if, and only if, A has finite rank.*

PROOF: If A has finite rank, then it follows by twofold application of Theorem 1, (*a*) (and the fact that A and A^* are isomorphic) that

$$r(A^*) = r(A) = r[L(A)] = r[L^2(A)];$$

and now we infer from the finiteness of $r(A^*)$ and the Rank Formula (*a*) that $A^* = L^2(A)$.

If A has infinite rank, then we deduce likewise from Theorem 1, (*b*) that

$$r[L^2(A)] = c^{r[L(A)]},\ r[L(A)] = c^{r(A)}$$

where c is the number of elements in the field F; and now we deduce from set theoretical principles that

$$r(A^*) = r(A) < c^{r(A)} < c^{c^{r(A)}} = r[L^2(A)];$$

and this implies $A^* < L^2(A)$.

REMARK 1: Because of the natural isomorphism between A and A^* we may identify the element t in A with the corresponding element t^* in A^*. Thus we obtain an identification of A and A^*. If $r(A)$ is finite, this actually is an identification of A and $L^2(A)$ so that in this case the adjoint of the adjoint may be considered as the original space. This puts into evidence the symmetry of the relation between A and $L(A)$ (see for further details the Appendix II).

The Galois Correspondence between A and $L(A)$

When constructing this fundamental and far-reaching correspondence we denote by XY for $X \leqslant A$ and $Y \leqslant L(A)$ the totality of elements xy for x in X and y in Y.

If T is a subset of A, then $E(T)$ is the set of all linear forms f such that $Tf = 0$.

If T is a subset of $L(A)$, then $S(T)$ is the set of all elements s in A such that $sT = 0$.

The reader will find it easy to verify the following important facts concerning the two operations E and S.

$E(T)$ is a subspace of $L(A)$.	$S(T)$ is a subspace of A.
$T' \leqslant T''$ implies $E(T'') \leqslant E(T')$.	$T' \leqslant T''$ implies $S(T'') \leqslant S(T')$.
$T \leqslant S[E(T)]$.	$T \leqslant E[S(T)]$.
$E(T) = ESE(T)$.	$S(T) = SES(T)$.
$TE(T) = 0$.	$S(T)T = 0$.

The last formulas distinguish as an important class of subspaces of A and $L(A)$ those of the form $S(T)$ and $E(T)$ respectively. It is an important problem to characterize them within the class of all subspaces; some discussion of this problem will be found below.

Proposition 1: *If M is a subspace of the linear manifold (F,A), then $E(M)$ is essentially the same as $L(A/M)$ and $L(A)/E(M)$ is essentially the same as $L(M)$.*

PROOF: If the linear form f belongs to $E(M)$, then $Mf = 0$. If X is a coset in A/M, then $X = M + x$ for every element x in X so that $Xf = (M + x)f = xf$ is a uniquely determined element in F. The mapping of X upon Xf is now easily seen to be a linear form f^* over the F-space A/M. If f and g are linear forms in $E(M)$ which give rise to the same linear form over A/M [so that $f^* = g^*$], then we have, for every x in A,

$$xf = (M + x)f = (M + x)f^* = (M + x)g^* = (M + x)g = xg;$$

and this implies $f = g$. Now it is easily seen that the mapping of f in $E(M)$ upon f^* in $L(A/M)$ constitutes an isomorphism of $E(M)$ into $L(A/M)$. If finally h is some linear form over A/M, then we define a linear form k in $E(M)$ by the rule:

$$xk = (M + x)h \qquad \text{for every } x \text{ in } A;$$

and it is clear that $k^* = h$. The mapping of f in $E(M)$ upon f^* in $L(A/M)$ constitutes therefore the desired isomorphism of $E(M)$ upon $L(A/M)$.

Every linear form f of A induces a linear form f^M over M, namely the one defined by the rule:

$$xf^M = xf \qquad \text{for every } x \text{ in } M.$$

It is clear that $(f' + f'')^M = f'^M + f''^M$ and $(fy)^M = f^My$ for f,f',f'' in $L(A)$ and y in F; and that $f^M = 0$ if, and only if, f belongs to $E(M)$. Thus the mapping of f in $L(A)$ upon f^M in $L(M)$ induces an isomorphism of $L(A)/E(M)$ into $L(M)$. Consider now some linear form h over the F-space M. Apply the Complementation Theorem of II.1 and find a subspace N of A such that $A = M \oplus N$. Then there exists one and only one linear form k over A such that

$$xk = \begin{cases} xh & \text{for } x \text{ in } M \\ 0 & \text{for } x \text{ in } N \end{cases}$$

and we see that $k^M = h$. Thus mapping f upon f^M induces the desired isomorphism of $L(A)/E(M)$ upon the whole space $L(M)$; and this completes the proof.

Corollary 3: *If M is a subspace of the F-space A, then*

$$r[L(A)] = r[L(M)] + r[L(A/M)]$$

and

$$r[E(M)] = \begin{cases} r(A/M) & \text{if } A/M \text{ has finite rank} \\ c^{r(A/M)} & \text{if } A/M \text{ has infinite rank and } c \text{ is the number of elements} \\ & \text{in the field } F. \end{cases}$$

This one deduces readily from the preceding proposition and Theorem 1, if one remembers that isomorphic linear manifolds have equal rank and that $r(A/B) + r(B) = r(A)$ for every subspace B of an F-space A (see II.2, Special Rank Formula).

Proposition 2: *If M is a subspace of the F-space A, then $M = S[E(M)]$.* PROOF: We have already pointed out that $M \leqslant S[E(M)]$. Suppose now that the element x in A does not belong to M. Form the subspace $M + Fx$. Note that $Fx \cap M = 0$, since x is not in the subspace M. Apply the Complementation Theorem of II.1 on the subspace $M \oplus Fx$ and find a subspace N of A such that $A = [M \oplus Fx] \oplus N$. Then it is possible to represent every element in A in one and only one way in the form

(*) $\quad a = a' + f(a)x + a''$ with a' in M, $f(a)$ in F and a'' in N.

The mapping h defined by the rule

$$ah = f(a) \qquad \text{for } a \text{ in } A \text{ and } f(a) \text{ from the representation (*)}$$

is a linear form over A which satisfies in particular $Mh = 0$ and $xh = 1$. Thus h belongs to $E(M)$ and hence x does not belong to $S[E(M)]$. Thus no element outside of M belongs to $S[E(M)]$; and this proves our contention $M = S[E(M)]$.

REMARK 2: As an illustration of the preceding results consider some hyperplane H of the F-space A. Then, by definition, $r(A/H) = 1$; and it follows from Corollary 3 that $r[E(H)] = 1$. Thus "hyperplanes satisfy just one equation," since all the equations satisfied by a hyperplane are multiples of one of them. It follows from Proposition 2 that $H = S[E(H)]$ so that hyperplanes are "defined by one equation."

Proposition 3: (*a*) *If the subspace T of $L(A)$ has finite rank, then $r(T) = r[A/S(T)]$.*

(*b*) *The following properties of the subspace T of $L(A)$ are equivalent.*

(I) *T has finite rank.*

(II) *$A/S(T)$ has finite rank.*

(III) *If the subspace X of $L(A)$ is part of T, then $X = E[S(X)]$.*

(*c*) *$X = E[S(X)]$ for every subspace X of $L(A)$ if, and only if, the F-space A has finite rank.*

We precede the proof of this proposition by a proof of the following lemma.

(3.*d*) *If* $f_1, \cdots, f_i$ *are finitely many elements in* $L(A)$, *then*

$$r\left[A/S\left(\sum_{j=1}^{i} f_jF\right)\right] \leqslant i.$$

PROOF: This lemma is certainly true for $i = 0$, since $S(0) = A$. Thus we may assume its validity for $i - 1$ in order to prove it for some positive i. The following fact is easily verified:

$$S\left(\sum_{j=1}^{i} f_jF\right) = S(f_1) \cap \cdots \cap S(f_i).$$

Next we note that

$$r\left[A/S\left(\sum_{j=1}^{i} f_jF\right)\right] = r\left[A/S\left(\sum_{j=1}^{i-1} f_jF\right)\right] + r\left[S\left(\sum_{j=1}^{i-1} f_jF\right)/S\left(\sum_{j=1}^{i} f_jF\right)\right],$$

as follows from the Rank Formulas of II.2, since

$$S\left(\sum_{j=1}^{i} f_jF\right) \leqslant S\left(\sum_{j=1}^{i-1} f_jF\right).$$

But it follows from the Isomorphism Laws [II.1] that

$$S\left(\sum_{j=1}^{i-1} f_jF\right)/S\left(\sum_{j=1}^{i} f_jF\right) = [S(f_1) \cap \cdots \cap S(f_{i-1})]/[S(f_1) \cap \cdots \cap S(f_i)]$$
$$\sim \left[S\left(\sum_{j=1}^{i-1} f_jF\right) + S(f_i)\right]/S(f_i);$$

and this latter F-space has at most rank 1, since $A/S(f)$ has for $f \neq 0$ exactly rank 1 as such an f maps A upon the totality of elements in F. This completes the induction and the proof of (3.*d*).

PROOF OF PROPOSITION 3:
If the subspace T of $L(A)$ has finite rank n, then $T = \sum_{i=1}^{n} t_iF$ where the elements $t_1, \cdots, t_n$ form a basis of T. It follows from Lemma 3.*d* that

(a^*) $r[A/S(T)] \leqslant r(T)$ *if the subspace* T *of* $L(A)$ *has finite rank.*

Assume now that the subspace T of $L(A)$ has finite rank. Then it follows from (a^*) that $r[A/S(T)]$ is finite; and thus we have shown that (I) implies (II).

Assume next that $A/S(T)$ has finite rank; and consider a subspace X of $L(A)$ such that $X \leqslant T$. Naturally $S(T) \leqslant S(X)$ so that $A/S(X)$ has finite rank too. We note that $X \leqslant E[S(X)]$; and it follows from Corollary 3 that

$$r(X) \leqslant r(E[S(X)]) = r[A/S(X)].$$

Thus $r(X)$ is finite; and it follows from (a^*) that $r[A/S(X)] \leqslant r(X)$. Hence $r(X) = r[A/S(X)] = r(E[S(X)])$. But X is a subspace of the subspace $E[S(X)]$ of finite rank; and now it follows from the Rank Formulas [II.2] that $X = E[S(X)]$. Thus we have shown that (III) is a consequence of (II). Incidentally we have completed the proof of (a). For we have shown that $A/S(T)$ has finite rank whenever T has finite rank; and we have shown $r(T) = r(E[S(T)])$ whenever $A/S(T)$ has finite rank.

Assume finally that the subspace T of $L(A)$ meets requirement (III). If X is a subspace of T, then we infer from Corollary 3 that the rank of $X = E[(S(X)]$ is not countably infinite (since $1 < c$ implies $\aleph < c^{\aleph}$; see Appendix S). But if the rank of T were infinite, then T would contain a subspace of countably infinite rank, since an infinite basis contains a countably infinite set of independent elements. But we just pointed out that T cannot contain a subspace of countably infinite rank. Hence $r(T)$ is finite. Thus we have shown that (III) implies (I); and this completes the proof of (b).

If we apply (b) on $T = L(A)$ and note that $S[L(A)] = S[E(O)] = 0$ by Proposition 2, then we see that (c) is an immediate consequence of (b).

When proving that (III) implies (I), we used a type of argument that has first been used by Dedekind when he proved the impossibility of obtaining a complete Galois correspondence for infinite algebraical extensions.

An alternative proof of the fact that $X = E[S(X)]$ for every subspace X of $L(A)$ provided $r(A)$ is finite may be constructed as follows. If $r(A)$ is finite, then it follows from Theorem 2 that we may identify A and $L^2(A)$. But then the S-operation from $L(A)$ to A may be identified with the E-operation from $L(A)$ to its adjoint space $L[L(A)]$ and the E-operation from A to $L(A)$ may be identified with the S-operation from the adjoint space $L[L(A)]$ to $L(A)$. Our contention is now an almost immediate consequence of Proposition 2.

It is customary to term *duality* a one-to-one mapping of one partially ordered set upon some other partially ordered set which inverts the partial ordering. The system of subspaces of a linear manifold is a partially ordered set with respect to inclusion. Now we may restate the main results of Propositions 2 and 3 in the following form.

Theorem 3: *If the F-space A has finite rank, then the mappings E and S* *constitute reciprocal dualities between the partially ordered sets of subspaces of A and $L(A)$ respectively.*

Applying Corollary 3 and remembering the complete symmetry between A, E, and $L(A)$, S we obtain readily the fundamental

Rank Formula: *If the F-space A has finite rank n, then*

$$n = r(M) + r[E(M)] \qquad \textit{for subspaces } M \textit{ of } A,$$
$$n = r(N) + r[S(N)] \qquad \textit{for subspaces } N \textit{ of } L(A).$$

(Note $r(A/M) = r(A) - r(M)$ and $r(A) = r[L(A)]$, since $r(A)$ is finite.)

APPENDIX I

Application to Systems of Linear Homogeneous Equations

Consider the system

$$\sum_{i=1}^{n} x_i a_{ik} = 0 \text{ for } k = 1, \cdots, m \tag{0}$$

of m linear homogeneous equations in the n unknowns x_i with coefficients a_{ik} in the field F. If we denote by S the system of all n-tuplets of numbers $x_1, \cdots, x_n$ in F which satisfy this system (0), then we make immediately the following observations:

If the n-tuplets $(x_1, \cdots, x_n)$ and $(y_1, \cdots, y_n)$ belong to S, so does their sum $(x_1 + y_1, \cdots, x_n + y_n)$; and if t is a number in F, then the n-tuplet $(tx_1, \cdots, tx_n)$ belongs to S likewise.

This we may restate in the language, introduced in II.1, in the following form: S is a subspace of the F-space (F,n) of all n-tuplets of elements in F; and we note immediately that (F,n) has rank n. The subspace S naturally has a rank not exceeding n; and this is usually expressed by saying that all the solutions of (0) may be derived from $r(S)$ of them [or even sometimes by saying that (0) has $r(S)$ solutions].

Suppose now that $b_1, \cdots, b_m$ are elements in F. Then it is fairly obvious that every solution of (0) is likewise a solution of the equation

$$\sum_{i=1}^{n} x_i \sum_{k=1}^{m} a_{ik} b_k = 0;$$

and these new equations are "derived" from the original ones. The system E of all the equations derived from the original ones may again be considered as a subspace of an F-space of rank n; only this time multiplication will be right multiplication. This will be connected best with the results of II.3 by making the following observations:

We denote by a_k the linear form over the F-space (F,n) which maps the n-tuplet $(x_1,\cdots, x_n)=x$ upon the n-tuplet $xa_k=\sum_{i=1}^{n} x_i a_{ik}$—that this is really a linear form is easily seen. What we termed E just now, is then nothing but the subspace $\sum_{i=1}^{m} a_i F$ of $L[(F,n)]$. What we termed S is then [in the notation of II.3] just the subspace $S(E)$ of (F,n). Since (F,n) has finite rank n, it follows from the results of II.3 that

$$E=E[S(E)] \text{ and } n=r(S)+r(E);$$

in other words:

The system of all equations derived from (0) is just the system of all equations satisfied by all solutions of (0), and n is the sum of the rank of the system of solutions of (0) plus the rank of the system of equations derivable from (0).

It is, of course, easy enough to obtain these results by the method of elimination, i.e., by eliminating one variable from the first equation and substituting the result in the remaining system of equations which leads us to a system of $m-1$ equations in $n-1$ unknowns.

The customary normalizations of a system of linear equations are now obtained by the following procedure. Select any one basis of the space (F,n) which contains a basis of the subspace S. If $s_1,\cdots, s_n$ is a basis of (F,n) which contains the basis $s_1,\cdots, s_k$ of S, then the linear forms $f_1,\cdots,f_k$ defined by the rule $s_j f_i=1$ or 0 according as j equals or does not equal i form a basis of E. Now one sees that by a change of coordinates the original system of linear equations has been transformed into the system of equations: $y_i=0$ for $i=1,\cdots, k$.

APPENDIX II

Paired Spaces

The symmetry between a linear manifold (of finite rank) and its adjoint becomes very much clearer, if we introduce the following notions which do not give any preference to either of the two manifolds.

Suppose that the following structures are given: a field F, an F-space A (which admits the elements in F as left-multipliers) and an F-space B (which admits the elements in F as right-multipliers). There is furthermore given a product ab with the following properties:

(*a*) For every element a in A and for every element b in B the product ab is a uniquely determined element in F.

(*b*) $(a' + a'')b = a'b + a''b$, $a(b' + b'') = ab' + ab''$ for every a,a',a'' in A and every b,b',b'' in B.

(*c*) $x(ab) = (xa)b$, $(ab)x = a(bx)$ for a in A, b in B, x in F.

The notations used in II.3 make it clear that an F-space A and its adjoint space $L(A)$ form such a pair of spaces; and that likewise the second adjoint $L^2(A)$ and the adjoint $L(A)$ form such a pair.

That these examples are quite typical, becomes clear from the following simple remarks:

(*d*) If b is an element in B, then the mapping of a in A upon ab constitutes a linear form over A; and if a is an element in A, then the mapping of b in B upon ab constitutes a linear form over B.

Whenever the F-spaces A and B are connected by some product ab which meets requirements (*a*), (*b*), (*c*), then we shall term them *paired F-spaces*. We shall denote this structure by A,B or by (B,F,A).

Theorem: *The following properties of paired F-spaces A,B of finite rank are equivalent.*

(I) $Ab = 0$ *implies* $b = 0$ *and* $aB = 0$ *implies* $a = 0$.

(II) $Ab = 0$ *implies* $b = 0$ *and every linear form over A is induced by an element in B.*

(III) *Every linear form over B is induced by an element in A and* $aB = 0$ *implies* $a = 0$.

(IV) *Every linear form over A is induced by an element in B and every linear form over B is induced by an element in A.*

PROOF: Assume first the validity of (I). If the elements b' and b'' induce the same linear form over A, then $ab' = ab''$ for every a in A. Hence $A(b' - b'') = 0$ and this implies $b' - b'' = 0$ or $b' = b''$ so that we may identify B with the space of induced linear forms. Hence B is a subspace of $L(A)$ such that [by (I)] $S(B) = 0$ and consequently we have [by II.3, Proposition 3] $B = E[S(B)] = E(0) = L(A)$. This shows that (II) is a consequence of (I); and that (III) is a consequence of (I), may be seen likewise.

Next we prove:

(*e*) *If every linear form over A is induced by an element in B, then* $a = 0$ *is a consequence of* $aB = 0$.

From $aB = 0$ and our hypothesis we infer that a belongs to $S[L(A)]$. But we have $S[L(A)] = S[E(0)] = 0$ by II.3, Proposition 2; and thus $a = 0$ is a consequence of $aB = 0$.

From (*e*) we infer that (I) is a consequence of (II); and that (I) is a consequence of (III), follows from reasons of symmetry. Thus (I), (II), (III) are equivalent.

If finally (IV) is true, then it follows from (*e*) that (III) is true. If conversely the equivalent properties (II) and (III) are valid, then so is [trivially] (IV). This completes the proof.

REMARK: Notice that the finiteness of the rank has been used only in part of the proof.

Corollary: *If the paired F-spaces A,B of finite rank meet the equivalent properties* (I) *to* (IV) [*of the preceding theorem*], *then each of them is the adjoint space of the other one and they have equal rank.*

This follows from a remark made in the course of the proof and from II.3, Theorem 1.

II.4. The Adjunct Space

If A is an F-space, then we obtain its adjoint space by considering linear forms which map A into F. It is natural to ask what happens if we map F into A; and this we propose to do now.

A linear anti-form f maps every element x in F upon a uniquely determined element xf in A, subject to the following rules:

(*a*) $(x + y)f = xf + yf$ for x,y in F.

(*b*) $(xy)f = x(yf)$ for x,y in F.

If f and g are linear anti-forms, then we define their sum $f + g$ by the rule:

(*c*) $x(f + g) = xf + xg$ for every x in F.

It is easily seen that sums of anti-forms are themselves anti-forms, and that the system $\mathbf{A} = \mathbf{A}(F,A)$ of all anti-forms is an abelian group with respect to addition.

If f is a linear anti-form, and if x is a number in F, then we define their product xf by the rule:

(*d*) $y(xf) = (yx)f$ for every element y in F.

A direct computation shows that the product xf is likewise a linear anti-form; and one verifies now easily that $\mathbf{A} = \mathbf{A}(F,A)$ is also an F-space which we call *the adjunct space of* A. We note that multiplication by elements in F is left-multiplication both in A and in its adjunct space.

Theorem: *Mapping the linear anti-form f upon the element $1f$ (where 1 is the field identity of F) we obtain an isomorphism of the F-space $\mathbf{A}$ upon the F-space A.*

PROOF: The mapping $f \to 1f$ is certainly single-valued. If $1f = 1g$, then we have for every x in F by (*b*)

$$xf = (x1)f = x(1f) = x(1g) = (x1)g = xg \text{ or } f = g;$$

and thus the mapping is one to one. If a is any element in A, then mapping x in F upon xa in A we obtain a linear anti-form f such that $1f = a$; and we have shown that our correspondence is a one-to-one mapping of **A** upon A. If f,g are anti-forms, then we have by (*c*)

$$1(f + g) = 1f + 1g$$

so that our mapping is additive. If x is in F and f is in **A**, then we have by (*d*)

$$1(xf) = (1x)f = (x1)f = x(1f)\,;$$

and thus we have verified that our mapping is isomorphic.

The isomorphism of our theorem is a natural isomorphism; and so linear manifolds and their adjunct spaces are essentially identical. This is the reason why the adjunct space gives little new information whereas the adjoint space was essentially distinct from the original space; and so its study yielded results. However, the preceding theorem will prove useful in later applications [V.2].

APPENDIX III

Fano's Postulate

In later parts of our investigation we shall have occasion to exclude the case where the field F [of coordinates] has characteristic 2 [$+1 = -1$]. A geometrical characterization of this fact may be obtained as follows.

The four points A, B, C, D in the linear manifold M are said to *form a quadrangle*, if they are coplanar [so that $A + B + C + D$ has rank 3] whereas no three of them are collinear [so that

$$A + B + C = A + B + D$$
$$= A + C + D = B + C + D$$

has rank 3]. Then it follows from the Rank Formulas of II.2 that the lines $A + B$ and $C + D$ meet in a point E_1, the lines $A + C$ and $B + D$ meet in a point E_2 and the lines $A + D$ and $B + C$ meet in a point E_3. These points E_i are called the diagonal points of the quadrangle; and it is not difficult to verify that $A, B, C, D, E_1, E_2, E_3$ are seven distinct points.

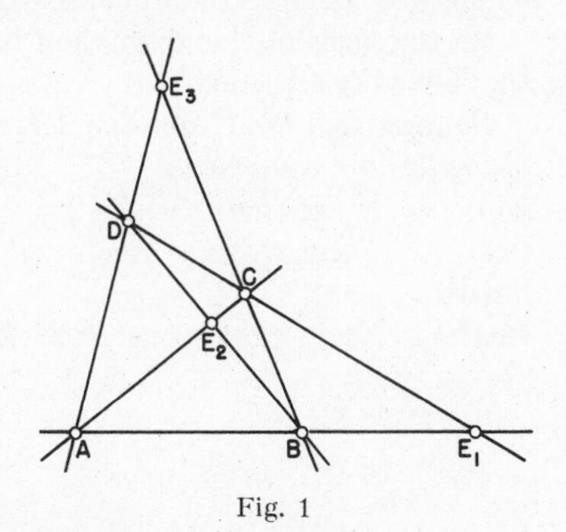

Fig. 1

The soc. Postulate of Fano excludes the possibility that the three diagonal points are collinear; and it is our objective to give an algebraical characterization of this geometrical property.

Suppose now that the points A,B,C,D in the F-space M form a quadrangle. Then $D = Fd$ belongs to the plane $A + B + C$; and consequently there exist elements a,b,c in A,B,C respectively such that $d = a + b + c$. If c were 0, then D would be on the line $A + B$ which is impossible; and likewise we see that a and b are not 0 either. Hence

$$A = Fa, \; B = Fb, \; C = Fc.$$

The points E_i are now determined as follows. Since

$$E_1 = (Fa + Fb) \cap (Fc + Fd)$$

and since $d = a + b + c$, it follows that $E_1 = F(a + b)$. (It is immediately clear that $F(a + b) \leqslant E_1$; and their equality is a consequence of the equality of their ranks.) Likewise we find that $E_2 = F(a + c)$ and that $E_3 = F(b + c)$.

Since the points E_i are all distinct, they are collinear if, and only if, $E_1 \leqslant E_2 + E_3$; and this is the case if, and only if, there exist numbers u,v in F such that

$$a + b = u(a + c) + v(b + c).$$

Since a,b,c are three independent elements in M, this last condition is equivalent to the validity of the following equations:

$$1 = u, \; 1 = v, \; 0 = u + v \text{ or } 0 = 1 + 1$$

which just asserts that the characteristic of the field F is 2.

On the basis of this discussion it is quite easy to verify the validity of the following assertion.

Suppose that the F-space M has rank not less than 3. Then the following properties are equivalent.

(I) *F has characteristic 2.*

(II) *There exists a quadrangle in M whose diagonal points are collinear.*

(III) *Every quadrangle in M has collinear diagonal points.*

CHAPTER III

Projectivities

The linear manifold (F,A) may be considered as the algebraical substratum of the projective geometry which consists of the totality of subspaces of the F-space A; and in particular what is called in classical language "the choice of a system of coordinates for the projective geometry" is nothing but a choice of some definite basis of the linear manifold A over F. By way of contrast we may say that in algebraical discussions of the linear manifold (F,A) the elements in A are the basic elements whereas in projective-geometrical discussions we have to consider the subspaces as the building stones. In Chapter II we have still emphasized the algebraical point of view; but from now on the point of view of projective geometry will move to the foreground.

The first problem arising in this context is the question as to the relations between algebraical and geometrical structure of the linear manifold (F,A). Typical in this direction is the question whether the field F [from which the "coordinates" are taken] is not only an algebraical, but also a projective-geometrical invariant. This question we are going to answer in the affirmative, provided the rank of (F,A) is at least 3; and it may be of historical interest that this question has been raised and answered only in the course of the last half- or quarter century.

If we want to relate algebraical and projective invariants, then we have to relate the algebraical isomorphisms of the linear manifold (F,A) [usually known under such names as linear and semi-linear transformations] with the geometrical isomorphisms, the soc. projectivities. This will lead us to certain theorems that are closely related to what used to be called the fundamental theorem of projective geometry. But since in classical discussions one restricted oneself usually to the study of autoprojectivities, mostly of a special nature, there will arise a variety of theorems taking the place of these classical results.

For a discussion of these problems in a more general framework the reader may consult R. Baer: A unified theory of projective spaces and finite abelian groups, *Transactions Amer. Math. Soc.*, vol. 52 (1942), pp. 283-343.

III.1. Representation of Projectivities by Semi-linear Transformations

A projectivity of the linear manifold (F,A) *upon the linear manifold* (G,B) is a mapping σ of the subspaces of A upon the subspaces of B with the following properties:

(*a*) If X is a subspace of A, then its image $X\sigma$ is a uniquely determined subspace of B.

(*b*) $X \leqslant Y$ if, and only if, $X\sigma \leqslant Y\sigma$.

(*c*) $X = Y$ if, and only if, $X\sigma = Y\sigma$.

(*d*) To every subspace Z of B there exists one [and only one] subspace Z' of A such that $Z'\sigma = Z$.

In short: a projectivity is a one-to-one, order preserving mapping of the partially ordered [by inclusion] set of all the subspaces of A upon the partially ordered set of all the subspaces of B. It is clear that the inverse of a projectivity is likewise a well determined projectivity.

The conditions (*a*) to (*d*) are obviously not independent. Thus (*c*) may be derived from (*b*); and there are some more sophisticated relations. It is very important to note that the linear manifolds A and B under consideration may be defined over different fields F and G respectively. It is one of our principal aims to derive the isomorphy of F and G whenever possible from the projective equivalence of A and B.

The reader ought to compare the concept of projectivity with the concept of isomorphism [II.1]. The former deals only with subspaces and we don't have to mention the field [of coordinates] whereas the elements and the field of coordinates are essential elements in the definition of isomorphy.

We begin by deriving a few simple and often used properties of projectivities.

(*e*) Projectivities preserve intersection and sum.

PROOF: If Φ is a set of subspaces of the linear manifold A, then the sum S of the subspaces in Φ is characterized [= uniquely determined] by the following two properties:

(′) $X \leqslant S$ for every X in Φ;

(″) If T is a subspace of A such that $X \leqslant T$ for every X in Φ, then $S \leqslant T$.

It is clear that the image $S\sigma$ meets the same requirements with respect to the subspaces $X\sigma$ with X in Φ; and thus $S\sigma$ is the sum of the subspaces in $\Phi\sigma$. The proof for the intersection is effected similarly.

(*f*) $0\sigma = 0$ and $A\sigma = B$.

This is obvious [see also (*e*)].

(g) Points are mapped on points.

For a point is a subspace P with the following two projectively invariant properties: $P \neq 0$; $0 < X \leqslant P$ implies $X = P$ [for X a subspace].

If T is an independent subset of the F-space A, then we term the set of points Ft with t in T likewise *an independent set of points.* We note that distinct elements t' and t'' in T lead to distinct points. The following simple lemma is a justification of this terminology and will also prove useful in the future.

Lemma 1: *The set T of elements in the F-space A is independent if, and only if, the following two conditions are satisfied by T:*

(α) *If t' and t'' are distinct elements in T, then t' does not belong to Ft''.*

(β) *If P is a point in the set T^* of points Ft with t in T, and if P' is the sum of all the other points in T^*, then $P \cap P' = 0$.*

PROOF: The necessity of condition (α) is quite obvious. If w is an element in $P \cap P'$ [using the notation of (β)], then $w = xt = \sum_{i=1}^{n} x_i t_i$ where the t, t_i are distinct elements in T and where the x, x_i belong to F. Now $w = 0$ is an immediate consequence of the independence of T. If conversely the conditions (α) and (β) are satisfied by T, then the points Ft with t in T are all distinct [by (α)]. If $0 = \sum_{i=1}^{n} f_i t_i$ where the t_i are distinct elements in T and where the f_i belong to F, then $f_i t_i$ belongs for every i to the intersection $Ft_i \cap \sum_{j \neq i} Ft_j$ which is 0 by (β). Hence $f_i = 0$, proving the independence of T.

REMARK: If t' and t'' are distinct elements, not 0, such that $Ft' = Ft''$, then the set T composed of t' and t'' has Property (β), but not Property (α). Thus this property is indispensable.

Proposition 1: *Projectivities preserve rank.*

PROOF: If σ is a projectivity of the F'-space A' upon the F''-space A'', and if S is a subspace of A', then we form first a basis T of S. This is an independent subset of A'. It follows from the Lemma 1 that the set T^* of points $F't$ with t in T contains as many elements as does T; and we note that S is the sum of the points in T^*. Consequently $S\sigma$ is the sum of the points in $T^*\sigma$. Form a set R such that $T^*\sigma$ is the set of points $F''r$ with r in R and such that distinct elements in R lead to distinct points in $T^*\sigma$. It follows from Lemma 1 that T^* has Property (β). Hence R has Properties (α) and (β); and it follows from Lemma 1 that R is independent and hence a basis of $S\sigma$. But R and T contain the same number of elements. Hence $r(S) = r(S\sigma)$.

It is, of course, impossible to show that projectively equivalent linear manifolds have the same field of coordinates. All we can hope for is that these fields are isomorphic; and this we shall be able to prove "almost always." For a proper analysis of this situation we need the following fundamental concept.

An algebraical isomorphism of the linear manifold (F,A) upon the linear manifold (G,B) is, in accordance with customary algebraical terminology, a one-to-one correspondence between the join of F and A and the join of G and B which preserves the basic operations: addition in A and in B, addition in F and in G, multiplication in F and in G and multiplication of the elements in F by elements in A [of the elements in G by elements in B]. In this fashion one is lead to the definition of a *semi-linear transformation of* (F,A) *upon* (G,B) [due to Corrado Segre] as a pair $\sigma = (\sigma',\sigma'')$ consisting of an isomorphism σ' of the additive group A upon the additive group B and an isomorphism σ'' of the field F upon the field G subject to the condition:

$$(fa)^{\sigma'} = f^{\sigma''}a^{\sigma'} \qquad \textit{for } f \textit{ in } F \textit{ and } a \textit{ in } A.$$

Note that we have used here the exponential way of writing the effect of a mapping as this is easier to visualize than the multiplicative one. Furthermore there will hardly ever be a danger of confusion, if we use the same symbol σ both for the isomorphism σ' of A upon B and the isomorphism σ'' of F upon G. Note that the above connecting formula then obtains the following simple form: $(fa)^\sigma = f^\sigma a^\sigma$. σ is called *linear*, if [$F = G$ and] $\sigma'' = 1$.

That the concept of a semi-linear transformation is more general than that of a linear transformation even if we restrict ourselves to just one F-space, may be seen from the following classical example: Let F be the field of all complex numbers, A the plane over F which consists of all the triplets (c_1,c_2,c_3) of complex numbers. Denote by $\bar{c}$ the conjugate complex number to c; and denote by σ the automorphism of A which maps (c_1,c_2,c_3) upon

$$(c_1,c_2,c_3)^\sigma = (\bar{c}_1,\bar{c}_2,\bar{c}_3).$$

The pair consisting of σ and the mapping $c \rightarrow \bar{c}$ is easily seen to be a semi-linear transformation. For a more detailed discussion of the "geometrical" differences between linear and semi-linear transformations, see III.2 below.

Proposition 2: *Semi-linear transformations induce projectivities.*

PROOF: If σ is a semi-linear transformation of the F-space A upon the G-space B, and if S is a subspace of the F-space A, then the set S^σ of all the elements s^σ with s in S is a subspace of B. [To verify this note that for g in G there is one and only one f in F such that $f^\sigma = g$ and that therefore

$gs^\sigma = f^\sigma s^\sigma = (fs)^\sigma$ belongs to S^σ whenever s is in S.] Hence the mapping of the subspace S of A upon the subspace S^σ of B is the desired "*induced*" projectivity of A upon B.

Proposition 3: *Suppose that A is an F-space and $1 < r(A)$.*

(*a*) *If the single-valued mapping σ of A into itself satisfies*

$$(a' + a'')^\sigma = a'^\sigma + a''^\sigma \qquad \textit{for } a', a'' \textit{ in } A$$

and

$$S^\sigma \leqslant S \qquad \textit{for every subspace } S \textit{ of } A,$$

then either $A^\sigma = 0$ [$\sigma = 0$] or else σ is a semi-linear transformation.

(*b*) *The mapping σ of A into itself is a semi-linear transformation satisfying $S^\sigma \leqslant S$ for every subspace S of A if, and only if, there exists a number $f \neq 0$ in F such that $a^\sigma = fa$ for every a in A.*

This proposition is important, since it provides a characterization of those transformations which induce the identity projectivity.

Proof: Suppose first that $f \neq 0$ is a number in F; and define $a^\sigma = fa$ for every element a in A. It is clear that this transformation σ is additive and satisfies $S = S^\sigma$ for every subspace S of A. If a is in A and x is in F, then we find

$$(xa)^\sigma = f(xa) = (fxf^{-1})fa = x^\sigma a^\sigma$$

where the mapping $x^\sigma = fxf^{-1}$ is an inner automorphism of the field F. Thus σ is a semi-linear transformation inducing the identity projectivity. [Note that σ is linear if, and only if, f commutes with every element in F.]

Assume next that the single-valued mapping σ of A into itself has the following properties:

$$(a' + a'')^\sigma = a'^\sigma + a''^\sigma \qquad \text{for } a', a'' \text{ in } A;$$

$$S^\sigma \leqslant S \qquad \text{for every subspace } S \text{ of the } F\text{-space } A.$$

It is clear that $0^\sigma = 0$. If $x \neq 0$ is an element in A, then it follows from the second condition that $(Fx)^\sigma \leqslant Fx$. Consequently x^σ belongs to Fx and there exists one and only one number $f(x)$ in F such that $x^\sigma = f(x)x$.

Assume now that x and y are independent elements in A. Then neither x nor y nor $x + y$ is 0; and we find therefore

$$f(x + y)x + f(x + y)y = f(x + y)[x + y] = (x + y)^\sigma = x^\sigma + y^\sigma$$
$$= f(x)x + f(y)y;$$

and it follows from the independence of x and y that

$$f(x) = f(x + y) = f(y).$$

Assume next that x and y are dependent elements in A neither of which is 0. From $1 < r(A)$ we infer the existence of an element z which is independent of both x and y; and it follows from what has already been shown that $f(x) = f(z) = f(y)$.

Thus we have shown that $f(x) = f(y)$ for any pair of elements x,y [neither of which is 0] in A. Denoting the common value of all these numbers $f(x)$ by f we have clearly $a^\sigma = fa$ for every a in A. If $f = 0$, then $\sigma = 0$; and if $f \neq 0$, then σ has been shown to be a semi-linear transformation. This completes the proof of our proposition.

Corollary 1: *If $Fx = Fy$, then there exists a semi-linear transformation σ such that $x^\sigma = y$ and $S^\sigma = S$ for every subspace S of A.*

This is an almost immediate consequence of Proposition 3.

Corollary 2: *Suppose that $1 < r(A)$. Then the semi-linear transformations σ and τ of the F-space A upon the G-space B induce the same projectivity if, and only if, there exists a number $g \neq 0$ in G such that $a^\sigma = ga^\tau$ for every a in A.*

Proof: σ and τ induce the same projectivity if, and only if, $\sigma\tau^{-1}$ induces the identity projectivity. It follows from Proposition 3 that this is equivalent to the existence of a number $f \neq 0$ in F such that $a^{\sigma\tau^{-1}} = fa$ for every a in A. But this last equation is satisfied if, and only if,

$$a^\sigma = (fa)^\tau = f^\tau a^\tau = ga^\tau$$

for every a in A where we let $g = f^\tau$.

The reader should find it easy to verify that Proposition 3 and Corollary 2 cease to be true, if we omit the hypothesis that $1 < r(A)$.

The First Fundamental Theorem of Projective Geometry

If the rank of the linear manifold A is at least 3, then every projectivity of A is induced by a semi-linear transformation.

[The group of theorems which is usually given the name "fundamental theorem of projective geometry" we shall discuss in III.2.]

The proof of this theorem will be effected in a number of steps. Assume that there are given an F-space A, a G-space B and a projectivity of A upon B which maps the subspace S of A upon the subspace S^* of B. We assume furthermore that $2 < r(A)$, though we shall use this hypothesis only towards the end of our proof.

It is a consequence of Proposition 1 that S and S^* have the same rank so that in particular $(Fx)^*$ for $x \neq 0$ is a point Gy. We prove:

(1) *If Fx and Fy are two distinct points in A, and if x' is an element in B such that $(Fx)^* = Gx'$, then there exists one and only one element $y' = h(x,x',y)$ in B such that $(Fy)^* = Gy'$ and $[F(x - y)]^* = G(x' - y')$.*

Proof: From $F(x - y) \leqslant Fx + Fy$ we deduce that

$$[F(x - y)]^* \leqslant (Fx)^* + (Fy)^*.$$

Since $F(x-y)$ and $[F(x-y)]^*$ are points, we have $[F(x-y)]^* = Gt$ where t necessarily belongs to $(Fx)^* + (Fy)^*$. From $(Fx)^* = Gx'$ we infer now the existence of a number g in G and of an element z in $(Fy)^*$ such that $t = gx' - z$. If g were 0, then t would be in $(Fy)^*$ so that $[F(x-y)]^* \leqslant (Fy)^*$. This would imply that $F(x-y) \leqslant Fy$ so that x would consequently belong to Fy. Hence $Fx = Fy$, an impossibility. Likewise we see that $z \neq 0$. Hence we may form $y' = g^{-1}z$. Then $y' \neq 0$ is in $(Fy)^*$ so that $(Fy)^* = Gy'$. Furthermore

$$[F(x-y)]^* = Gt = Gg^{-1}t = G(x'-y');$$

and so y' meets all our requirements.

Assume now that y'' is some element in B such that $(Fy)^* = Gy''$ and $[F(x-y)]^* = G(x'-y'')$. Then we have

$$Gy' = Gy'' \quad \text{and} \quad G(x'-y') = G(x'-y'').$$

Consequently there exist numbers u,v in G neither of which is 0 such that $y' = uy''$ and

$$v(x'-y'') = x'-y' = x'-uy''.$$

But Gx' and Gy'' are distinct points; and so x' and y'' are independent elements in B. Hence we deduce from $vx' - vy'' = x' - uy''$ that $v = 1$ and $v = u$; and we have consequently $y' = y''$, as we claimed.

The single-valued function $h(x,x',y)$ is defined for every pair of independent elements x and y in A and every x' in B such that $(Fx)^* = Gx'$. It depends, of course, on the projectivity S^* too. It will be convenient—and in accord with previous definitions—to let $h(x,x',0) = 0$, since $0^* = 0 = G0$.

(2) *$h(x,x',y) = y'$ if, and only if, $h(y,y',x) = x'$.*

PROOF: This is an immediate consequence of the uniqueness part of (1), since the defining equations for both $h(x,x',y) = y'$ and $h(y,y',x) = x'$ are the same, namely

$$(Fx)^* = Gx', \qquad [F(x-y)]^* = G(x'-y'), \qquad (Fy)^* = Gy'.$$

(3) *If x,y,z are three independent elements in A, then*

$$F(y-z) = [Fy + Fz] \cap [F(x-y) + F(x-z)].$$

PROOF: It is obvious that

$$F(y-z) \leqslant [Fy + Fz] \cap [F(x-y) + F(x-z)] = J.$$

If conversely j is an element in J, then we have by definition

$$j = ay + bz = d(x-y) + e(x-z) \qquad \text{with } a,b,d,e \text{ in } F.$$

Since x,y,z are independent, this implies $a = -d$, $b = -e$, $0 = d + e$ and hence $a = -b$ so that $j = a(y-z)$ belongs to $F(y-z)$. Thus $J \leqslant F(y-z)$; and this proves the desired equation.

(4) *If x,y,z are three independent elements in A, then $h(x,x',y) = y'$ and $h(x,x',z) = z'$ imply $h(y,y',z) = z'$.*

Proof: By hypothesis we are assured of the validity of the following equations:

$$(Fx)^* = Gx', \qquad (Fy)^* = Gy', \qquad (Fz)^* = Gz',$$
$$[F(x-y)]^* = G(x'-y'), \qquad [F(x-z)]^* = G(x'-z').$$

Thus the independence of the three elements x',y',z' is a consequence of

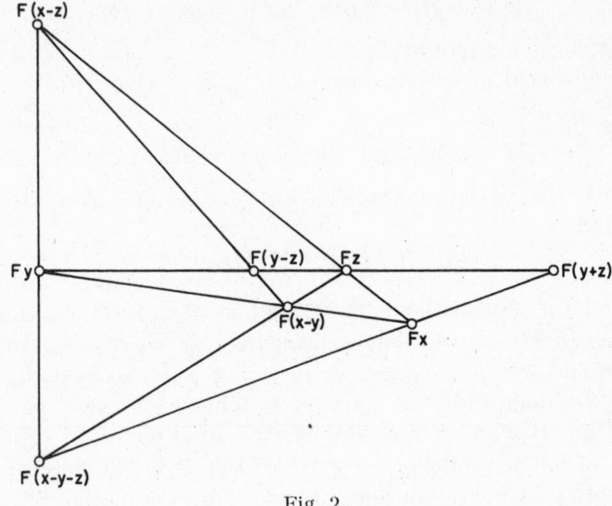

Fig. 2

the independence of the three elements x,y,z [see Lemma 1]. Hence we may apply (3) both on the elements x,y,z in A and on the elements x',y',z' in B. Consequently we find that

$$\begin{aligned}[F(y-z)]^* &= ([Fy + Fz] \cap [F(x-y) + F(x-z)])^* \\ &= [(Fy)^* + (Fz)^*] \cap [(F(x-y))^* + (F(x-z))^*] \\ &= [Gy' + Gz'] \cap [G(x'-y') + G(x'-z')] = G(y'-z').\end{aligned}$$

But the validity of the three equations

$$(Fy)^* = Gy', \qquad [F(y-z)]^* = G(y'-z'), \qquad (Fz)^* = Gz'$$

is equivalent to the statement $h(y,y',z) = z'$ by (1); and this we intended to prove.

(5) *If x,y,z are three independent elements in A, then*

$$F(x-y-z) = [F(x-y) + Fz] \cap [F(x-z) + Fy] \text{ and}$$
$$F(y+z) = [Fy + Fz] \cap [F(x-y-z) + Fx].$$

The proof may be left to the reader, as it is quite similar to that of (3) [see Figure 2]. Another way of arguing would be to show that the intersections on the right have rank not more than 1 and contain the points on the left.

(6) *If Fx is a point, if $(Fx)^* = Gx'$, and if $Fx \cap (Fy + Fz) = 0$, then*

$$h(x,x',y + z) = h(x,x',y) + h(x,x',z).$$

PROOF: We distinguish several cases.

Case 1: The elements x,y,z in A are independent.

We let $y' = h(x,x',y)$, $z' = h(x,x',z)$; and we note the validity of the following equations:

$$(Fx)^* = Gx', \qquad (Fy)^* = Gy', \qquad (Fz)^* = Gz';$$
$$[F(x - y)]^* = G(x' - y'), \qquad [F(x - z)]^* = G(x' - z').$$

The independence of the three elements x',y',z' is again a consequence of the independence of the three elements x,y,z. Thus we may apply (5) both on x,y,z and on x',y',z'. Consequently we have

$$\begin{aligned}[F(x - y - z)]^* &= ([F(x - y) + Fz] \cap [F(x - z) + Fy])^* \\ &= ([F(x - y)]^* + (Fz)^*) \cap ([F(x - z)]^* + (Fy)^*) \\ &= [G(x' - y') + Gz'] \cap [G(x' - z') + Gy'] = G(x' - y' - z');\end{aligned}$$

and using this result we find similarly that

$$\begin{aligned}[F(y + z)]^* &= ([Fy + Fz] \cap [F(x - y - z) + Fx])^* \\ &= [(Fy)^* + (Fz)^*] \cap [(F(x - y - z))^* + (Fx)^*] \\ &= [Gy' + Gz'] \cap [G(x' - y' - z') + Gx'] = G(y' + z').\end{aligned}$$

But the three equations $(Fx)^* = Gx'$, $[F(x - y - z)]^* = G(x' - y' - z')$, $[F(y + z)]^* = G(y' + z')$ are the three defining equations of $h(x,x',y + z)$ which is uniquely determined by (1); and thus we have $h(x,x',y + z) = y' + z'$ which proves our claim in this first case.

Case 2: The elements x,y,z are not independent.

If y or z is 0, then there is nothing to prove, since $h(x,x',0)$ is 0, by definition. Thus we may assume that neither y nor z is 0; and then it follows from our hypotheses that Fx and $Fy = Fz$ are two distinct points. But we made the fundamental hypothesis that the rank of A is at least three; and consequently there exists a point Fw such that the three points Fx, $Fy = Fz$, Fw are independent.

Then either $y + z = 0$ or else x, $y + z$, w are three independent elements. The elements x, $w + y$, z are certainly independent, since $Fy = Fz$ etc; and the elements x,w,y are independent. Applying the result of Case 1 various times and $h(x,x',0) = 0$ in case $y + z = 0$, we obtain now:

$$\begin{aligned}h(x,x',w) + h(x,x',y + z) &= h(x,x',w + y + z) = h(x,x',w + y) + h(x,x',z) \\ &= h(x,x',w) + h(x,x',y) + h(x,x',z);\end{aligned}$$

and this implies again the desired equation

$$h(x,x',y+z)=h(x,x',y)+h(x,x',z).$$

This completes the proof of (6).

Now we have made all the necessary preparations for the construction of the desired semi-linear transformation. Since the rank of A is at least three, we may select [in a variety of ways] three independent elements u,v,w in A. Since $(Fu)^*$ is a point, we may select [in a variety of ways] an element u' such that $(Fu)^* = Gu'$. We let $v' = h(u,u',v)$ and $w' = h(u,u',w)$; and we note that [as a consequence of (2) and (4)] we have

(7) *$h(x,x',y) = y'$ if x and y are any two [distinct] elements of the three elements u,v,w.*

We have, of course, $f(x,x',0) = 0$ for $x = u,v,w$. Suppose now that $t \neq 0$ is some element in A. Then Ft is a point which is different from at least two of the points Fu,Fv,Fw.

(8) *If Fx and Fy are two [distinct] points of the three points Fu,Fv,Fw, and if Ft is different from Fx and Fy, then $h(x,x',t) = h(y,y',t)$.*

Proof: Assume for convenience of language that $x = u$ and $y = v$. If Ft is not on the line $Fu + Fv$, then u,v,t are three independent elements. Let $h(u,u',t) = t'$. Then we have $h(u,u',v) = v'$ and $h(u,u',t) = t'$; and it follows from (4) that $h(v,v',t) = t'$, as we claimed. If Ft is on the line $Fu + Fv$, then it follows from our hypotheses that Ft is neither on the line $Fu + Fw$ nor on the line $Fw + Fv$; and the same argument as before [together with (7)] shows that $h(u,u',t) = h(w,w',t) = h(v,v',t)$; and this completes the proof.

Thus we have shown that, for every t in A, of the three functional values $h(u,u',t)$, $h(v,v',t)$, $h(w,w',t)$ at least two are defined; and that all that are defined have the same value which we denote by t^σ. Thus we can say:

(9) *If x is one of the three values u,v,w, then $h(x,x',t)$ is either undefined or else it is the uniquely determined value t^σ.*

Thus we have found a single-valued mapping σ of the elements in A upon elements in B. This mapping has the following properties.

(10) *$(Ft)^* = Gt^\sigma$ for every t in A.*

This is an obvious consequence of the definition of the function h in (1).

(11) *$(a+b)^\sigma = a^\sigma + b^\sigma$ for a,b in A.*

Proof: At least one of the points Fu,Fv,Fw is not on $Fa + Fb$, since u,v,w are independent. If Fu is not on $Fa + Fb$, then it follows from the definition of σ and from (6) that

$$(a+b)^\sigma = h(u,u',a+b) = h(u,u',a) + h(u,u',b) = a^\sigma + b^\sigma,$$

as we claimed.

(12) *σ is an isomorphism of [the additive group] A upon [the additive group] B.*

PROOF: If $t^\sigma = 0$, then we infer from (10) that $0 = Gt^\sigma = (Ft)^*$; and this implies $t = 0$. Thus σ is certainly an isomorphism of A into B [$a^\sigma = b^\sigma$ implies $(a - b)^\sigma = 0$, etc]. Suppose now that s is some element in B. Since $0^\sigma = 0$, we may assume that $s \neq 0$. Thus Gs is a point in the G-space B; and it follows from the properties of projectivities [see (d)] that there exists one and only one point T in the F-space A such that $T^* = Gs$. This point T is different of at least one of the points Fu, Fv, Fw. Thus we may assume without loss in generality that $Fu \neq T$. Then $Gu' \neq Gs$. Consider the point $G(u' + s)$ on the line

$$Gu' + Gs = (Fu)^* + T^* = [Fu + T]^*.$$

As before we deduce the existence of one and only one point P on $Fu + T$ such that $P^* = G(u' + s)$. Since $T^* = Gs \neq G(u' + s) = P^*$, we have likewise $T \neq P$; and now one sees easily that the point P on $Fu + T$ has the form $P = F(u + t)$ with t in T [$= Ft$]. We deduce from (10) and (11) that

$$G(u' + s) = P^* = [F(u + t)]^* = G(u + t)^\sigma = G(u^\sigma + t^\sigma) = G(u' + t^\sigma).$$

Hence $s - t^\sigma$ belongs to the intersection of P^* and $Gs = T^* = (Ft)^* = Gt^\sigma$; and this intersection is 0, since P and T are different points. Consequently $s = t^\sigma$; and we have shown that $A^\sigma = B$.

(13) *$S^* = S^\sigma$ for every subspace S of the F-space A.*

This is an almost immediate consequence of (10) and (12).

If t is an element, not 0, in A, and if $f \neq 0$ is an element in F, then we have $Gt^\sigma = (Ft)^* = [F(ft)]^* = G(ft)^\sigma$ by (10); and consequently there exists one and only one number $g(f,t) \neq 0$ in G such that $(ft)^\sigma = g(f,t)t^\sigma$. If we let $g(0,t) = 0$, then the defining formula

$$(ft)^\sigma = g(f,t)t^\sigma$$

holds for every f in F and for every $t \neq 0$ in A.

(14) *$g(f,x) = g(f,y)$ for every f in F and every x,y, not 0, in A.*

PROOF: We assume first that Fx and Fy are distinct points. Then we find that

$$\begin{aligned} g(f,x + y)x^\sigma + g(f,x + y)y^\sigma &= g(f,x + y)(x + y)^\sigma = [f(x + y)]^\sigma \\ &= [fx + fy]^\sigma = (fx)^\sigma + (fy)^\sigma \\ &= g(f,x)x^\sigma + g(f,y)y^\sigma. \end{aligned}$$

Since Fx and Fy are distinct points, so are $Gx^\sigma = (Fx)^*$ and $Gy^\sigma = (Fy)^*$. Hence x^σ and y^σ are independent ; and it follows that

$$g(f,x) = g(f,x + y) = g(f,y).$$

If the points Fx and Fy are not distinct, then there exists a point $Fz \neq Fx = Fy$. From what we have already shown, it follows that $g(f,x) = g(f,z) = g(f,y)$; and thus we have shown (14) in all generality.

We denote now the common value of all the $g(f,x)$ for $x \neq 0$ in A by f^σ. Thus σ is defined as a single-valued mapping of F into G such that

(15) *$(ft)^\sigma = f^\sigma t^\sigma$ for f in F and t in A.*

We show next that

(16) *σ is an isomorphism of the field F upon the field G.*

PROOF: If x and y are elements in F, and if t is some element, not 0, in A, then we deduce from (15) that

$$(x + y)^\sigma t^\sigma = [(x + y)t]^\sigma = [xt + yt]^\sigma = (xt)^\sigma + (yt)^\sigma$$
$$= x^\sigma t^\sigma + y^\sigma t^\sigma = (x^\sigma + y^\sigma)t^\sigma;$$

and $(x + y)^\sigma = x^\sigma + y^\sigma$ is a consequence of $t^\sigma \neq 0$.

Likewise we see that

$$(xy)^\sigma t^\sigma = [(xy)t]^\sigma = [x(yt)]^\sigma = x^\sigma(yt)^\sigma = x^\sigma(y^\sigma t^\sigma) = (x^\sigma y^\sigma)t^\sigma$$

so that $(xy)^\sigma = x^\sigma y^\sigma$; and now it is readily seen that σ is an isomorphism of the field F into the field G. If finally z is some element in G, $t \neq 0$ some element in A, then there exists because of (12) an element s in A such that $s^\sigma = zt^\sigma$. It follows from (13) that

$$(Fs)^* = Gs^\sigma = Gzt^\sigma = Gt^\sigma = (Ft)^* \text{ and hence } Fs = Ft.$$

Consequently there exists an element k in F such that $s = kt$; and we deduce from (15) that $zt^\sigma = s^\sigma = (kt)^\sigma = k^\sigma t^\sigma$; and $z = k^\sigma$ is a consequence of $t^\sigma \neq 0$. Thus $F^\sigma = G$; and this completes the proof.

It is a consequence of (12), (15), (16) that σ is a semi-linear transformation of the F-space A upon the G-space B; and it follows from (13) that this semi-linear transformation induces the given projectivity. This completes the proof of our theorem.

REMARK 1: The following considerations show the indispensability of the condition $2 < r(A)$. If A is a linear manifold of rank 2 [a line], then the only subspaces of A, apart from 0 and A, are points; and different points are never included in each other. Consequently every one-to-one correspondence mapping the set of all the points in A upon the set of all the points in the linear manifold B of rank 2 leads to a projectivity of A upon B by mapping 0 upon 0 and A upon B. It is clear that not all these projectivities may be derived from semi-linear transformations. An example illustrating this would be the following one: Let A be the line with real coefficients and B the line with complex coefficients. Both contain a continuum of points and thus the above construction may be applied. On the other hand the fields of real and complex numbers are not isomorphic; and so it is impossible to construct a semi-linear transformation of A upon B. Another example of this type is the following one: Let L be the line with rational coefficients. This line carries a countable infinity of points. There exists a continuum of permutations of these points; and consequently L possesses a continuum of auto-projectivities. On the other hand every semi-linear transformation of L upon itself is actually linear; and their number is countably infinite. Consequently there exist auto-projectivities of L which are not induced by semi-linear transformations. Similar examples of finite and infinite lines can be constructed with great ease.

REMARK 2: The impossibility of proving that every projectivity can be induced by a linear transformation will be discussed in more detail in the next section. At present we want to point out only that projectively equivalent spaces may be defined over different, though isomorphic fields in which case no linear transformation between them could exist. Furthermore the reader ought to recall the example of a semi-linear transformation of the complex plane onto itself which we gave earlier. This transformation induces a projectivity; and it is easy to see that no linear transformation induces this projectivity.

Note that the expression "projective equivalence of linear manifolds" is just another way of saying that there exists a projectivity mapping the one upon the other.

Projective Structure Theorem: *The F-space A and the G-space B are projectively equivalent if, and only if, they satisfy the following conditions:*
(*a*) $r(A) = r(B)$.
(*b*) $r(A) = r(B) = 1$; *or* $r(A) = r(B) = 2$ *and F and G contain the same number of elements*; *or* $2 < r(A) = r(B)$ *and F and G are isomorphic fields.*

PROOF: That projectively equivalent linear manifolds have equal rank, is a consequence of Proposition 1. This shows the necessity of condition (*a*).

It is trivial that any two points are projectively equivalent. If the F-space A is a line $[r(A) = 2]$, then A possesses a basis u,v; and every point, not Fv, may be represented in one and only one way in the form $F(u + xv)$ with x in F. If c is the number of elements in F, then this implies that A possesses just $c + 1$ points. Hence A and the line B possess the same number of points if, and only if, F and G contain the same number of elements; and now it is clear that lines are projectively equivalent if, and only if, the corresponding fields contain the same number of elements [see also Remark 1].

If finally the F-space and the G-space B have equal rank greater than 2, then every projectivity of A upon B is induced by a semi-linear transformation of A upon B; and thus projective equivalence implies isomorphy of the fields F and G. If conversely F and G are isomorphic, then we form a basis A_0 of A and a basis B_0 of B. From $r(A) = r(B)$ we deduce the existence of a one-to-one correspondence τ mapping A_0 upon B_0; and from the isomorphy of F and G we deduce the existence of some definite isomorphism σ of F upon G. Mapping now the "general" element $\sum_{a \in A_0} f(a)a$ in A upon $\sum_{a \in A_0} f(a)^\sigma a^\tau$ in B we obtain a semi-linear transformation of A upon B; and this transformation induces a projectivity of A upon B [by Proposition 2]. This completes the proof.

THE GROUP OF AUTO-PROJECTIVITIES of a linear manifold A of rank not less than 3 can now be represented as follows. Denote by Π the group of auto-projectivities of A; and denote by Λ the group of semi-linear transformations of A. If σ is in Λ, then σ induces by Proposition 2 an auto-projectivity σ^* of A. The mapping of σ upon σ^* is clearly homomorphic $[\sigma^*\tau^* = (\sigma\tau)^*]$ and it is a mapping of the group Λ upon the full group $\Pi = \Lambda^*$ [by the First Fundamental Theorem]. The mappings σ in Λ with $\sigma^* = 1$ are by Proposition 3 just the "multiplications" $a^\sigma = fa$ for every a in A; they form a normal subgroup N of Λ. Thus we see that Π *is essentially the same as* Λ/N.

HISTORICAL NOTE: The first satisfactory proof of a theorem of the type of our First Fundamental Theorem of Projective Geometry has been given by E. Kamke.

APPENDIX I

Projective Construction of the Homothetic Group

The linear manifold (F,A) determines a projective geometry, namely that of the subspaces of A. It is the content of the First Fundamental Theorem of Projective Geometry [III.1] that the structure of this projective geometry determines completely the algebraical structure of the linear manifold. Consequently it must be possible to reconstruct the linear manifold (F,A) from the projective geometry of its subspaces. This problem can be attacked on two essentially different levels: a deeper one where the projective geometry is given abstractly, where nothing is supposed to be known concerning the underlying linear manifold and where the construction of this underlying linear manifold is the principal objective; and the considerably simpler problem where the projective geometry is given as the projective geometry of subspaces of some definite linear manifold and where we want to prove the essential identity of the linear manifold with some structure derived constructively from the projective geometry. The former of these problems will occupy us in Chapter VII whereas in the present appendix we shall be concerned with the latter problem only. But many of the constructions and arguments will have their counterpart [and will serve as motivation] for operations undertaken in Chapter VII.

What we shall construct now is not so much the linear manifold itself, but a certain group of transformations which is closely connected with it. If $f \neq 0$ is an element in F and w is some element in A, then the transformation σ defined by the rule

$$x^{\sigma} = fx + w \qquad \text{for } x \text{ in } A$$

is called *a homothetic transformation of the linear manifold* (F,A); and it is clear that the totality of homothetic transformations is a group: *the homothetic group* H *of* (F,A). This group has two subgroups of particular interest for us: the group T of all the *translations* $[x^{\sigma} = x + w]$ and the group Δ of all the *dilatations* $[x^{\sigma} = fx]$. By mapping w in A upon the translation $x^{\sigma} = x + w$, we obtain a [natural] isomorphism of the additive group of A upon the multiplicative group of translations; and by mapping $f \neq 0$ in F upon the dilatation $x^{\sigma} = f^{-1}x$ we obtain a [natural] isomorphism of the multiplicative group of F upon the multiplicative group of dilatations. [The reader ought to check these simple remarks.]

It is clear that every homothetic transformation η of (F,A) may be written in one and only one way in the form $\eta = \delta\tau$ where δ is a dilatation and τ a translation. Suppose now that β is the translation $x^{\beta} = x + b$; and that $x^{\sigma} = fx + w$ is a homothetic transformation. Then one verifies by direct computation that

$$x^{\sigma^{-1}\beta\sigma} = x + fb \qquad \text{for } x \text{ in } A.$$

Thus the group T of translations is a normal subgroup of the homothetic group H; and the inner automorphisms of H induce in T a group of automorphisms which is essentially the same as the group Δ of dilatations and as the multiplicative group of F.

From the preceding remarks it becomes apparent that the structure of the linear manifold (F,A) is known completely once we know the structure of the extension H of T by Δ; and thus it will be our objective to give a projective construction of the homothetic group with its subgroups of translations and dilatations.

We shall construct the desired groups as groups of projectively defined transformations of projectively defined objectives. Naturally we have to define the objectives first. This will be done in two steps.

A BASIC SEXTET is an *ordered* set of six points R,S,T; R',S',T' subject to the following requirements:

$$\text{(I.1)}\quad \begin{cases} \text{The points } R,S,T \text{ are independent [span a plane].} \\ \text{If } X,Y,Z \text{ is a permutation of the three letters } R,S,T, \text{ then} \\ \quad X+Y=Y+Z'=Z'+X. \\ R'+S'=S'+T'=T'+R'. \end{cases}$$

It is clear that the three points R',S',T' span a line L', and that this line L' meets the three sides of the triangle R,S,T just in the points T',R',S'.

Such a basic sextet can exist only if the rank of the linear manifold

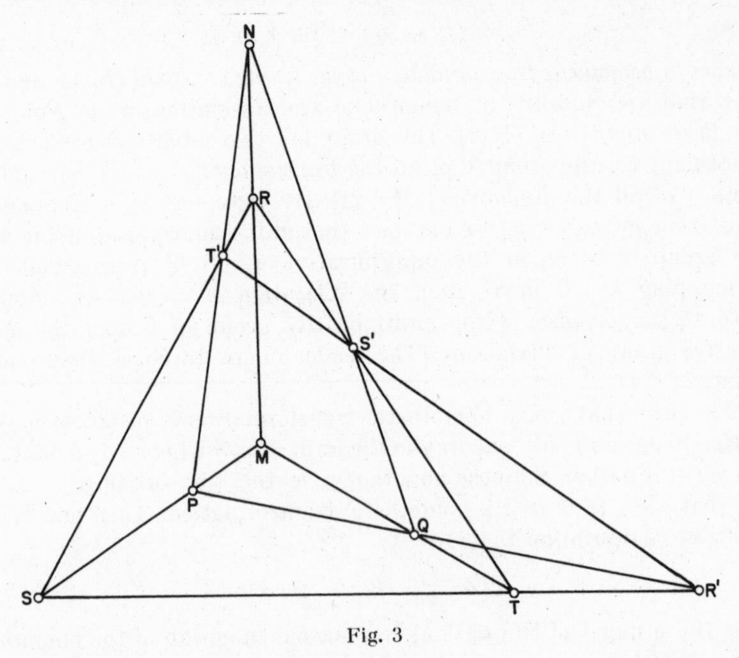

Fig. 3

(F,A) is at least 3. Consequently we shall assume $2 < r(A)$ throughout this appendix.

Proposition 1: (*a*) *If r,s,t are three independent elements in A, then the points Fr,Fs,Ft; $F(s-t)$, $F(t-r)$, $F(r-s)$ form a basic sextet which we call the basic sextet determined by the triplet r,s,t of independent elements.*

(*b*) *Every basic sextet is determined by some triplet of independent elements.*

(*c*) *The triplets r,s,t and u,v,w of independent elements determine the same basic sextet if, and only if, there exists a number $f \neq 0$ in F such that $u = fr$, $v = fs$, $w = ft$.*

PROOF: The verification of (*a*) and (*c*) is simple enough to be left to the reader. To prove (*b*) consider some basic sextet $R,S,T;\ R',S',T'$ and select at random some element r such that $R = Fr$. Since the point R is on the line $S + T'$, there exist uniquely determined elements s and t' in S and T' respectively such that $r = s + t'$. Since R,S,T' are three distinct points, it is impossible that s or t' is 0. Hence $S = Fs$ and $T' = Ft' = F(r - s)$. Similarly we find an element t such that $T = Ft$ and $S' = F(t - r)$. Next we note that R' is the intersection of the lines $S' + T'$ and $S + T$. Hence

$$R' = (S + T) \cap (S' + T') = (Fs + Ft) \cap (F[t - r] + F[r - s]).$$

But the points R,S,T span a plane; and thus r,s,t are three independent elements. Since the intersection of the lines $S + T$ and $S' + T'$ is a point, namely R', and since this intersection clearly contains the point $F(s - t)$, it follows that $R' = F(s - t)$. Thus the given basic sextet is determined by the triplet r,s,t of independent elements, as we claimed.

AN ADMISSIBLE QUARTET [WITH RESPECT TO A GIVEN BASIC SEXTET $R,S,T;\ R',S',T'$] is an *ordered* set $[M,N,P,Q]$ of subspaces of the linear manifold A which meet the following requirements.

$$\text{(I.2)}\quad \begin{cases} r(X) \leqslant 1 \text{ for } X = M,N,P,Q. \\ M + R = R + N = N + M,\ P + R' = R' + Q = Q + P, \\ M + S = S + P = P + M,\ Q + S' = S' + N = N + Q, \\ M + T = T + Q = Q + M,\ N + T' = T' + P = P + N. \end{cases}$$

It is quite easy to see that at most one of the "coordinates" of an admissible quartet can be 0; and that the following quartets are the only quartets with one coordinate equal to 0:

$$[0,R,S,T],\ [R,0,T',S'],\ [S,T',0,R'],\ [T,S',R',0].$$

Proposition 2: *If the basic sextet $R,S,T;\ R',S',T'$ is determined by the triplet r,s,t of independent elements in A, then a one-to-one mapping of A upon the totality of admissible quartets is obtained by mapping the element x in A upon the quartet*

$$[Fx,F(x - r),F(x - s),F(x - t)] = x^*.$$

PROOF: It is an almost immediate consequence of the independence of r,s,t that x^* is, for every x in A, an admissible quartet with respect to the basic sextet Fr, Fs, Ft; $F(s-t)$, $F(t-r)$, $F(r - s)$. If $x^* = y^* = [M,N,P,Q]$, then we infer from the definition of z^* that $x - y$ belongs to $M \cap N \cap P \cap Q$, since, for instance, $x - y = (x - r) - (y - r)$. Since the rank of M,N,P,Q does not exceed 1, the intersection $M \cap N \cap P \cap Q = 0$, as otherwise $M = N = P = Q$ would be a point and this point would be equal to R,S and T which is clearly impossible. But the vanishing of this intersection implies that $x - y = 0$. Hence $x = y$ is a consequence of $x^* = y^*$ so that our mapping is a one-to-one mapping of A into the totality of admissible quartets.

Consider now an admissible quartet $J = [M, N, P, Q]$. If one of its coordinates is 0, then $J = 0^*$, or r^* or s^* or t^*, as has already been pointed out before; and thus we may assume that none of the coordinates of J is 0. Now we distinguish two cases.

Case 1: M equals one of the points R, S, T.

We may assume without loss in generality that $M = R = Fr$. Then we infer from (I.2) that N is part of $M + R = Fr$. But N is a point; and so it follows that $M = N = R = Fr$. Consequently we have $Fr + Fs = Fs + P = P + Fr$; and this implies the existence of an element p in P and an element u in F such that $s = ur + p$ where neither u nor p can be 0. Then $P = Fp = F(ur - s)$; and likewise we show the existence of a number $v \neq 0$ in F such that $Q = F(vr - t)$. We deduce from (I.2) that

$$F(s - t) = R' \leqslant P + Q = F(ur - s) + F(vr - t)\,;$$

and now we deduce from the independence of r,s,t that $u = v$. If u were 1, then it would follow from (I.2) that

$$Fr = N \leqslant P + T' = F(r - s)$$

which contradicts the independence of r and s; and now one verifies easily that $J = (ur)^*$.

Case 2: The point M is different from the three points R, S, T.

Since the three lines $R + S, S + T, T + R$ are all distinct, the point M can be on at most one of these three lines; and thus we may assume that M is neither on $Fr + Fs$ nor on $Fs + Ft$.

Since $R + M = M + N = N + R$ is a line, it follows that R,M,N are three distinct collinear points. Consequently there exist uniquely determined elements m and n in M and N respectively such that $r = m + n$ and such that neither m nor n is 0. Hence

$$M = Fm \text{ and } N = Fn = F(m - r).$$

If the lines

$$F(m - s) + Fs = Fm + Fs = M + S = S + P = P + M$$
$$F(m - s) + F(s - r) = F(s - r) + F(m - r) = T' + N = N + P = P + T'$$

were equal, then this line would equal $R + S$ and would carry M which is impossible. Thus these two lines meet in a point; and it is easy to see that their common point is $F(m - s) = P$. Likewise we show that $F(m - t) = Q$; and thus we have shown that $m^* = J$. This completes the proof.

REMARK 1: If the basic sextet R,S,T; R',S',T' is determined by the triplet r,s,t of independent elements in A, then we have shown in the preceding proposition that the mapping of x in A upon

$$x^* = [Fx, F(x-r), F(x-s), F(x-t)]$$

is a one-to-one mapping of A upon the totality of admissible quartets. If r^0,s^0,t^0 is another triplet of independent elements in A which determines the given basic sextet, then this would lead us to the mapping of x in A upon the admissible quartet

$$x^0 = [Fx, F(x-r^0), F(x-s^0), F(x-t^0)].$$

We may deduce from Proposition 1, (c) the existence of a number $g \neq 0$ in F such that $u^0 = gu$ for $u = r,s,t$; and now one verifies that

$$(gx)^0 = x^* \qquad \text{for every } x \text{ in } A.$$

Thus we see that different triplets of independent elements which determine the same basic sextet lead to different mappings of A upon the totality of admissible quartets; but any two of these mappings differ by a "trivial" semi-linear transformation only.

The reader will find it a useful exercise to determine $(x+y)^*$ in terms of x^* and y^*. We shall, however, have no occasion to make use of such a result.

Consider now some fixed basic sextet R,S,T; R',S',T' which we denote by B. The totality of quartets admissible with respect to B we shall denote by $V(B)$. The coordinates of the admissible quartets we shall denote according to the following systematic rule:

$$J = [J_O, J_R, J_S, J_T] \text{ for every } J \text{ in } V(B).$$

Now we define *a homothetic transformation of* $V(B)$ as a permutation σ of $V(B)$ which satisfies

$$\text{(H)} \quad \begin{aligned} &(J_O + K_O) \cap (J_R + K_R) \cap (J_S + K_S) \cap (J_T + K_T) \\ &= (J_O^\sigma + K_O^\sigma) \cap (J_R^\sigma + K_R^\sigma) \cap (J_S^\sigma + K_S^\sigma) \cap (J_T^\sigma + K_T^\sigma) \end{aligned}$$

for every J and K in $V(B)$.

The homothetic transformation σ is *a translation of* $V(B)$, if it satisfies

$$\text{(T)} \quad \begin{aligned} &(J_O + J_O^\sigma) \cap (J_R + J_R^\sigma) \cap (J_S + J_S^\sigma) \cap (J_T + J_T^\sigma) \\ &= (K_O + K_O^\sigma) \cap (K_R + K_R^\sigma) \cap (K_S + K_S^\sigma) \cap (K_T + K_T^\sigma) \end{aligned}$$

for every J and K in $V(B)$.

The homothetic transformation σ is *a dilatation of* $V(B)$, if it satisfies

$$\text{(D)} \qquad J_O^\sigma = J_O \text{ for every } J \text{ in } V(B).$$

It is quite obvious that the homothetic transformations form a group and that the dilatations form a subgroup of this group. It is not quite so obvious that the translations form a normal subgroup of the group of dilatations; but we shall obtain this result as a by-product from later results.

Consider now some triplet r,s,t of independent elements in A which determines the given basic sextet B. Mapping x in A upon

$$x^* = [Fx, F(x-r), F(x-s), F(x-t)]$$

in $V(B)$ we obtain a one-to-one mapping of A upon $V(B)$. If σ is a permutation of A, then there exists one and only one permutation σ^* of $V(B)$ such that

$$(x^\sigma)^* = x^{*\sigma^*} \text{ for every } x \text{ in } A \text{ ;}$$

and the mapping of σ upon σ^* is an isomorphy of the group of all permutations of A upon the group of all permutations of $V(B)$.

Proposition 3: *The isomorphism $\sigma \to \sigma^*$ maps the group of homothetic transformations of (F,A) upon the group of homothetic transformations of $V(B)$, the group of dilatations of (F,A) upon the group of dilatations of $V(B)$ and the group of translations of (F,A) upon the totality of translations of $V(B)$.*

Since isomorphisms map normal subgroups upon normal subgroups, this theorem shows incidentally that the translations of $V(B)$ form a normal subgroup of the group of homothetic transformations of $V(B)$.

It will be convenient to precede the proof of this proposition by a proof of the following fact.

Lemma : *If x and y are elements in A, then*

$$F(x-y) = (x^*_O + y^*_O) \cap (x^*_R + y^*_R) \cap (x^*_S + y^*_S) \cap (x^*_T + y^*_T).$$

PROOF: The right hand intersection of four subspaces of A is equal to $D = (Fx + Fy) \cap [F(x-r) + F(y-r)] \cap \cdots$. It is clear that $F(x-y)$ is part of D. Since r,s,t are three independent elements, they cannot all belong to $Fx + Fy$; and thus we may assume without loss in generality that r is not in $Fx + Fy$. Every element d in D has the form

$$d = x'x + y'y = x''(x-r) + y''(y-r)$$

where x',x'',y',y'' are suitable numbers in F. It follows that

$$(x''-x')x + (y''-y')y = (x''+y'')r.$$

Since r is not in $Fx + Fy$, this implies $x'' + y'' = 0$ so that $d = x''(x-y)$ belongs to $F(x-y)$. Hence $D = F(x-y)$, as we claimed. [One may also use a rank argument instead of the last argument.]

An immediate consequence of this lemma is the following proposition.

Corollary: *Suppose that σ is a permutation of A and that σ^* is the corresponding permutation of $V(B)$.*

(*a*) σ^* *has Property* (H) *if, and only if,* $F(x-y) = F(x^\sigma - y^\sigma)$ *for every* x,y *in* A.

(*b*) σ^* *has Property* (T) *if, and only if,* $F(x-x^\sigma) = F(y-y^\sigma)$ *for every* x *and* y *in* A.

(*c*) σ^* *has Property* (D) *if, and only if,* $Fx = Fx^\sigma$ *for every* x *in* A.

If σ is a translation of A, then $x^\sigma = x + b$ for every x in A so that $x - y = x^\sigma - y^\sigma$ and $x - x^\sigma = -b = y - y^\sigma$. Hence σ^* is a translation of $V(B)$ whenever σ is a translation of (F,A). If σ is a dilatation of A, then $x^\sigma = fx$ for $f \neq 0$ in F so that $x^\sigma - y^\sigma = f(x-y)$. Hence σ^* is a dilatation of $V(B)$ whenever σ is a dilatation of (F,A); and that σ^* is a homethetic transformation of $V(B)$ whenever σ is a homothetic transformation of (F,A), is seen likewise [or deduced from the preceding remarks].

(1) *The homothetic transformation* η *of* $V(B)$ *is a dilatation if, and only if,* $0^* = 0^{*\eta}$.

Proof: If η is a dilatation, then $0^{*\eta} = [0,\cdots]$. But the only admissible quartet with first coordinate equal to 0 is 0^* so that $0^{*\eta} = 0^*$. If conversely $0^* = 0^{*\eta}$, then we remember first the existence of one and only one permutation σ of A such that $\sigma^* = \eta$. Clearly $0^\sigma = 0$. Since η satisfies (H), it follows from Corollary (*a*) that $Fx = F(x-0) = F(x^\sigma - 0^\sigma) = Fx^\sigma$; and now it follows from Corollary (*c*) that $\sigma^* = \eta$ is a dilatation, as we claimed.

(2) *If* τ' *and* τ'' *are translations of* $V(B)$ *such that* $0^{*\tau'} = 0^{*\tau''}$, *then* $\tau' = \tau''$.

Proof: There exist permutations σ' and σ'' of A such that $\sigma'^* = \tau'$ and $\sigma''^* = \tau''$. From $(x^\sigma)^* = x^{*\sigma^*}$ and our condition it follows now that $0^{\sigma'} = 0^{\sigma''} = w$. Applying Corollary (*a*), (*b*) with $y = 0$ we obtain

$$(2^*)\qquad \begin{cases} F(x^{\sigma'} - w) = Fx = F(x^{\sigma''} - w) \text{ for every } x \text{ in } A, \\ F(x^{\sigma'} - x) = Fw = F(x^{\sigma''} - x) \text{ for every } x \text{ in } A. \end{cases}$$

If $w = 0$, then $x^{\sigma'} = x = x^{\sigma''}$ for every x; and thus we may assume that $w \neq 0$.

Suppose first that x is not in Fw. Then we deduce from the equations (2*) the existence of numbers f,f',f'' in F such that

$$x^{\sigma'} - w = f(x^{\sigma''} - w), \qquad f'w = x^{\sigma'} - x, \qquad f''w = x^{\sigma''} - x.$$

Eliminating $x^{\sigma'}$ and $x^{\sigma''}$ from these equations we find that

$$f'w + x - w = f[f''w + x - w] \text{ or } (f-1)x = (f' - 1 - ff'' + f)w.$$

But x is not in Fw. Hence $f = 1$ and consequently $x^{\sigma'} = x^{\sigma''}$.

Assume next that x is in Fw. If $x = 0$, then we know already that $0^{\sigma'} = w = 0^{\sigma''}$. Hence we assume $x \neq 0$ and this implies $Fx = Fw$. There exist elements y,z such that x,y,z are three independent elements. Then neither y nor z is in $Fx = Fw$; and it follows from what we have shown in the preceding paragraph of this proof that

$$y^{\sigma'} = y^{\sigma''} = y' \text{ and } z^{\sigma'} = z^{\sigma''} = z'.$$

Applying Corollary (a) we find that

$$F(x^{\sigma'} - y') = F(x - y) = F(x^{\sigma''} - y'),$$
$$F(x^{\sigma'} - z') = F(x - z) = F(x^{\sigma''} - z').$$

Consequently there exist numbers h',h'',k',k'' in F such that

$$y' - x^{\sigma''} = h''(x - y), \qquad z' - x^{\sigma''} = k''(x - z),$$
$$x^{\sigma'} - y' = h'(x - y), \qquad x^{\sigma'} - z' = k'(x - z).$$

If follows by addition that

$$(h' + h'')(x - y) = x^{\sigma'} - x^{\sigma''} = (k' + k'')(x - z);$$

and it follows from the independence of x,y,z that $h' + h'' = k' + k'' = 0$. But then $x^{\sigma'} = x^{\sigma''}$ for every x so that $\tau' = \sigma'^* = \sigma''^* = \tau''$; and this completes the proof of (2).

(3) *If δ is a dilatation of $V(B)$ which leaves invariant at least one admissible quartet different from 0^*, then $\delta = 1$.*

Proof: There exists a permutation σ of A such that $\delta = \sigma^*$; and it follows from our hypotheses [and (1)] that σ leaves invariant 0 and an element different from 0. Next we show:

(3*) If x and y are independent elements in A such that $x^{\sigma} = x$, then $y^{\sigma} = y$.

It follows from Corollary (a), (c) that $F(x - y) = F(x - y^{\sigma})$ and $Fy = Fy^{\sigma}$. Consequently there exist numbers f,g in F such that $y^{\sigma} - x = f(y - x)$ and $y^{\sigma} = gy$. Hence $gy - x = f(y - x)$; and it follows from the independence of x and y that $g = f = 1$ or $y^{\sigma} = y$. This proves (3*).

From (3*) and our hypotheses concerning δ and σ one deduces immediately that $\sigma = 1$ and $\delta = 1$, since there exist at least three independent elements in A.

(4) *The permutation σ of A is a dilatation of (F,A) if, and only if, σ^* is a dilatation of $V(B)$.*

Proof: We have already pointed out that σ^* is a dilatation whenever σ is a dilatation. Assume now that σ^* is a dilatation. Then it follows from Corollary (c) that $Fr = Fr^{\sigma}$. Hence there exists a number $f \neq 0$ in F such that $r = fr^{\sigma}$. The mapping $x^{\tau} = fx$ for x in A is a dilatation of A; and consequently τ^* is a dilatation of $V(B)$. The dilatations of $V(B)$ form a group; and so $\sigma^*\tau^* = (\sigma\tau)^*$ is a dilatation of $V(B)$. From $r^{\sigma\tau} = fr^\sigma = r$ it follows that the dilatation $(\sigma\tau)^*$ of $V(B)$ leaves invariant at least one admissible quartet, not 0^*. We infer $\sigma^*\tau^* = (\sigma\tau)^* = 1$ from (3). Hence $\sigma\tau = 1$ so that σ is the dilatation $x^\sigma = f^{-1}x$ of (F,A), as we claimed.

(5) *The permutation σ of A is a translation of (F,A) if, and only if, σ^* is a translation of $V(B)$.*

PROOF: We have already pointed out that σ^* is a translation whenever σ is a translation. Assume now that σ^* is a translation. Let $0^\sigma = w$. Then $x^\tau = x + w$ is a translation of (F,A); and consequently τ^* is a translation of $V(B)$. We have

$$0^{*\sigma^*} = (0^\sigma)^* = w^* = (0^\tau)^* = 0^{*\tau^*};$$

and it follows from (2) that $\sigma^* = \tau^*$. But this implies $\sigma = \tau$ so that σ is the translation $x^\tau = x + w$, as we wanted to show.

(6) *The permutation σ of A is a homothetic transformation of (F,A) if, and only if, σ^* is a homothetic transformation of $V(B)$.*

PROOF: We have already pointed out that σ^* is a homothetic transformation whenever σ is a homothetic transformation. Assume now that σ^* is a homothetic transformation. Let $0^\sigma = w$ and denote by τ the translation $x^\tau = x - w$ of (F,A). Then τ^* is a translation of $V(B)$ and therefore a homothetic transformation of $V(B)$. The homothetic transformations of $V(B)$ form a group. Hence $(\sigma\tau)^* = \sigma^*\tau^*$ is a homothetic transformation. One notices that $0^{\sigma\tau} = 0$; and it follows from (1) that $(\sigma\tau)^*$ is a dilatation. We infer from (4) that $\sigma\tau$ is a dilatation. Since τ and $\sigma\tau$ are therefore homothetic transformations of (F,A), σ itself is a homothetic transformation, as we wanted to show.

Remembering the remarks we made preceding the enunciation of Proposition 3 one sees now that Proposition 3 is an immediate consequence of the statements (3), (4), (5) which we just proved.

REMARK 2: It should be noted that the somewhat more involved statement (2) was necessitated by the fact that we were not assured at that time that the translations of $V(B)$ form a group.

REMARK 3: On the basis of the preceding results one may deduce from a projectivity of the linear manifold (F,A) upon the linear manifold (G,B) an isomorphism of the homothetic group of (F,A) upon (G,B) which maps translations upon translations, dilatations upon dilatations and which may be induced by a semi-linear transformation. We omit the details of this discussion; and note only that in this way one may obtain a second proof of the First Fundamental Theorem of Projective Geometry. The reader should note that some of the arguments in this appendix are closely related to those of III.1.

III.2. The Group of Collineations

Throughout this section we shall assume that *A is an F-space whose rank is at least* 3. Then we have shown in § 1 that every auto-projectivity of A is induced [may be represented] by a semi-linear transformation. Those-auto-projectivities of A which are represented by linear transformations of A we shall term *collineations* [the customary terminology varies somewhat]. It is clear that the collineations form a subgroup Π_0 of the group Π of all auto-projectivities of A; and using the fact that every projectivity may be represented by a semi-linear transformation one verifies easily that Π_0 is a normal subgroup of Π.

Lemma 1: *The semi-linear transformation $\sigma = (\sigma', \sigma'')$ induces a collineation if, and only if, σ'' is an inner automorphism of F.*

PROOF: If σ induces a collineation, then there exists a linear transformation [isomorphism of A upon itself] τ such that σ and τ induce the same projectivity. It follows that $\sigma\tau^{-1}$ leaves invariant every subspace of A; and we deduce from III.1, Proposition 3 the existence of a number $f \neq 0$ in F such that $a^{\sigma\tau^{-1}} = fa$ for every a in A. Remembering that τ is a linear transformation of A we deduce

$$a^{\sigma'} = fa^{\tau} \text{ for every } a \text{ in } A.$$

Consequently we find that

$$x^{\sigma''}a^{\sigma'} = (xa)^{\sigma'} = f(xa)^{\tau} = fxa^{\tau} = fxf^{-1}fa^{\tau} = fxf^{-1}a^{\sigma'}$$

so that $x^{\sigma''} = fxf^{-1}$ for every x in F. Thus σ'' is an inner automorphism of the field F.

If conversely σ'' is an inner automorphism of the field F, then there exists a number $v \neq 0$ in F such that $x^{\sigma''} = v^{-1}xv$ for every x in F. Define the mapping ρ of A by the formula:

$$a^{\rho} = va^{\sigma'} \qquad \text{for every } a \text{ in } A.$$

Then we verify that ρ is an additive automorphism of A which induces the same auto-projectivity of A as does σ; and we have furthermore

$$(xa)^{\rho} = v(xa)^{\sigma'} = vx^{\sigma''}a^{\sigma'} = v(v^{-1}xv)a^{\sigma'} = xva^{\sigma'} = xa^{\rho}$$

so that ρ is a linear transformation. Hence σ induces a collineation.

Corollary 1: *Every projectivity of the F-space A [upon itself] is a collineation if, and only if, every automorphism of the field F is an inner automorphism.*

This is an almost obvious consequence of Lemma 1 and the First Fundamental Theorem of Projective Geometry [III.1].

A commutative field has this property if, and only if, the identity is its only automorphism. Examples of such fields are the field of rational numbers [and all the soc. prime fields], the field of real numbers [and all the real closed fields which it contains]. On the other hand the field of complex numbers possesses many automorphisms; the simplest example consists in mapping every number upon its conjugate complex one. By means of set theoretical methods it is possible to show that their number is $2^{2^{\aleph_0}}$.

The simplest example of a non-commutative field with our property is the field of quaternions. More generally it can be shown: if F is finite over its center, and if the identity is the only automorphism of its center, then every automorphism of F is inner. [See e. g. Albert (1).]

A Survey of the Groups of a Linear Manifold

In III.1 we had introduced the group Λ of all semi-linear transformations of the linear manifold (F,A) of rank not less than 3. Every semi-linear transformation σ induces a well determined auto-projectivity σ^* of A; and mapping σ onto σ^* constitutes a homomorphism of Λ upon the group Π of auto-projectivities of A. The kernel of this homomorphism we denoted by N; and we pointed out that it consists just of the multiplications $a^\sigma = fa$ [where $f \neq 0$ is some number in F].

Suppose now that σ is some semi-linear transformation and that $a^\nu = fa$ is some multiplication in N. Then we have

$$x^{\sigma^{-1}\nu\sigma} = (fx^{\sigma^{-1}})^\sigma = f^\sigma x \qquad \text{for every } x \text{ in } A.$$

Consequently σ and ν commute with each other if, and only if, $f = f^\sigma$. Since every number $f \neq 0$ appears in one and only one multiplication in N, it follows that σ commutes with every multiplication if, and only if, $f = f^\sigma$ for every f in F. This is equivalent to saying that the automorphism of F which constitutes a component of σ is the identity; in other words: that σ is a linear transformation. Now it is customary in group theory to term the totality of elements in a group G which commute with a certain subset S of G the *centralizer* of S in G. Using this concept we may restate our result as follows:

The group T *of linear transformations is the centralizer of* N *in* Λ.

Denote next by Λ_0 the subgroup of those semi-linear transformations whose component automorphism is an inner automorphism of F. It follows from Lemma 1 that the semi-linear transformation σ belongs to Λ_0 if, and only if, the induced auto-projectivity σ^* is a collineation. This implies among other things that the group of multiplications N is part of Λ_0. Likewise T is part of Λ_0. On the other hand it follows from Lemma 1 that there exists to every semi-linear transformation σ in Λ_0 a linear transformation σ' such that σ and σ' induce the same collineation. But then it follows from III.1, Proposition 3 that $\sigma'\sigma^{-1}$ is a multiplication; and thus we have shown that

$$\Lambda_0 = \mathrm{NT}.$$

Consider now some basis B of A and denote by $\Gamma = \Gamma(B)$ the totality of semi-linear transformations σ of A which satisfy:

$$b^\sigma = b \qquad \text{for every } b \text{ in } B.$$

Clearly Γ is a subgroup of Λ. If we note that every basis may be mapped on every other basis by one and only one linear transformation of A, then we verify easily that

$$\Lambda = \mathrm{T}\Gamma \text{ and } 1 = \mathrm{T} \cap \Gamma.$$

One verifies furthermore that every transformation in Γ is completely determined by its component automorphism of the field F and that every automorphism of F appears in one and only one transformation in Γ. Thus mapping σ in Γ upon its component automorphism of F constitutes *an isomorphism of* Γ *upon the group of all automorphisms of the field* F.

There does not seem to exist a geometrical criterion which permits to decide whether a given auto-projectivity is a collineation. But it is possible to give a geometrical characterization of the whole group of collineations as a subgroup of the group of auto-projectivities; and this we are going to do next.

Lemma 2: *A projectivity is a collineation, if it possesses a full line of fixed points.*

PROOF: Suppose that the projectivity under consideration is induced by the semi-linear transformation σ; and that it leaves invariant every point on the subspace L of rank 2. Since σ induces a semi-linear transformation on the linear manifold L which leaves invariant every subspace of L, we may deduce from III.1, Proposition 3 the existence of a number $f \neq 0$ such that $t^\sigma = ft$ for every t in L. If $t \neq 0$ is in L, and if x is in F, then we find accordingly

$$x^\sigma ft = x^\sigma t^\sigma = (xt)^\sigma = f(xt) = (fx)t \text{ or } x^\sigma f = fx \text{ or } x^\sigma = fxf^{-1}.$$

Thus σ is the inner automorphism of the field F which is induced by f; and it follows from Lemma 1 that the projectivity induced by σ is a collineation.

DEFINITION 1: *The projectivity* ν *is a perspectivity, if there exists a subspace* H *of* A *such that* $r(A/H) = 1$ *and every subspace of* H *is left invariant by* ν.

Since $2 < r(A)$, we have $2 \leqslant r(H)$. It follows from Lemma 2 that perspectivities are collineations. It is clear that it suffices to require that every point on H is left invariant by ν; and thus we shall refer to H as to *the hyperplane of fixed points of the perspectivity* ν.

Lemma 3: *If the elements a and b in A are both not contained in the subspace S of A, then there exists a linear transformation σ of A which induces a perspectivity and which satisfies $a^\sigma = b$, $s^\sigma = s$ for every s in S.*

PROOF: From the Complementation Theorem we deduce the existence of a subspace T such that $A = [S + Fa + Fb] \oplus T$. Now we distinguish two cases.

Case 1: $S + Fa = S + Fb$.

Then there exists one and only one linear transformation σ such that $a^\sigma = b$ and $x^\sigma = x$ for x in $S + T$. Since $S + T$ is a hyperplane [and $A = (S + T) \oplus Fa = (S + T) \oplus Fb$], σ induces a perspectivity with $S + T$ for its hyperplane of fixed points.

Case 2: $S + Fa \neq S + Fb$.

Since neither a nor b is in S, we have $A = S \oplus T \oplus Fa \oplus Fb$; and there exists one and only one linear transformation σ such that $a^\sigma = b$, $b^\sigma = a$, $x^\sigma = x$ for x in $S + T$. But then $H = S + T + F(a + b)$ is a hyperplane all of whose elements are left invariant by the linear transformation σ. Consequently σ induces a perspectivity with H for its hyperplane of fixed points.

Proposition 1: *If ν is a collineation and if S is a subspace of finite rank of A, then there exists a product π of not more than r(S) perspectivities of A such that $X^\nu = X^\pi$ for every subspace X of S.*

PROOF: There exists a linear transformation σ which induces the collineation ν. Denote furthermore by $s_1, \cdots, s_n$ a basis of S so that $n = r(S)$. We deduce from Lemma 3 the existence of a linear transformation σ_1 such that $s_1 = s_1^{\sigma\sigma_1}$ and such that σ_1 induces a perspectivity. It is clear that $\rho_1 = \sigma\sigma_1$ is a linear transformation which leaves invariant every element in $S_1 = Fs_1$.

Now we make the inductive hypothesis that for some i with $0 < i < n$ we have constructed linear transformations $\sigma_1, \cdots, \sigma_i$ such that each of them induces a perspectivity and such that $\rho_i = \sigma\sigma_1 \cdots \sigma_i$ leaves invariant every element in the subspace $S_i = \sum_{j=1}^{i} Fs_j$. Then neither s_{i+1} nor $s_{i+1}^{\rho_i}$ belongs to S_i; and we infer from Lemma 3 the existence of a linear transformation σ_{i+1} which induces a perspectivity, leaves invariant every element in S_i and maps $s_{i+1}^{\rho_i}$ upon s_{i+1}. Then $\rho_{i+1} = \rho_i\sigma_{i+1}$ is a linear transformation which leaves invariant every element in S_i and the element s_{i+1} so that every element in $S_{i+1} = S_i + Fs_{i+1}$ is left invariant by ρ_{i+1}. This completes the induction.

Consequently there exist linear transformations $\sigma_1, \cdots, \sigma_n$ each of which induces a perspectivity such that $\sigma\sigma_1 \cdots \sigma_n$ leaves invariant every element in S. The projectivity ρ induced by $\sigma_n^{-1} \cdots \sigma_1^{-1}$ is consequently a product of [at most] n perspectivities and satisfies $X^\nu = X^\pi$ for every subspace X of S.

Theorem 1: *A projectivity is a collineation if, and only if, it is a product of* [*at most* 3] *projectivities each of which possesses a full line of fixed points.*
PROOF: The sufficiency of our condition is an immediate consequence of Lemma 2. If conversely σ is a collineation, and if L is a line, then we deduce from Proposition 1 the existence of perspectivities σ_1 and σ_2 such that $P^\sigma = P^{\sigma_1\sigma_2}$ for every point P on L. Then $\sigma = \rho\sigma_1\sigma_2$ where the projectivity ρ leaves invariant every point on the line L and where the perspectivities σ_i have at least a line of fixed points each.

Theorem 2: *A projectivity of a linear manifold of finite rank is a collineation if, and only if, it is a product of perspectivities.*

This is an immediate consequence of Lemma 2 and Proposition 1.

For linear manifolds of infinite rank we may give to Proposition 1 a similar interpretation by saying that a projectivity is a collineation if, and only if, it may be "approximated" by products of perspectivities. We leave the details to the topologically interested reader.

III.3. The Second Fundamental Theorem of Projective Geometry

We assume in the course of the present section that A is an F-space of finite rank $r(A) = n$. We shall assume that $1 < n$, since otherwise our investigation would not have any content [and some assertions would cease to be valid]. It is, however, not neccessary to assume that $2 < n$, since we are going to investigate collineations which are *defined* by the fact that they are induced by linear transformations.

It will be convenient to term *simplex* a set of $r(A) + 1$ points no $r(A)$ of which are contained in a hyperplane of A. This is equivalent to saying that any $r(A)$ of the $r(A) + 1$ points of a simplex are independent. If, for instance, A is a plane, then a simplex consists of four points no three of which are collinear.

Lemma 1: *If the points P_i [$i = 0, 1, \cdots, r(A)$] form a simplex, then there exists a basis $b_1, \cdots, b_n$ of A such that $P_i = Fb_i$ for $i = 1, \cdots, n$ and*

$$P_0 = F\sum_{j=1}^{n} b_j.$$

PROOF: Let b_0 be any element such that $P_0 = Fb_0$. Since the points $P_1, \cdots, P_n$ are independent, and since n is the finite rank of A, we have $A = \sum_{j=1}^{n} P_j$. Consequently there exist elements b_j in P_j such that $b_0 = b_1 + \cdots + b_n$. If b_i were 0, then b_0 and hence P_0 would be part of the hyperplane $\sum_{i \neq j} P_j$ which is impossible. Consequently $P_i = Fb_i$; and from the independence of the points $P_1, \cdots, P_n$ it follows now that $b_1, \cdots, b_n$ is a basis of A.

Proposition 1: *If P'_i and P''_i are simplices, then there exists a collineation σ such that $P''_i = P'^{\sigma}_n$ for $i = 0, 1, \cdots, n$.*

PROOF: We infer from Lemma 1 the existence of bases $p'_1, \cdots, p'_n$ and $p''_1, \cdots, p''_n$ of A such that $P'_i = Fp'_i$, $P''_i = Fp''_i$ for $i = 1, \cdots, n$ and $P'_0 = F\sum_{i=1}^{n} p'_i$, $P''_0 = F\sum_{i=1}^{n} p''_i$. There exists furthermore one and only one linear transformation σ of A such that $p'^{\sigma}_i = p''_i$ for $i = 1, \cdots, n$; and it is clear that the collineation induced by σ meets all the requirements.

Proposition 2: *The identity is the only collineation leaving invariant every point of a simplex if, and only if, F is a commutative field.*

PROOF: Denote by P_i some simplex of A. Then there exists, by Lemma 1, a basis $b_1, \cdots, b_n$ of A such that $P_i = Fb_i$ for $i = 1, \cdots, n$ and $P_0 = F\sum_{i=1}^{n} b_i$.

Assume now that the identity is the only collineation which leaves invariant every point P_i of our simplex. Suppose that $f \neq 0$ is a number in F. Then there exists one and only one linear transformation σ of A such that $b^{\sigma}_i = fb_i$ for $i = 1, \cdots, n$. This linear transformation σ induces a collineation which satisfies $P^{\sigma}_i = P_i$ for $i = 0, 1, \cdots, n$. Hence σ induces the identity projectivity. Consider now some element z in F. Then it follows that

$$F(b_1 + zb_2) = [F(b_1 + zb_2)]^{\sigma} = F(fb_1 + zfb_2) = F(b_1 + f^{-1}zfb_2).$$

Because b_1 and b_2 are independent, it follows now that $z = f^{-1}zf$ or $fz = zf$; and this is true for every f and every z. Hence F is commutative.

Assume now conversely that the field F be commutative. Consider some collineation σ which leaves invariant each of the points P_i of our simplex. This collineation σ is induced by a linear transformation which we denote by σ too. From $P^{\sigma}_i = P_i$ for $i = 0, 1, \cdots, n$ we deduce the existence of numbers $f_i \neq 0$ in F such that

$$b_i^\sigma = f_i b_i \text{ for } i = 1, \cdots, n, \quad \left[\sum_{i=1}^{n} b_i\right]^\sigma = f_0 \sum_{i=1}^{n} b_i.$$

But this implies

$$\sum_{i=1}^{n} f_0 b_i = \sum_{i=1}^{n} b_i^\sigma = \sum_{i=1}^{n} f_i b_i;$$

and we infer from the independence of the b_i that $f_0 = f_1 = \cdots = f_n$. If a is any element in A, then

$$a^\sigma = \left[\sum_{i=1}^{n} a_i b_i\right]^\sigma = \sum_{i=1}^{n} a_i f_0 b_i = f_0 \sum_{i=1}^{n} a_i b_i = f_0 a$$

since F is commutative. Hence σ leaves invariant every point in A so that σ is the identity projectivity.

REMARK 1: Mapping the point (x_0,x_1,x_2) of the complex plane upon its conjugate complex point $(\bar{x}_0,\bar{x}_1,\bar{x}_2)$ we obtain a semi-linear transformation which leaves invariant every point of the simplex (1,1,1), (1,0,0), (0,1,0), (0,0,1); but which does not leave invariant the point represented by $(i,1,0)$ where $i^2 = -1$. Thus the Proposition 2 ceases to be true, if we omit the restriction to collineations.

REMARK 2: If the rank of A is at least 3, then we may consider the group of all projectivities which leave invariant every point of some given simplex. This group may be seen [by application of the First Fundamental Theorem of Projective Geometry of III.1] to be isomorphic to the group of automorphisms of the field F. See III.2, Survey of the groups of a linear manifold.

REMARK 3: If F is not commutative, then there exist collineations, not the identity, which leave invariant every point in a given simplex [Proposition 2]. It follows from III.2, Theorem 2 that this collineation is a product of perspectivities. Thus a product of perspectivities may leave invariant every point in a simplex without being the identity.

THE SECOND FUNDAMENTAL THEOREM OF PROJECTIVE GEOMETRY may now be stated as follows:

Every simplex may be mapped upon every other simplex by one and only one collineation if, and only if, the field of coordinates is commutative.

The proof is an immediate consequence of Propositions 1 and 2.

REMARK 4: It is possible to extend the results of this section to linear manifolds of infinite rank. If one wants to do this, one has to investigate the effect of a linear transformation on the subspaces of finite rank. Furthermore one has to find a substitute for the concept of simplex which in its present form looses its meaning in the case of infinite rank. We leave this discussion to the interested reader.

APPENDIX II

The Theorem of Pappus

In the Second Fundamental Theorem of Projective Geometry [§ 3] we have met a first instance where the commutativity of the field F [of coor-

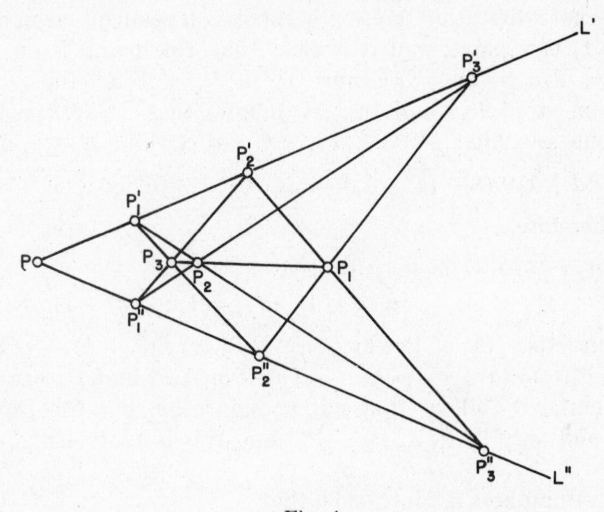

Fig. 4

dinates] appears important. In the present appendix we give Hilbert's celebrated characterization of commutativity of F by means of geometric configurations.

We are considering configurations of ten coplanar points which are obtained as follows. There are given two distinct lines L' and L'' meeting in a point P and on each of the lines three distinct points P_i' on L' and and P_i'' on L'' such that $P \neq P_i'$ and $P \neq P_i''$ for $i = 1,2,3$. One verifies easily that the three intersections

$$P_i = (P_j' + P_k'') \cap (P_j'' + P_k')$$

where i,j,k is a permutation of 1,2,3 are well determined points; and we say that *this configuration* $[P_i,P_i',P_i'']$ *has the Property of Pappus, if* P_1,P_2,P_3 *are collinear points.*

We select at random an element a such that $P = Fa$. From

$$P \leqslant L' = P'_1 + P'_2$$

we infer the existence of elements b and b' in P'_1 and P'_2 respectively such that $a = b' - b$; and b and b' are different from 0, since P is different from P'_i. Hence $P'_1 = Fb$ and $P'_2 = F(a + b)$. Since P'_3 is on $L' = P + P'_1$, we have $P'_3 = F(a + d'b)$ with $d' \neq 0, 1$, if we only note that P'_3 is different from P, P'_1, P'_2. Likewise we find an element c in A and an element $d'' \neq 0, 1$ in F such that $P''_1 = Fc$, $P''_2 = F(a + c)$, $P''_3 = F(a + d''c)$.

The configuration $[P_i, P'_i, P''_i]$ *has the Property of Pappus if, and only if,* $d'd'' = d''d'$.

PROOF: We note first that a,b,c are three independent elements. Thus $F(a + b + c)$ is a point; and it is clear that this point is on the lines $P'_2 + P''_1 = F(a + b) + Fc$ and $P''_2 + P'_1 = F(a + c) + Fb$. But these lines meet in P_3; and thus it follows that $P_3 = F(a + b + c)$. Similarly one sees that $P_2 = F(a + d'b + d''c)$. Finally we note that

$$(d' - 1)^{-1}d' + (d'' - 1)^{-1} = (d' - 1)^{-1} + (d'' - 1)^{-1}d''$$

and that therefore

$$(d' - 1)^{-1}d'(a + b) + (d'' - 1)^{-1}(a + d''c)$$
$$= (d' - 1)^{-1}(a + d'b) + (d'' - 1)^{-1}d''(a + c)$$

is the intersection P_1 of the lines $P'_2 + P''_3 = F(a + b) + F(a + d''c)$ and $P''_2 + P'_3 = F(a + c) + F(a + d'b)$. Since P_2 and P_3 are certainly different points, it follows that our configuration has the Property of Pappus if, and only if, $P_1 \leqslant P_2 + P_3$; and this is equivalent to the property:

There exist numbers h,k in F such that

$$(1) \qquad \begin{aligned} &(d' - 1)^{-1}d'(a + b) + (d'' - 1)^{-1}(a + d''c) \\ &= h(a + d'b + d''c) + k(a + b + c). \end{aligned}$$

From the independence of a,b,c it follows that the numbers h,k satisfy equation (1) if, and only if, they satisfy the following equations:

$$(2) \qquad \begin{cases} h + k = (d' - 1)^{-1}d' + (d'' - 1)^{-1}, \\ hd' + k = (d' - 1)^{-1}d', \\ hd'' + k = (d'' - 1)^{-1}d''. \end{cases}$$

Thus the existence of solutions h,k of (2) is necessary and sufficient for the validity of the Property of Pappus.

If there exist numbers h,k which satisfy (2), then subtraction of the first equation (2) from the other two equations (2) gives:

$$h(d' - 1) = -(d'' - 1)^{-1}, \qquad h(d'' - 1) = 1 - (d' - 1)^{-1}d' = -(d' - 1)^{-1}.$$

Consequently

$$-(d''-1)^{-1}(d'-1)^{-1}=h=-(d'-1)^{-1}(d''-1)^{-1}$$

or $(d'-1)(d''-1)=(d''-1)(d'-1)$; and this implies $d'd''=d''d'$.

If conversely $d'd''=d''d'$, then one sees easily that

$$h=-(d''-1)^{-1}(d'-1)^{-1}, \qquad k=(d'-1)^{-1}d''(d''-1)^{-1}$$

are solutions of the equations (2). Thus we have shown that $d'd''=d''d'$ is necessary and sufficient for the validity of the Property of Pappus, as we claimed.

Finally we say that the *Postulate of Pappus* is satisfied by the linear manifold (F,A) [of rank at least 3], if every configuration $[P_i,P_i',P_i'']$ has the Property of Pappus. If a,b,c are three independent elements, and if neither of the numbers d',d'' is 0 or 1, then the points $P_1'=Fb$, $P_2'=F(a+b)$, $P_3'=F(a+d'b)$ and $P_1''=Fc$, $P_2''=F(a+c)$, $P_3''=F(a+d''c)$ lead as before to a configuration of the type considered here. Consequently we have the fundamental result:

The Postulate of Pappus is satisfied by the linear manifold (F,A) of rank not less than 3 if, and only if, the field F is commutative.

III.4. The Projective Geometry of a Line in Space; Cross Ratios

A line [= linear manifold of rank 2] has no geometrical structure, if considered as an isolated or absolute phenomenon, since then it is nothing but a set of points with the number of points on the line as the only invariant [see III.1, Remark 1 and the Projective Structure Theorem]. But the line receives a definite geometrical structure, if imbedded into a linear manifold of higher rank.

If (F,A) is a linear manifold of rank not less than 3, and if L' and L'' are lines in (F,A), then geometrical equivalence of a configuration on L' with one on L'' in the absolute sense is defined by any one-to-one correspondence between the points on L' and those on L'' whereas geometrical equivalence relative to the containing manifold (F,A) is defined by those correspondences which are induced by auto-projectivities of (F,A). Thus there arises the problem of characterizing those correspondences which are induced by projectivities. To do this we have to introduce the concept of cross ratio; and we emphasize that this concept, so central in classical discussions of projective geometry, is needed only for a treatment of the rather special problem under consideration.

Definition 1: *Suppose that P,Q,R,S are four distinct points on the line*

L. Then the number s in F belongs to the cross ratio $\begin{bmatrix} P & Q \\ S & R \end{bmatrix}$ *if, and only if, there exist elements p,q in A such that*

$$P = Fp, \qquad Q = Fq, \qquad R = F(p + q), \qquad S = F(p + sq).$$

The justification for the "square" symbol indicating the cross ratio [which is due to F. Levi] will be found in the symmetry rules that we are going to derive later. The justification for defining the cross ratio not as a number but as set of numbers will be found in the following proposition.

Proposition 1: *The cross ratio of four distinct collinear points is a complete class of conjugate numbers.*

Here we term as usual the numbers h and k in F *conjugate,* if there exists a number $c \neq 0$ in F such that $h = c^{-1}kc$; and we shall designate the totality of numbers conjugate to the number k in F by the symbol $\langle k \rangle$.

PROOF: Consider four distinct collinear points P,Q,R,S. Then

$$P + Q = P + R = P + S = Q + R = Q + S = R + S = L$$

is a line in our linear manifold (F,A). We have to show first that there exists at least one number in the cross ratio $\begin{bmatrix} P & Q \\ S & R \end{bmatrix}$.

To do this consider any element r such that $R = Fr$. Since r belongs to $P \oplus Q$, there exist uniquely determined elements p,q in P and Q respectively such that $r = p + q$; and p and q are different from 0, since R is different from Q and P. Hence $P = Fp, Q = Fq, R = F(p + q)$. Since p is on $Q + S = P + Q$, there exist elements q' and w in Q and S respectively such that $p = q' + w$; and q' and w are different from 0, since P is different from S and Q. Hence $Q = Fq'$ and $S = Fw$. From $Fq = Q = Fq'$ we deduce the existence of a number $s \neq 0$ such that $q' = - sq$; and now it follows that $S = Fw = F(p + sq)$. Hence s belongs to the cross ratio $\begin{bmatrix} P & Q \\ S & R \end{bmatrix}$; and we note that $s \neq 0,\ 1$.

Suppose now that s is some definite number in the cross ratio. Then there exist elements p,q in A such that

$$P = Fp, \qquad Q = Fq, \qquad R = F(p + q), \qquad S = F(p + sq).$$

Consider now a number s' in F conjugate to s. Then there exists a number $t \neq 0$ in F such that $s = t^{-1}s't$; and we find

$$P = F(tp), \qquad Q = F(tq), \qquad R = F(tp + tq), \qquad S = F(tp + s'(tq))$$

so that s' belongs to the cross ratio $\begin{bmatrix} P & Q \\ S & R \end{bmatrix}$. Assume conversely that s' belongs to the cross ratio. Then there exist elements p',q' in A such that

$$P = Fp', \qquad Q = Fq', \qquad R = F(p' + q'), \qquad S = F(p' + s'q').$$

There exist numbers h,k,m,n in F neither of which is 0 such that

$$p' = hp, \qquad q' = kq, \qquad p' + q' = m(p + q), \qquad p' + s'q' = n(p + sq).$$

Hence $m(p + q) = hp + kq$ and $n(p + sq) = hp + s'kq$. From the independence of p and q we deduce now the validity of $m = h = k$ and $n = h$, $ns = s'k$ or $s' = nsn^{-1}$ so that s' is conjugate to s. Thus the cross ratio $\begin{bmatrix} P & Q \\ S & R \end{bmatrix}$ is the totality of numbers conjugate to s, as we intended to prove.

REMARK 1: We have shown incidentally the obvious fact that 0, 1 do not belong to the cross ratio $\begin{bmatrix} P & Q \\ S & R \end{bmatrix}$, if we insist that the points P,Q,R,S are all distinct. If we would permit $S = P$, then our definition would give the value 0 to the cross ratio; and if we permit $S = R$, then our definition would give the value 1 to the cross ratio. These two possibilities we shall consequently allow occasionally. But $S = Q$ cannot be admitted unless we want to give the value ∞ to the cross ratio; and the points P,Q,R have to be considered distinct in any case. By and large it will prove most convenient to insist that P,Q,R,S are four distinct collinear points.

REMARK 2: From $P = Fp$, $Q = Fq$, $R = F(p + q)$, $S = F(p + sq)$ we deduce also $S = F(q + s^{-1}p)$, provided $s \neq 0$; and this shows that $\begin{bmatrix} Q & P \\ S & R \end{bmatrix}$ is the totality of inverses of elements in $\begin{bmatrix} P & Q \\ S & R \end{bmatrix}$. This fact we may express in the form

$$\begin{bmatrix} P & Q \\ S & R \end{bmatrix}^{-1} = \begin{bmatrix} Q & P \\ S & R \end{bmatrix};$$

and one verifies similarly that

$$\begin{bmatrix} P & Q \\ S & R \end{bmatrix}^{-1} = \begin{bmatrix} P & Q \\ R & S \end{bmatrix}.$$

Next we note that $1 - t^{-1}st = t^{-1}(1 - s)t$. Thus $1 - s$ ranges with s over a whole class of conjugate numbers; and with every class C of conjugate numbers $1 - C$ is a well defined class of conjugate numbers in F. If we note now that

$$S = F(p + sq), \qquad P = F(-p), \qquad Q = F[(p + sq) + (-p)],$$
$$R = F[(p + sq) + (1 - s)(-p)],$$

then we find that

$$\begin{bmatrix} S & P \\ R & Q \end{bmatrix} = 1 - \begin{bmatrix} P & Q \\ S & R \end{bmatrix}.$$

The values of the cross ratios which we obtain by all the 24 possible permutations of the four points P,Q,R,S are now easily computed. In particular one checks the validity of *the symmetry formulas*

$$\begin{bmatrix} P & Q \\ S & R \end{bmatrix} = \begin{bmatrix} S & R \\ P & Q \end{bmatrix} = \begin{bmatrix} Q & P \\ R & S \end{bmatrix} = \begin{bmatrix} R & S \\ Q & P \end{bmatrix}.$$

REMARK 3: One verifies without difficulty that every class of conjugate numbers in F, not 0 or 1, is the cross ratio of some quadruplet of collinear points.

The following formulas will prove useful in our discussion; and they may also remind the reader of known expressions for cross ratios in classical instances.

Lemma 1: *Suppose that p and q are two independent elements in A.*

(*a*) *If u,v,w are three distinct numbers in F, then*

$$\begin{bmatrix} Fq & F(p + uq) \\ F(p + wq) & F(p + vq) \end{bmatrix} = \langle (v - u)(w - u)^{-1} \rangle.$$

(*b*) *If h,k,m,n are four distinct numbers in F, then*

$$\begin{bmatrix} F(p + hq) & F(p + kq) \\ F(p + nq) & F(p + mq) \end{bmatrix} = \langle (h - n)(k - n)^{-1}(k - m)(h - m)^{-1} \rangle.$$

PROOF: Formula (*a*) one verifies by realising that

$$F(p + vq) = F[(v - u)q + (p + uq)],$$
$$F(p + wq) = F[(v - u)q + (v - u)(w - u)^{-1}(p + uq)].$$

To prove (*b*) we verify first the following equality.

(*c*) If x,y,z are distinct numbers in F, then

$$F(p + xq) = F[- (p + yq) + (y - x)(z - x)^{-1}(p + zq)].$$

This identity follows from the following identities:

$$(y - x)(z - x)^{-1} - 1 = (y - x - z + x)(z - x)^{-1} = (y - z)(z - x)^{-1},$$
$$(y - x)(z - x)^{-1}z - y = (y - x)(z - x)^{-1}z - z - (y - z)$$
$$= [(y - x)(z - x)^{-1} - 1]z - (y - z)$$
$$= (y - z)(z - x)^{-1}z - (y - z) = (y - z)(z - x)^{-1}[z - (z - x)]$$
$$= (y - z)(z - x)^{-1}x.$$

From (*c*) one deduces (*b*) by remarking that

$$F(p + mq) = F[- (p + hq) + (h - m)(k - m)^{-1}(p + kq)]$$

and

$$F(p + nq) = F[- (p + hq) + (h - n)(k - n)^{-1}(p + kq)]$$
$$= F[- (p + hq) + (h - n)(k - n)^{-1}(k - m)(h - m)^{-1}(h - m)(k - m)^{-1}(p + kq)].$$

So far our discussion of cross ratios has been very similar to the classical discussion of this concept, in spite of the fact that the cross ratio is a class of conjugate numbers. We come now to a result which is rather different from classical results. If P,Q,R are three distinct collinear points, and if $\langle s \rangle$ is a class of conjugate numbers in F, then we may always find a point S on the line $P + Q$ such that $\begin{bmatrix} P & Q \\ S & R \end{bmatrix} = \langle s \rangle$, since we can always find elements p,q in A such that $P = Fp$, $Q = Fq$, $R = F(p + q)$, and since we may then choose $S = F(p + sq)$. But the point S is, in general, not uniquely determined by P,Q,R and the cross ratio, as may be seen from the following proposition.

Proposition 2: *The following properties of the four distinct points P,Q,R,S on the line L are equivalent.*

(I) *If X is a point on L such that* $\begin{bmatrix} P & Q \\ S & R \end{bmatrix} = \begin{bmatrix} P & Q \\ X & R \end{bmatrix}$, *then $S = X$.*

(II) $\begin{bmatrix} P & Q \\ S & R \end{bmatrix}$ *is a one element set.*

(III) $\begin{bmatrix} P & Q \\ S & R \end{bmatrix}$ *is part of the center of F.*

PROOF: The equivalence of properties (III) and (II) is evident once one remembers that an element in F belongs to the center of F if, and only if, it commutes with every element in F. To prove the equivalence of (I) and (II) we remark first that there exist elements p,q in A and an element s in F such that

$$P = Fp, \quad Q = Fq, \quad R = F(p + q), \quad S = F(p + sq)\,; \begin{bmatrix} P & Q \\ S & R \end{bmatrix} = \langle s \rangle.$$

Assume now the validity of (I) and consider some element $f \neq 0$ in F. If we let $X = F(p + f^{-1}sfq)$, then

$$\begin{bmatrix} P & Q \\ X & R \end{bmatrix} = \langle s \rangle = \begin{bmatrix} P & Q \\ S & R \end{bmatrix};$$

and it follows from (I) that $F(p + f^{-1}sfq) = X = S = F(p + sq)$; and this implies $f^{-1}sf = s$ because of the independence of the elements p,q. Thus $\langle s \rangle$ is a one element set; and we have shown that (II) is a consequence of (I). If conversely (II) is satisfied, and if X is a point on L such that $\begin{bmatrix} P & Q \\ S & R \end{bmatrix} = \begin{bmatrix} P & Q \\ X & R \end{bmatrix}$, then $X = F(p + xq)$, since X is different from Q. Hence $\langle s \rangle = \langle x \rangle$; and this implies $s = x$ by (II) so that $S = X$. Thus (I) is a consequence of (II); and this completes the proof.

REMARK 4: It is an immediate consequence of Proposition 2 that commutativity of the field F is necessary and sufficient for the fact that three collinear points and the cross ratio determine uniquely the fourth point; see Theorem 2 below.

REMARK 5: If F does not have characteristic 2, then -1 is a center element different from 0 and 1. It is customary to term the four distinct collinear points P,Q,R,S *a harmonic set*, if $\begin{bmatrix} P & Q \\ S & R \end{bmatrix} = -1$. In this case the four points may be represented in the form:

$$P = Fp, \qquad Q = Fq, \qquad R = F(p+q), \qquad S = F(p-q).$$

It is now easy to verify that the following property is necessary and sufficient for the four points P,Q,R,S to form a harmonic set:

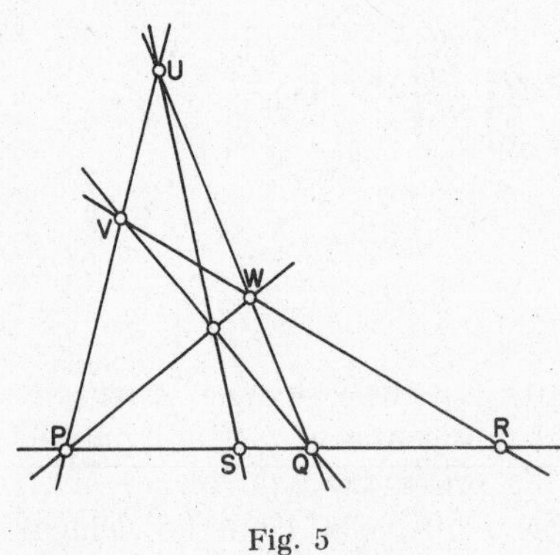

Fig. 5

If U is a point not on the line $P+Q$, if V and W are points such that

$$U+V = V+P = P+U,$$
$$U+W = W+Q = Q+U,$$
$$V+W = W+R = R+V,$$

then the three lines $U+S$, $W+P$, $V+Q$ meet in a point.

Since we do not want to exclude fields of characteristic 2 from our discussion, it will be impossible [and usually unnecessary] to give harmonic sets the preferential treatment so often accorded to them.

Theorem 1: *Suppose that P,Q,R,S and P',Q',R',S' are quadruples of distinct collinear points.*

(*a*) *There exists a projectivity π of the linear manifold (F,A) such that $P^\pi = P'$, $Q^\pi = Q'$, $R^\pi = R'$, $S^\pi = S'$ if, and only if, there exists an automorphism α of the field F such that* $\begin{bmatrix} P & Q \\ S & R \end{bmatrix}^\alpha = \begin{bmatrix} P' & Q' \\ S' & R' \end{bmatrix}$.

(*b*) *There exists a collineation σ of (F,A) such that $P^\sigma = P'$, $Q^\sigma = Q'$, $R^\sigma = R'$, $S^\sigma = S'$ if, and only if,* $\begin{bmatrix} P & Q \\ S & R \end{bmatrix} = \begin{bmatrix} P' & Q' \\ S' & R' \end{bmatrix}$.

PROOF: We note first that we may represent the points of the first quadruple in the form:

$$P = Fp, \quad Q = Fq, \quad R = F(p+q), \quad S = F(p+sq), \quad \begin{bmatrix} P & Q \\ S & R \end{bmatrix} = \langle s \rangle.$$

It follows from the First Fundamental Theorem of Projective Geometry

[III.1] that every projectivity π is induced by a semi-linear transformation α. But then

$$P^\pi = Fp^\alpha, \qquad Q^\pi = Fq^\alpha, \qquad R^\pi = F(p^\alpha + q^\alpha), \qquad S^\pi = F(p^\alpha + s^\alpha q^\alpha),$$

$$\begin{bmatrix} P^\pi & Q^\pi \\ S^\pi & R^\pi \end{bmatrix} = \langle s^\alpha \rangle = \langle s \rangle^\alpha = \begin{bmatrix} P & Q \\ S & R \end{bmatrix}^\alpha;$$

and this proves the necessity of our condition.

If there exists an automorphism α such that $\begin{bmatrix} P' & Q' \\ S' & R' \end{bmatrix} = \begin{bmatrix} P & Q \\ S & R \end{bmatrix}^\alpha$, then $\begin{bmatrix} P' & Q' \\ S' & R' \end{bmatrix} = \langle s^\alpha \rangle$. Consequently there exist elements p',q' in A such that

$$P' = Fp',\ Q' = Fq',\ R' = F(p' + q'),\ S' = F(p' + s^\alpha q').$$

Since p,q as well as p',q' are pairs of independent elements, one constructs easily a semi-linear transformation (α,β) of (F,A) such that $p^\beta = p'$, $q^\beta = q'$, $(xy)^\beta = x^\alpha y^\beta$ for x in F and y in A [use the fact that p,q as well as p',q' are part of some basis and that any two bases contain the same number of elements; see Chapter II]. This semi-linear transformation induces a projectivity π which maps P,Q,R,S upon P',Q',R',S' respectively. This completes the proof of (a). The proof of (b) is quite similar to the proof of (a), once one remembers that collineations are induced by linear transformations [$\alpha = 1$ in the above notation].

The preceding theorem shows that cross ratios characterize in a way the projective structure of quadruples of collinear points. Our next objective is the characterization of those mappings which preserve some or all cross ratios. It is a fairly obvious consequence of Theorem 1 that it will suffice to consider permutations of the points on some line which possess at least three fixed points.

Suppose now that σ is a permutation of the points on the line L and that the three distinct points U,V,W are fixed points of σ. Then we show as usual the existence of [independent] elements p,q such that $U = Fp$, $V = Fq$, $W = F(p + q)$. Every point, not V, on L may be represented in one and only one way in the form $F(p + xq)$ for x in F so that the elements p,q define a one-to-one correspondence between the points, not V, on L and the elements in F. Consequently there exists a well determined permutation σ' of the numbers in F such that

$$[F(p + xq)]^\sigma = F(p + x^{\sigma'} q).$$

It is clear that $0 = 0^{\sigma'}$, $1 = 1^{\sigma'}$; and that every permutation σ' of the numbers in F which leaves invariant 0 and 1 comes from a permutation of the points on L with fixed points U,V,W.

Proposition 3: *Assume that p,q are two independent elements in A, that σ is a permutation of the points on the line $Fp + Fq = L$ with fixed points Fp, Fq, $F(p + q)$ and that the corresponding permutation of the numbers in F is denoted by σ too so that*

$$[F(p + xq)]^\sigma = F(p + x^\sigma q) \text{ for every } x \text{ in } F.$$

If z is a number neither 0 nor 1 in the center of F, then the following properties of z and σ are equivalent.

(I) *If P,Q,R,S are four distinct points on L such that* $\begin{bmatrix} P & Q \\ S & R \end{bmatrix} = z$, *then* $\begin{bmatrix} P^\sigma & Q^\sigma \\ S^\sigma & R^\sigma \end{bmatrix} = z$.

(II) *The permutation σ of F satisfies $z = z^\sigma$ and $x^\sigma + y^\sigma = (x + y)^\sigma$, $(xyx)^\sigma = x^\sigma y^\sigma x^\sigma$ for every x,y in F.*

(III) *σ is an automorphism or anti-automorphism of F which leaves invariant z.*

We note that the center element z is the only element in its class of conjugate elements and that we may consequently identify z with its class $\langle z \rangle$. Furthermore we remind the reader that an *anti-automorphism* of the field F is a permutation of the elements in F which preserves addition, but inverts the order of multiplication $[(xy)^\sigma = y^\sigma x^\sigma]$.

Proof: Assume first the validity of (I). Then $z^\sigma = z$, since

$$\begin{bmatrix} Fp & Fq \\ F(p + zq) & F(p + q) \end{bmatrix} = z,$$

and since therefore

$$\langle z^\sigma \rangle = \begin{bmatrix} Fp & Fq \\ F(p + z^\sigma q) & F(p + q) \end{bmatrix} = \begin{bmatrix} [Fp]^\sigma & [Fq]^\sigma \\ [F(p + zq)]^\sigma & [F(p + q)]^\sigma \end{bmatrix} = z.$$

Next we deduce from Lemma 1, (*a*) that we have for every $x \neq 0$ in F

$$\begin{bmatrix} Fq & Fp \\ F(p + xq) & F(p + zxq) \end{bmatrix} = z;$$

and consequently it follows from (I) and Lemma 1, (*a*) as before that

$$z = \begin{bmatrix} Fq & Fp \\ F(p + x^\sigma q) & F(p + zx)^\sigma q) \end{bmatrix} = \langle (zx)^\sigma (x^\sigma)^{-1} \rangle.$$

But this implies clearly

(I.1) $$(zx)^\sigma = zx^\sigma \text{ for every } x \text{ in } F.$$

[Note that our proof holds for $x \neq 0$ only; but our formula is true for $x = 0$ too, since $0 = 0^\sigma$.]

If u and v are two different numbers in F, then we let $w = z^{-1}(v - u) + u$ which is possible, since $z \neq 0$. Since $z \neq 1$, w is different from u and v. Since z is in the center of F, we deduce from Lemma 1, (*a*) that

$$\begin{bmatrix} Fq & F(p + uq) \\ F(p + wq) & F(p + vq) \end{bmatrix} = \langle (v - u)(w - u)^{-1} \rangle = z.$$

As before we deduce from (i) and Lemma 1, (*a*) [which may be applied, since the permutation σ maps distinct elements upon distinct elements] that

$$z = \begin{bmatrix} Fq & F(p + u^\sigma q) \\ F(p + w^\sigma q) & F(p + v^\sigma q) \end{bmatrix} = \langle (v^\sigma - u^\sigma)(w^\sigma - u^\sigma)^{-1} \rangle.$$

Consequently

$$z(w^\sigma - u^\sigma) = v^\sigma - u^\sigma \qquad \text{or} \qquad zw^\sigma = v^\sigma - u^\sigma + zu^\sigma.$$

From (i,1) and the definition of w we deduce that

$$zw^\sigma = (zw)^\sigma = (v - u + zu)^\sigma;$$

and thus we have shown the validity of

(i.2) $\qquad (v - u + zu)^\sigma = v^\sigma - u^\sigma + zu^\sigma$ for every u,v in F.

[Actually we have verified this formula for $u \neq v$ only; but in case $u = v$, our formula is equivalent with (i.1).]

If x and y are numbers in F, then we may form $u = (z - 1)^{-1}y$, since $z \neq 1$. It follows from (i.2) that

$$(x + y)^\sigma = (x + zu - u)^\sigma = x^\sigma - u^\sigma + zu^\sigma = x^\sigma + (z - 1)u^\sigma.$$

Letting $v = 0$ in (i.2) we obtain $[(z - 1)u]^\sigma = (z - 1)u^\sigma$; and now it follows that

$$x^\sigma + (z - 1)u^\sigma = x^\sigma + [(z - 1)u]^\sigma = x^\sigma + y^\sigma.$$

Thus we have shown:

(i.3) $\qquad (x + y)^\sigma = x^\sigma + y^\sigma$ for every x and y in F.

Next we prove that

(i.4) $\qquad [k(zk - h)^{-1}h]^\sigma = k^\sigma(zk^\sigma - h^\sigma)^{-1}h^\sigma, \qquad$ provided $h \neq zk$.

This formula is trivially true, if k or $h = 0$. Thus we assume that neither h nor k is 0. If $h = k$, then (i.4) is a consequence of $(z - 1)u^\sigma = [(z - 1)u]^\sigma$ which equation is a consequence of (i.1) and (i.3). Thus we assume that $h \neq k$. Now we let $m = (z - 1)k(zk - h)^{-1}h$. If h were equal to m, then we would have $(z-1)k = zk - h$ or $k = h$ which is impossible; and $k \neq m$ is shown likewise. Now we infer from Lemma 1 (*b*) that

$$\begin{bmatrix} F(p+hq) & F(p+kq) \\ Fp & F(p+mq) \end{bmatrix} = \langle hk^{-1}(k-m)(h-m)^{-1} \rangle.$$

But $k^{-1}(k-m) = 1 - k^{-1}m = 1 - (z-1)(zk-h)^{-1}h$

$$= (zk-h)^{-1}[zk - h - (z-1)h] = (zk-h)^{-1}z(k-h)$$

and consequently

$$\begin{aligned} z(h-m) &= z[1 - (z-1)k(zk-h)^{-1}]h = z[(zk-h) - (z-1)k](zk-h)^{-1}h \\ &= z[k-h](zk-h)^{-1}h = (zk - h + h - zh)(zk-h)^{-1}h \\ &= h + (1-z)h(zk-h)^{-1}h = h[1 + (1-z)(zk-h)^{-1}h] \\ &= h(zk-h)^{-1}[zk - h + (1-z)h] = h(zk-h)^{-1}z(k-h) \\ &= hk^{-1}(k-m) \end{aligned}$$

or

$$\begin{bmatrix} F(p+hq) & F(p+kq) \\ Fp & F(p+mq) \end{bmatrix} = z.$$

Now we apply (I) and Lemma 1, (*b*) to find that

$$h^\sigma(k^\sigma)^{-1}(k^\sigma - m^\sigma)(h^\sigma - m^\sigma)^{-1} = z \quad \text{or} \quad h^\sigma(k^\sigma)^{-1}(k^\sigma - m^\sigma) = z(h^\sigma - m^\sigma).$$

By a direct computation it follows that $m^\sigma = (z-1)k^\sigma(zk^\sigma - h^\sigma)^{-1}h^\sigma$; and now (I.4) is an immediate consequence of the formula $[(z-1)u]^\sigma = (z-1)u^\sigma$ which we have used before and which is contained in (I.1) and (I.3).

(I.5) $\qquad (ab^{-1}a)^\sigma = a^\sigma(b^\sigma)^{-1}a^\sigma$ for $b \neq 0$ and a in F.

We let $k = a$ and $h = zk - b$. Then $h \neq zk$, since $b \neq 0$. Thus we may apply (I.4) and find that [$zk - h = b$ and therefore]

$$\begin{aligned} [ab^{-1}(za-b)]^\sigma &= [k(zk-h)^{-1}h]^\sigma = k^\sigma\,(zk^\sigma - h^\sigma)^{-1}h^\sigma \\ &= k^\sigma[(zk-h)^\sigma]^{-1}h^\sigma = a^\sigma(b^\sigma)^{-1}(za-b)^\sigma \end{aligned} \quad \text{by (I.1) and (I.3).}$$

From (I.3) we deduce now $(ab^{-1}za)^\sigma = a^\sigma(b^\sigma)^{-1}(za)^\sigma$; and (I.5) is a consequence of (I.1).

If we let in (I.5) in particular $a = 1$ and remember that $1^\sigma = 1$, then we find that $(b^{-1})^\sigma = (b^\sigma)^{-1}$; and now one deduces from (I.5) the validity of $(aba)^\sigma = a^\sigma b^\sigma a^\sigma$. Thus (II) is a consequence of (I).

Assume now the validity of (II). Considering a and $b = a^{-1}$ we find that $a^\sigma = a^\sigma(a^{-1})^\sigma a^\sigma$ or $(a^{-1})^\sigma = (a^\sigma)^{-1}$. Letting $b = 1$ and remembering $1^\sigma = 1$ we find that $(a^2)^\sigma = (a^\sigma)^2$; and letting $a = x + y$ we deduce from this last formula:

$$\begin{aligned} (a^\sigma)^2 + a^\sigma b^\sigma + b^\sigma a^\sigma + (b^\sigma)^2 &= [(a+b)^\sigma]^2 = [(a+b)^2]^\sigma \\ &= [a^2\ b + ab + a + b^2]^\sigma = (a^2)^\sigma + (ab)^\sigma + (ba)^\sigma + (b^2)^\sigma \\ &= (a^\sigma)^2 + (ab)^\sigma + (ba)^\sigma + (b^\sigma)^2 \end{aligned}$$

or

$$a^\sigma b^\sigma + b^\sigma a^\sigma = (ab)^\sigma + (ba)^\sigma.$$

Next we note that

$$(ba)^\sigma = [(ab)(ab)^{-1}(ba)]^\sigma = [a(b(ab)^{-1}b)a]^\sigma = a^\sigma[b(a)b^{-1}b]^\sigma a^\sigma$$
$$= a^\sigma b^\sigma[(ab)^{-1}]^\sigma b^\sigma a^\sigma = a^\sigma b^\sigma[(ab)^\sigma]^{-1}b^\sigma a^\sigma.$$

Hence

$$a^\sigma b^\sigma + b^\sigma a^\sigma = (ab)^\sigma + (ba)^\sigma = (ab)^\sigma + a^\sigma b^\sigma[(ab)^\sigma]^{-1}b^\sigma a^\sigma$$

and consequently

$$[(ab)^\sigma - a^\sigma b^\sigma][(ab)^\sigma]^{-1}[(ab)^\sigma - b^\sigma a^\sigma] = 0$$

so that we have shown:

(II.1) If a,b are elements in F, then $(ab)^\sigma = \begin{cases} a^\sigma b^\sigma \\ b^\sigma a^\sigma. \end{cases}$

Next we prove:

(II.2) If a,b,c are elements in F, then

(*a*) either $(ab)^\sigma = a^\sigma b^\sigma$ and $(ac)^\sigma = a^\sigma c^\sigma$ or else $(ab)^\sigma = b^\sigma a^\sigma$ and $(ac)^\sigma = c^\sigma a^\sigma$; and

(*b*) either $(ba)^\sigma = b^\sigma a^\sigma$ and $(ca)^\sigma = c^\sigma a^\sigma$ or else $(ba)^\sigma = a^\sigma b^\sigma$ and $(ca)^\sigma = a^\sigma c^\sigma$.

Suppose (*a*) were false. Then we may assume without loss in generality that $(ab)^\sigma \neq a^\sigma b^\sigma$. It follows from (II.1) that $b^\sigma a^\sigma = (ab)^\sigma \neq a^\sigma b^\sigma$; and this implies because of the invalidity of (*a*) that $(ac)^\sigma \neq c^\sigma a^\sigma$. Hence $a^\sigma c^\sigma = (ac)^\sigma$ by (II.1). Likewise

$$b^\sigma a^\sigma + a^\sigma c^\sigma = (ab)^\sigma + (ac)^\sigma = (ab + ac)^\sigma = [a(b + c)]^\sigma$$
$$= \begin{cases} a^\sigma(b + c)^\sigma = a^\sigma b^\sigma + a^\sigma c^\sigma \\ (b + c)^\sigma a^\sigma = b^\sigma a^\sigma + c^\sigma a^\sigma \end{cases}$$

and both these possibilities lead to a contradiction. Hence (*a*) is true; and the validity of (*b*) is shown likewise.

If σ were neither an automorphism nor an anti-automorphism of F, then there would exist elements d,e,f,g in F such that $(de)^\sigma \neq d^\sigma e^\sigma$ and $(fg)^\sigma \neq g^\sigma f^\sigma$. We infer from (II.1) and (II.2) that

$$(de)^\sigma = e^\sigma d^\sigma \text{ and } (dx)^\sigma = x^\sigma d^\sigma,\ (xe)^\sigma = e^\sigma x^\sigma \text{ for every } x \text{ in } F,$$
$$(fg)^\sigma = f^\sigma g^\sigma \text{ and } (fy)^\sigma = f^\sigma y^\sigma,\ (yg)^\sigma = y^\sigma g^\sigma \text{ for every } y \text{ in } F.$$

In particular we have therefore $(dg)^\sigma = d^\sigma g^\sigma = g^\sigma d^\sigma$ and $(fe)^\sigma = e^\sigma f^\sigma = f^\sigma e^\sigma$. From (II.1) we deduce now that

$$e^\sigma d^\sigma + (fe)^\sigma + (dg)^\sigma + f^\sigma g^\sigma = (de)^\sigma + (fe)^\sigma + (dg)^\sigma + (fg)^\sigma$$
$$= [(d + f)(e + g)]^\sigma$$
$$= \begin{cases} (d + f)^\sigma(e + g)^\sigma = d^\sigma e^\sigma + (fe)^\sigma + (dg)^\sigma + f^\sigma g^\sigma \\ (e + g)^\sigma(d + f)^\sigma = e^\sigma d^\sigma + (fe)^\sigma + (dg)^\sigma + g^\sigma f^\sigma. \end{cases}$$

But both these possibilities lead to a contradiction. Thus it is impossible that σ is neither an automorphism nor an anti-automorphism. Hence (III) is a consequence of (II).

Assume finally the validity of (III). Then σ certainly satisfies the following conditions:

$$(x + y)^\sigma = x^\sigma + y^\sigma, \qquad (zx)^\sigma = zx^\sigma \text{ for } x,y \text{ in } F,$$

since z is a center element of F such that $z = z^\sigma$. Consider now four distinct points P,Q,R,S on the line L such that $\begin{bmatrix} P & Q \\ S & R \end{bmatrix} = z$. We distinguish two cases.

Case 1: One of the points P,Q,R,S is the point Fq.

In Remark 2 we have shown that the cross ratio does not change its value, if the four points undergo a certain transitive group of permutations. Hence we may assume without loss in generality that

$$P = Fq, \qquad Q = F(p + uq), \qquad R = F(p + vq), \qquad S = F(p + wq).$$

It follows from Lemma 1, (a) that

$$(v - u)\,(w - u)^{-1} = z \qquad \text{or} \qquad v - u = z(w - u).$$

From the properties of σ already stated it follows that

$$v^\sigma - u^\sigma = (v - u)^\sigma = [z(w - u)]^\sigma = z(w - u)^\sigma = z(w^\sigma - u^\sigma)$$

or $(v^\sigma - u^\sigma)(w^\sigma - u^\sigma)^{-1} = z$; and now it follows from Lemma 1, (a) that

$$\begin{bmatrix} P^\sigma & Q^\sigma \\ S^\sigma & R^\sigma \end{bmatrix} = \begin{bmatrix} Fq & F(p + u^\sigma q) \\ F(p + w^\sigma q) & F(p + v^\sigma q) \end{bmatrix} = \langle (v^\sigma - u^\sigma)(w^\sigma - u^\sigma)^{-1} \rangle = z.$$

Case 2: None of the points P,Q,R,S is the point Fq.

Then our points have the form $P = F(p + hq)$, $Q = F(p + kq)$, $R = F(p + mq)$, $S = F(p + nq)$; and it follows from Lemma 1, (b) that

$$(*) \qquad (h - n)(k - n)^{-1}(k - m)(h - m)^{-1} = z;$$

and that

$$(**) \qquad \begin{bmatrix} P^\sigma & Q^\sigma \\ S^\sigma & R^\sigma \end{bmatrix} = \langle (h^\sigma - n^\sigma)(k^\sigma - n^\sigma)^{-1}(k^\sigma - m^\sigma)(h^\sigma - m^\sigma)^{-1} \rangle.$$

If σ happens to be an automorphism of the field F, then we deduce from (*), (**) and the invariance of z immediately that

$$\begin{bmatrix} P^\sigma & Q^\sigma \\ S^\sigma & R^\sigma \end{bmatrix} = z.$$

Thus we assume finally that σ is an anti-automorphism of F. We note first that

$$(+)\left\{\begin{aligned} &1-[(h-n)(k-n)^{-1}-1][(h-n)(h-m)^{-1}-1]\\ &\quad=(h-n)(k-n)^{-1}+(h-n)(h-m)^{-1}-(h-n)(k-n)^{-1}(h-n)(h-m)^{-1}\\ &\quad=(h-n)[(k-n)^{-1}(h-m)+1-(k-n)^{-1}(h-n)](h-m)^{-1}\\ &\quad=(h-n)(k-n)^{-1}[(h-m)+(k-n)-(h-n)](h-m)^{-1}\\ &\quad=(h-n)(k-n)^{-1}(k-m)(h-m)^{-1}.\end{aligned}\right.$$

Since this expression is by (*) the center element z, and since from $z=1-uv$ we may deduce $z=u^{-1}[1-uv]u=1-vu$, we find that

$$\begin{aligned} z&=(h-n)(k-n)^{-1}(k-m)(h-m)^{-1}\\ &=1-[(h-n)(k-n)^{-1}-1][(h-n)(h-m)^{-1}-1]\\ &=1-[(h-n)(h-m)^{-1}-1][(h-n)(k-n)^{-1}-1]\\ &=1-[(n-h)(m-h)^{-1}-1][(n-h)(n-k)^{-1}-1] \qquad \text{[by (+)]}\\ &=(n-h)(m-h)^{-1}(m-k)(n-k)^{-1}\\ &=(h-n)(h-m)^{-1}(k-m)(k-n)^{-1}\\ &=(h-m)^{-1}(k-m)(k-n)^{-1}(h-n)\end{aligned}$$

since z is in the center and since therefore transformation with $(h-n)$ does not change the value of z. Now it follows from (**) and the invariance of z under σ that

$$\begin{bmatrix} P^\sigma & Q^\sigma \\ S^\sigma & R^\sigma \end{bmatrix}=\langle[(h-m)^{-1}(k-m)(k-n)^{-1}(h-n)]^\sigma\rangle=z;$$

and thus we have shown that (I) is a consequence of (III). This completes the proof.

REMARK 6: When proving that (III) is a consequence of (II), we actually showed the following fact:

A permutation σ of the field F is an automorphism or anti-automorphism of F if, and only if, $1=1^\sigma$, $(x+y)^\sigma=x^\sigma+y^\sigma$, $(xyx)^\sigma=x^\sigma y^\sigma x^\sigma$.

These permutations σ of F shall be termed *semi-automorphisms of F*.

The proof we gave is a reproduction of a proof due to Hua [see L.-K. Hua: On the Automorphism of a Sfield, *Proceedings of the National Acad. of Sciences*, vol. 35 (1949), 386-389]. For a more recent investigation of the general problem of semi-automorphisms the reader is advised to consult N. Jacobson–C. E. Rickart: Jordan Homomorphisms of Rings, *Trans. Amer. Math. Soc.*, vol. 69 (1950), pp. 479–502.

REMARK 7: If one considers in particular the case $z=-1$ [which is impossible if, and only if, F has characteristic 2], then one investigates the permutations of the points of a line which preserve harmonic sets [Remark 5]; and one aims at a generalization of what is known as von

Staudt's Theorem; see in this context Proposition 4 below and G. Ancochea: Le théorème de von Staudt en géométrie projective quaternionienne, *Journal für die reine und angewandte Mathematik*, vol. 184 (1942), pp. 193-198, where further references may be found.

For a convenient enunciation of further results we have to introduce a concept which generalizes the relation between point permutations and number permutations appearing in Proposition 3. Consider a one-to-one correspondence σ mapping the totality of points on the line L upon the totality of points on the line L^σ. Then we term the permutation τ of the numbers in the field F a *σ-admissible permutation of F*, if there exist independent elements p,q in L and independent elements p',q' in L^σ such that

$$(Fp)^\sigma = Fp',\ [F(p+q)]^\sigma = F(p'+q'),\ (Fq)^\sigma = Fq',$$
$$[F(p+xq)]^\sigma = F(p'+x^\tau q') \qquad \text{for every } x \text{ in } F.$$

We note that $0 = 0^\tau, 1 = 1^\tau$ is an immediate consequence of the preceding equations; and that the field permutation σ appearing in Proposition 3 is admissible with respect to the point permutation σ. It is easy to see that to every one-to-one mapping σ of the points on L upon the points on L^σ there exists at least one σ-admissible permutation of the field F; and that permutations of the field F which leave invariant 0 and 1 lead [in many ways] to one-to-one mappings of the points on one line onto the points on some other line.

Lemma 2: *If σ is a one-to-one correspondence mapping the points on the line L upon the totality of points on the line L^σ, and if the semi-automorphism α of the field F is σ-admissible, then*

$$\begin{bmatrix} P^\sigma & Q^\sigma \\ S^\sigma & R^\sigma \end{bmatrix} = \begin{bmatrix} P & Q \\ S & R \end{bmatrix}^\alpha \quad \textit{for any four points } P,Q,R,S \textit{ on } L.$$

Proof: Since σ is α-admissible, there exist elements p,q,p',q' such that

$$L = Fp + Fq, \qquad L^\sigma = Fp' + Fq',$$
$$(Fp)^\sigma = Fp', \qquad [F(p+q)]^\sigma = F(p'+q'), \qquad (Fq)^\sigma = Fq',$$
$$[F(p+xq)]^\sigma = F(p'+x^\alpha q') \text{ for every } x \text{ in } F.$$

From now on we are going to use arguments quite similar to those used already when deriving (i) from (iii) in the proof of Proposition 3.

Consider four distinct points P,Q,R,S on L. We distinguish two cases.

Case 1: One of the four points is Fq.

We may assume without loss in generality that

$$P = Fq \text{ and } Q = F(p+uq),\ R = F(p+vq),\ S = F(p+wq).$$

Then u,v,w are three distinct numbers in F; and we deduce from Lemma 1, (*a*) that

$$\begin{bmatrix} P & Q \\ S & R \end{bmatrix} = \langle (v-u)(w-u)^{-1} \rangle, \qquad \begin{bmatrix} P^\sigma & Q^\sigma \\ S^\sigma & R^\sigma \end{bmatrix} = \langle (v^\alpha - u^\alpha)(w^\alpha - u^\alpha)^{-1} \rangle.$$

But α is an automorphism or anti-automorphism of F; and the numbers xy^{-1} and $y^{-1}x = y^{-1}(xy^{-1})y$ are conjugate in F. Hence

$$\langle (v^\alpha - u^\alpha)(w^\alpha - u^\alpha)^{-1} \rangle = \langle [(v - u)(w - u)^{-1}]^\alpha \rangle$$

or

$$\begin{bmatrix} P^\sigma & Q^\sigma \\ S^\sigma & R^\sigma \end{bmatrix} = \begin{bmatrix} P & Q \\ S & R \end{bmatrix}^\alpha$$

Case 2: None of the four points P,Q,R,S equals Fq.

In this case we have $P = F(p + hq)$, $Q = F(p + kq)$, $R = F(p + mq)$, $S = F(p + nq)$ where h,k,m,n are four different numbers in F. It follows from Lemma 1, (b) that

$$\begin{bmatrix} P & Q \\ S & R \end{bmatrix} = \langle (h - n)(k - n)^{-1}(k - m)(h - m)^{-1} \rangle,$$

$$\begin{bmatrix} P^\sigma & Q^\sigma \\ S^\sigma & R^\sigma \end{bmatrix} = \langle (h^\alpha - n^\alpha)(k^\alpha - n^\alpha)^{-1}(k^\alpha - m^\alpha)(h^\alpha - m^\alpha)^{-1} \rangle.$$

During the last part of the proof of Proposition 3 we verified an identity:

$$(+) \qquad (h - n)(k - n)^{-1}(k - m)(h - m)^{-1}$$
$$= 1 - [(h - n)(k - n)^{-1} - 1][(h - n)(h - m)^{-1} - 1] = t;$$

and this number t is conjugate in F to

$$\begin{aligned} t' &= 1 - [(h - n)(h - m)^{-1} - 1][(h - n)(k - n)^{-1} - 1] \\ &= 1 - [(n - h)(m - h)^{-1} - 1][(n - h)(n - k)^{-1} - 1] \\ &= (n - h)(m - h)^{-1}(m - k)(n - k)^{-1} \end{aligned}$$

by a second application of the identity (+); and this number t' is conjugate in F to

$$t'' = (m - h)^{-1}(m - k)(n - k)^{-1}(n - h) = (h - m)^{-1}(k - m)(k - n)^{-1}(h - n)$$

which is obtained from t by inversion of the order of the factors. We may state this result as follows:

$$(++) \quad \langle (h - n)(k - n)^{-1}(k - m)(h - m)^{-1} \rangle = \langle (h - m)^{-1}(k - m)(k - n)^{-1}(h - n) \rangle.$$

If α is an automorphism, then we may apply it on the first expression in (++); and if α is an anti-automorphism, then we apply it on the second expression in (++). In this way we see that

$$\begin{bmatrix} P & Q \\ S & R \end{bmatrix}^\alpha = \langle (h^\alpha - n^\alpha)(k^\alpha - n^\alpha)^{-1}(k^\alpha - m^\alpha)(h^\alpha - m^\alpha)^{-1} \rangle = \begin{bmatrix} P^\sigma & Q^\sigma \\ S^\sigma & R^\sigma \end{bmatrix};$$

and this completes the proof of the lemma.

Proposition 4: *Suppose that σ is a one-to-one correspondence mapping the points on the line L upon the totality of points on the line L^σ; and suppose that the center of F contains at least three elements. Then σ preserves all the* *cross ratios if, and only if, the σ-admissible permutations of the field F are semi-automorphisms which leave invariant every class of conjugate elements in F.*

PROOF: The sufficiency of our condition is an immediate consequence of Lemma 2. Assume conversely that σ preserves all the cross ratios and consider a σ-admissible permutation τ of the elements in the field F. Then there exist pairs of independent elements p,q and p',q' in L and L^σ respectively such that

$$(Fp)^\sigma = Fp', \qquad [F(p+q)]^\sigma = F(p'+q'), \qquad (Fq)^\sigma = Fq',$$
$$[F(p+xq)]^\sigma = F(p'+x^\tau q') \text{ for every } x \text{ in } F.$$

There exists clearly a linear transformation η of the F-space L^σ upon the F-space L such that $(xp'+yq')^\eta = xp+yq$ for x,y in F. This linear transformation η induces a one-to-one correspondence σ' of the totality of points on L^σ upon the totality of points on L; and it is easy to see [either by application of Lemma 1 or by application of Lemma 2] that σ' preserves cross ratios. Consequently $\sigma\sigma'$ is a cross ratio preserving permutation of the points on the line L; and this permutation has clearly the following further properties:

$$(Fp)^{\sigma\sigma'} = Fp, \qquad [F(p+q)]^{\sigma\sigma'} = F(p+q), \qquad (Fq)^{\sigma\sigma'} = Fq,$$
$$[F(p+xq)]^{\sigma\sigma'} = F(p+x^\tau q) \text{ for } x \text{ in } F.$$

By hypothesis there exists in the center of F a number $z \neq 0, 1$. Since the permutation $\sigma\sigma'$ preserves all cross ratios, it satisfies Proposition 3, (I); and consequently it follows from Proposition 3 that τ is a semi-automorphism. That this semi-automorphism τ leaves invariant every class of conjugate elements in F, is an immediate consequence of the fact that

$$\langle x\rangle = \begin{bmatrix} Fp & Fq \\ F(p+xq) & F(p+q) \end{bmatrix} = \begin{bmatrix} Fp & Fq \\ F(p+x^\tau q) & F(p+q) \end{bmatrix} = \langle x^\tau \rangle$$

holds for every x in F. This completes the proof.

REMARK 8: If σ is a semi-automorphism of the field F which leaves invariant every class of conjugate elements in F, then σ leaves invariant every element in the center Z of F. If σ is an automorphism of the field F which leaves invariant every element in the center Z of F, and if F is finite over Z, then it is a celebrated theorem that σ is an inner automorphism of F [see, for instance, Artin-Nesbitt-Thrall (1), p. 66, Corollary 7.2D]. But if F is infinite over Z, then there exist automorphisms of F which leave invariant every class of conjugate elements in F and which are *not* inner automorphisms of F. [An example may be found in G. Köthe: Schiefkörper unendlichen Ranges über dem Zentrum, *Math.*

Ann., vol. 105 (1931), pp. 15-39.] If F is finite over Z, and if σ is an anti-automorphism of F, then σ leaves invariant every element in Z whereas the elements x and x^σ, for x in F, always satisfy the same equations over Z. It follows that x and x^σ are conjugate elements in F [see, for instance, Artin-Nesbitt-Thrall (1), p. 67, Theorem 7.2 E]. An example of such an anti-automorphism may be obtained by mapping every real quaternion $x_0 + x_1 i + x_2 j + x_3 k$ upon its "conjugate" $x_0 - x_1 i - x_2 j - x_3 k$. This shows that semi-automorphisms which leave invariant every class of conjugate elements may be inner automorphisms, automorphisms which are not inner and anti-automorphisms which are not automorphisms; and it is clear that cross ratios do not suffice for distinguishing between these various kinds of mappings of the points on a line.

Theorem 2: *The field F is commutative if, and only if, the identity is the only permutation of the points on a line which preserves cross ratios and possesses three fixed points.*

This theorem is a kind of complement to the Second Fundamental Theorem of Projective Geometry [III.3]; and it is also an improvement upon Remark 4.

PROOF: Assume first that the identity is the only permutation of the points on a line which preserves cross ratios and possesses three fixed points. If $g \neq 0$ is a number in F, and if p,q are two independent elements in A, then we may define a permutation σ of the points on the line $L = Fp + Fq$ by the rule:

$$(Fq)^\sigma = Fq, \qquad [F(p + xq)]^\sigma = F(p + g^{-1}xgq) \text{ for } x \text{ in } F.$$

It is clear that σ possesses the three distinct fixed points, Fp, Fq, $F(p+q)$; and it follows from Lemma 2 that σ preserves cross ratios, since the mapping $x \to g^{-1}xg$ is an inner automorphism of the field F. It follows from our hypothesis that $\sigma = 1$. Hence $p + g^{-1}xgq$ belongs, for every x in F, to $F(p + xq)$. This implies $x = g^{-1}xg$ for every x in F, since p and q are independent elements. Thus g belongs to the center of F. Since we have shown that every element in F belongs to the center of F, we have shown the commutativity of F.

Assume conversely the commutativity of the field F. If F consists of the two elements 0 and 1 only, then lines carry three points only; and the only permutation of the points on a line with three fixed points is naturally the identity. Thus we may assume that F contains at least three elements. Consider now a cross ratio preserving permutation σ of the points on the line L which possesses three fixed points. These fixed points may be assumed to have the form Fp, $F(p + q)$, Fq. Then there exists a σ-admissible permutation τ of the numbers in F such that

$$[F(p +)]xq^\sigma = F(p + x^\tau q)$$

for every x in F. It follows from Proposition 4 that τ is a semi-automorphism of F which leaves invariant every class of conjugate elements in F. But F is a commutative field so that $\tau = 1$. Hence $\sigma = 1$, as we wanted to show.

REMARK 9: It is easy to see the equivalence of the conditions of Theorem 2 with the following property:

The one-to-one correspondences σ' and σ'' of the totality of points on the line L upon the totality of points on the line L' satisfy $\sigma' = \sigma''$ if, and only if,

(a) There exist three distinct points P_i $[i = 1,2,3]$ on L such that $P_i^{\sigma'} = P_i^{\sigma''}$ for $i = 1,2,3$; and

(b) $\begin{bmatrix} P^{\sigma'} & Q^{\sigma'} \\ S^{\sigma'} & R^{\sigma'} \end{bmatrix} = \begin{bmatrix} P^{\sigma''} & Q^{\sigma''} \\ S^{\sigma''} & P^{\sigma''} \end{bmatrix}$ for any distinct points P,Q,R,S on L.

Proposition 5: *Suppose that the characteristic of the field F is not 2 [so that $+1 \neq -1$]. Then the following properties of the one-to-one mapping σ of the points on the line L onto the points on the line L^σ are equivalent.*

(I) $\begin{bmatrix} P & Q \\ S & R \end{bmatrix} = \begin{bmatrix} P' & Q' \\ S' & R' \end{bmatrix}$, *if, and only if,* $\begin{bmatrix} P^\sigma & Q^\sigma \\ S^\sigma & R^\sigma \end{bmatrix} = \begin{bmatrix} P'^\sigma & Q'^\sigma \\ S'^\sigma & R'^\sigma \end{bmatrix}$.

(II) *Suppose that P,Q,R,S is a harmonic set. Then P',Q',R',S' is a harmonic set if, and only if,*

$$\begin{bmatrix} P^\sigma & Q^\sigma \\ S^\sigma & R^\sigma \end{bmatrix} = \begin{bmatrix} P'^\sigma & Q'^\sigma \\ S'^\sigma & R'^\sigma \end{bmatrix}$$

(III) *σ-admissible permutations of F are semi-automorphisms.*

PROOF: It is trivial that (I) implies (II), since harmonic sets have cross ratio -1. Assume next the validity of (II). Consider any harmonic set P,Q,R,S; and suppose that W is a point on L^σ such that

$$\begin{bmatrix} P^\sigma & Q^\sigma \\ S^\sigma & R^\sigma \end{bmatrix} = \begin{bmatrix} P^\sigma & Q^\sigma \\ W & R^\sigma \end{bmatrix}$$

Then there exists one and only one point V on the line L such that $W = V^\sigma$; and it follows from (II) that P,Q,R,V is also a harmonic set. It follows from Proposition 2 that $V = S$; and this implies $W = S^\sigma$. Applying Proposition 2 again it follows now that the cross ratio $\begin{bmatrix} P^\sigma & Q^\sigma \\ S^\sigma & R^\sigma \end{bmatrix}$ is a center element z of F. With P,Q,R,S also P,Q,S,R is a harmonic set; and it follows from (II) that

$$z = \begin{bmatrix} P^\sigma & Q^\sigma \\ S^\sigma & R^\sigma \end{bmatrix} = \begin{bmatrix} P^\sigma & Q^\sigma \\ R^\sigma & S^\sigma \end{bmatrix} = z^{-1} \text{ or } z = \pm 1.$$

Since σ maps the four distinct points P,Q,R,S upon four distinct points, it is impossible that $z = 1$; and thus we have shown that σ maps harmonic sets upon harmonic sets. By an argument quite similar to the one used in the proof of Proposition 4 we may deduce now from Proposition 3 that σ-admissible permutations of F are semi-automorphisms; and thus (iii) is a consequence of (ii). That (i) is a consequence of (iii), is a fairly immediate consequence of Lemma 2.

Lemma 3: *Suppose that σ is a one-to-one mapping of the points on the line L upon the points on the line L^σ.*

(*a*) *σ is induced in L by an auto-projectivity of the F-space A if, and only if, the σ-admissible permutations of F are automorphisms.*

(*b*) *σ is induced in L by a collineation of the F-space A if, and only if, the σ-admissible permutations of F are inner automorphisms.*

The proof is easily constructed once one remembers that projectivities are induced by semi-linear transformations [III.1, First Fundamental Theorem of Projective Geometry] and that a semi-linear transformation induces a collineation if, and only if, the component automorphism of F is inner [III.2, Lemma 1].

If the σ-admissible permutations of the field F are anti-automorphisms but not automorphisms, then σ cannot be induced by an auto-projectivity. It is, however, possible to deduce these mappings in some fashion from auto-dualities by means of the theory developed in the next chapter. We leave this to the interested reader.

It is clear from the discussion of Remark 8 that cross ratios will not suffice to characterize those mappings of the points on a line which are induced by auto-projectivities. Thus we need a certain generalization of the concept of cross ratio.

If P,Q,R are three distinct points on the line L, then we may represent them in the form: $P = Fp$, $Q = Fq$, $R = Fr$ subject to the condition that $p + q + r = 0$. If X,Y,Z are points on L none of which equals Q, then these points may be represented in one and only one way in the form $X = F(p + xq)$, $Y = F(p + yq)$, $Z = F(p + zq)$. If we select another admissible representation: $P = Fp'$, $Q = Fq'$, $R = Fr'$ subject to the condition $p' + q' + r' = 0$, then there exists a number $g \neq 0$ in F such that $p' = gp$, $q' = gq$, $r' = gr$; and we find that

$$X = F(p' + gxg^{-1}q'),\ Y = F(p' + gyg^{-1}q'),\ Z = F(p' + gzg^{-1}q').$$

Hence the second representation leads us to the conjugate triplet

$$g(x,y,z)g^{-1} = (gxg^{-1},gyg^{-1},gzg^{-1})$$

of numbers in F. It is easy to see that by changing the admissible representations we obtain a full class of conjugate triplets of numbers in F. Using the notation $\langle (x,y,x) \rangle$ for the totality of triplets $g(x,y,z)g^{-1}$ with $g \neq 0$ in F, we may now define

DEFINITION 2: $$\begin{bmatrix} Fp & Fq & F(p+q) \\ F(p+xq) & F(p+yq) & F(p+zq) \end{bmatrix} = \langle (x,y,z) \rangle.$$

We note that $\begin{bmatrix} P & Q & R \\ X & Y & Z \end{bmatrix}$ is defined whenever the points P,Q,R,X,Y,Z are collinear, the points P,Q,R are different from each other and none of the points X,Y,Z equals Q. If furthermore

$$\begin{bmatrix} P & Q & R \\ X & Y & Z \end{bmatrix} = \langle (x,y,z) \rangle,$$

then

$$\begin{bmatrix} P & Q \\ X & R \end{bmatrix} = \langle x \rangle, \begin{bmatrix} P & Q \\ Y & R \end{bmatrix} = \langle y \rangle, \begin{bmatrix} P & Q \\ Z & R \end{bmatrix} = \langle z \rangle;$$

but the converse will not always be true except if x,y,z belong all three to the center of F.

Proposition 6: *Suppose that the center of the field F contains at least three elements, and that σ is a one-to-one mapping of the points on the line L upon the points on the line L^σ. Then σ is induced by an auto-projectivity of the F-space A if, and only if, there exists an automorphism α of the field F such that*

$$\begin{bmatrix} P^\sigma & Q^\sigma & R^\sigma \\ X^\sigma & Y^\sigma & Z^\sigma \end{bmatrix} = \begin{bmatrix} P & Q & R \\ X & Y & Z \end{bmatrix}^\alpha \textit{ for all [admissible] } P,Q,R,X,Y,Z \textit{ on } L.$$

PROOF: It suffices to prove the sufficiency of the condition. If the condition is satisfied, then there exists a σ-admissible permutation τ of F which satisfies

$$\langle (x^\tau,y^\tau,z^\tau) \rangle = \langle (x^\alpha,y^\alpha,z^\alpha) \rangle \text{ for } x,y,z \text{ in } A.$$

One notes furthermore easily that $\tau\alpha^{-1}$ is a σ'-admissible permutation of F where σ' is an [easily constructed] permutation of the points on L which preserves cross ratios and has at least three fixed points. It follows from Proposition 3 that $\tau\alpha^{-1}$ is a semi-automorphism β of the field F such that (x,y,z) and $(x^\beta,y^\beta,z^\beta)$ are conjugate in F. Consequently there exists to every pair x,y of elements in F an element $g \neq 0$ in F such that

$$x^\beta = g^{-1}xg,\ y^\beta = g^{-1}yg,\ (xy)^\beta = g^{-1}xyg.$$

But this implies

$$(xy)^\beta = g^{-1}xg\, g^{-1}yg = x^\beta y^\beta.$$

Hence β is an automorphism of F. Since α is an automorphism of F, τ is an automorphism of F; and now it follows from Lemma 3 that σ is induced by an auto-projectivity of the F-space A.

The discussion of Remark 8 makes it evident that this new concept, though sufficient to characterize the mappings induced by auto-projectivities, will not suffice to characterize those induced by collineations, again with the exception of certain special situations. What distinguishes the mapping induced by collineations from the others, is the fact that the former admit of geometrical construction; and this we are going to show now.

Suppose that the lines L' and L'' belong to the same plane of the F-space A; and that P is some point in this plane which is neither on L' nor on

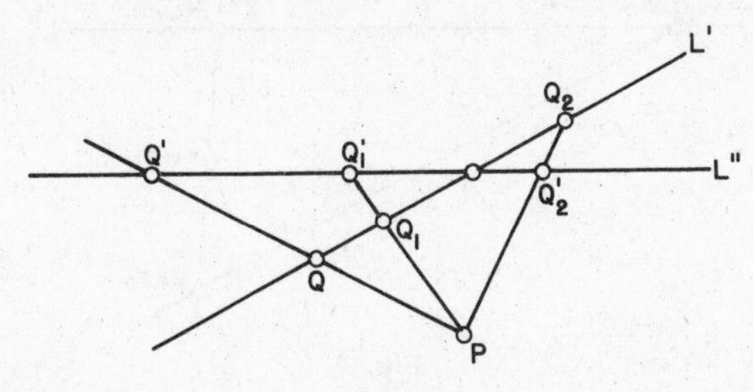

Fig. 6

L''. Then $L' \oplus P = L'' \oplus P$ is a plane carrying the lines L' and L''. If Q' is a point on L', then $Q'' = L'' \cap (P + Q')$ is a point on L'' [as follows from II.2, Rank Formula]. Likewise $L' \cap (P + Q'')$ is a point on L' for every point Q'' on L''; and these two operations are reciprocals of each other, since one deduces from Dedekind's Law that

$$\begin{aligned} L' \cap (P + [L'' \cap (P + Q')]) &= L' \cap (P + L'') \cap (P + Q') \\ &= L' \cap (P + Q') = Q', \end{aligned}$$

if we only note that $L' \leqslant P + L''$. This implies in particular that the mapping of the point Q' on L' upon the point $L'' \cap (P + Q')$ on L'' is one-to-one and exhaustive; and this mapping we shall call *the projection of the line L' from the point P upon the line L''*. [Note that this projection is the identity mapping, if $L' = L''$.]

Suppose now that $L_0, L_1, \cdots, L_k$ are lines in the F-space A and that L_i, L_{i+1} are always coplanar lines. If P_i is some point neither on L_i nor on L_{i+1}, such that $L_i + P_i = P_i + L_{i+1}$, then the projection σ_i of the line L_i from the point P_i upon the line L_{i+1} is well defined. We may form the product $\sigma = \sigma_0\sigma_1 \cdots \sigma_{k+1}$ which is a one-to-one mapping of the totality of points on L_0 upon the totality of points on L_k. Any such mapping σ of the points on L_0 upon the points on L_k shall be called *a linear mapping of the line L_0 upon the line L_k*. [Various names have been given in the literature to what we called here, for purely algebraical reasons, a linear mapping.]

Proposition 7: *The mapping σ of the points on the line L' upon the points on the line L'' is a linear mapping if, and only if, σ is induced by a collineation of the F-space A.*

PROOF: We want to show first that linear mappings are induced by collineations. Since every linear mapping is a product of projections, and since every product of collineations is itself a collineation, it suffices to prove that every projection is induced by a collineation. Consider therefore two different coplanar lines L' and L'' and a point P on the plane

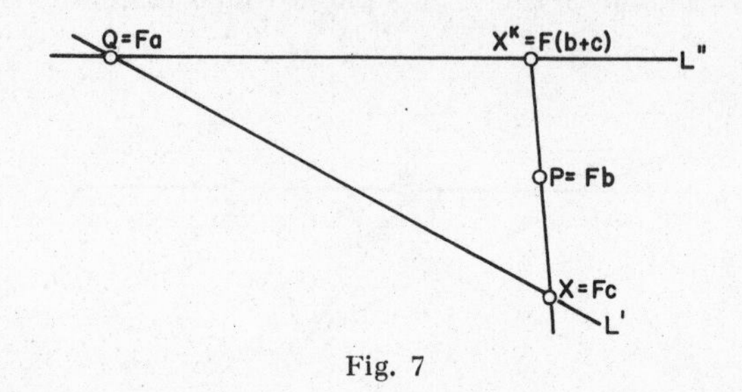

Fig. 7

$L' + L''$ which is neither on L' nor on L''. Then $L' + L'' = L'' + P = P + L'$ is a plane; and the lines L' and L'' meet in a point Q. Consider some point X on L' such that $X \neq Q$; and denote by $\varkappa$ the projection of L' from P onto L''. Then $X, P, X^\varkappa$ are three distinct collinear points; and as usual we can find elements b, c such that $P = Fb$, $X = Fc$, $X^\varkappa = F(b + c)$; and we let a be some element such that $Q = Fa$. Then there exists a linear transformation τ of the F-space A such that $a = a^\tau$, $b = b^\tau$, $c^\tau = b + c$ which leaves invariant every element in some space complementary to $L' + L''$. Every point on L' has the form $F(ua + vc)$. From

$$(ua + vc)^\tau = ua + v(b + c) = (ua + vc) + vb$$

we infer that $[F(ua + vc)]^\tau$ is the intersection $[F(ua + vc)]^\varkappa$ of the lines $L'' = Fa + F(b + c)$ and $F(ua + vc) + P$. Thus $\varkappa$ is induced by τ; and this shows, as we remarked before, that every linear mapping is induced by a collineation of the F-space A.

Suppose next that the mapping σ of the line L' upon the line L'' is induced by a collineation of the F-space A. Every collineation is induced by a linear transformation; and thus we may assume that σ is induced by the linear transformation η of the F-space A.

Case 1: $L' \neq L''$.

Since L' carries at least three points, there exist two different points Fu and Fv on L' such that neither $(Fu)^\sigma = Fu^\eta$ nor $(Fv)^\sigma = Fv^\eta$ is on L'. At most one of the points Fu and Fv is on L''; and thus we may assume without loss in generality that Fv is not on L''. If we let $u^{\eta} = u'$, $v^{\eta} = v'$, then it follows from our choice of u and v that the triplets u,v,u' and

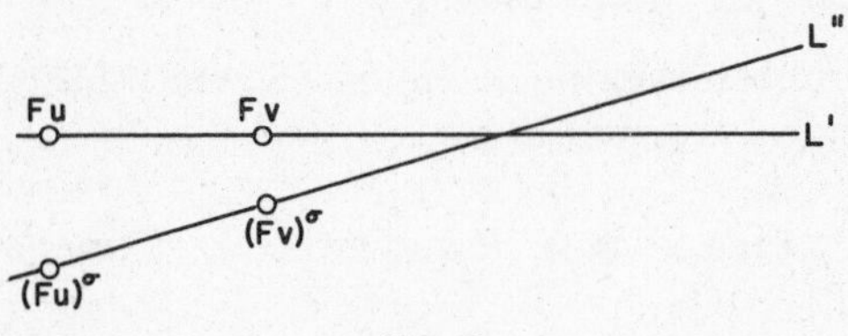

Fig. 8

v,u',v' are independent. Now we define linear transformations ν,ω of the planes $Fu + Fv + Fu'$ and $Fv + Fu' + Fv'$ respectively by the rules

$$u^{\nu} = u',\ v^{\nu} = v,\ u'^{\nu} = 2u' - u; \qquad v^{\omega} = v',\ u'^{\omega} = u',\ v'^{\omega} = 2v' - v.$$

One verifies that ν induces a projection of the line $Fu + Fv = L'$ from the point $F(u' - u)$ upon the line $Fu' + Fv$; that ω induces a projection of the line $Fu' + Fv$ from the point $F(v' - v)$ upon the line $Fu' + Fv' = L''$; and that $\nu\omega$ induces a linear transformation of L' upon L'' which coincides with the linear transformation η. Thus we have represented σ as a product of two projections.

Case 2: $L' = L''$.

Since the F-space A has rank not less than 3, there exists a line L different from $L' = L''$. There exists a projection σ' of L' upon L; and we have shown in the first part of the proof that σ' is induced by a collineation and therefore by a linear transformation η' of the F-space A. The linear transformation $\eta\eta'$ of the F-space A induces the mapping $\sigma\sigma'$ of the line L' upon the line $L \neq L'$, since σ maps L' upon $L'' = L'$. From what we have shown in Case 1 it follows that $\sigma\sigma'$ is the product of two projections. Since the inverse of a projection is likewise a projection, we have shown that σ is the product of three projections. This completes the proof.

APPENDIX III

Projective Ordering of a Space

Suppose that an algebraical ordering of the field F is given. By this we mean that *a domain of positivity* P has been distinguished in F. Such a subset P meets the following requirements:

(*a*) 0 is not in P.

(*b*) If x and y are in P, then $x + y$ and xy are in P.

(*c*) If $x \neq 0$ is not in P, then $-x$ is in P.

The numbers in P are referred to as positive numbers. [The reader should verify the fact that the definition

$$x < y \text{ if, and only if, } y - x \text{ is in } P$$

leads to an algebraical ordering of the field F which has the properties we expect an algebraical ordering to have.]

One proves:

(*d*) If $x \neq 0$, then x^2 is in P and either x, x^{-1} are both in P or x, x^{-1} are both not in P.

We leave the simple verification to the reader. From (*d*) one deduces easily.

(*e*) If y is in P, then $x^{-1}yx$ is in P for every $x \neq 0$ in F.

Consider now four distinct points P, Q, R, S on a line L in the F-space A. Then their cross ratio $\begin{bmatrix} P & Q \\ S & R \end{bmatrix}$ is a complete class of conjugate numbers in F none of which is 0 [or 1]. Thus it follows from (*e*) that either all the numbers in the cross ratio belong to P or none of them is in P. In the first case we term the cross ratio *positive* and in the second case *negative*. In a similar fashion we may see that either all the numbers in the cross ratio are *smaller than* 1 or all of them are *greater than* 1. [For $f^{-1}xf < 1$ if, and only if, $1 - f^{-1}xf = f^{-1}(1 - x)f$ belongs to P; and this is equivalent to saying that $1 - x$ is in P or that $x < 1$.] Now we may define:

The pair P, Q separates the pair R, S if, and only if, the cross ratio $\begin{bmatrix} P & Q \\ S & R \end{bmatrix}$ *is both positive and smaller than* 1.

That this soc. cyclical ordering of the points on L has all the properties we expect of it, the reader will easily verify using the symmetry properties of cross ratios. [For an account of the basic properties of cyclical order see, for instance, Coxeter (1), p. 22-24.] This ordering is furthermore invariant under all projections and collineations as follows from III.4, Proposition 7 and III.4, Lemma 2.

REMARK: The theory of commutative ordered fields is well developed [Artin-Schreier Theory; see, for instance, van der Waerden: Moderne Algebra, vol. 1]. But as regards non-commutative ordered fields there is known not much more than their existence [see, for instance, Reidemeister (1)] and the equivalence of orderability and formal reality recently established by I. Szele.

Another approach to the problem of order in projective spaces is given by E. Sperner: Beziehungen zwischen geometrischer und algebraischer Anordnung, *Sitzungsberichte der Heidelberger Akademie der Wissenschaften*; Math.-nat. Kl., 1949 (10), and E. Sperner: Die Ordnungsfunktionen einer Geometrie, *Math. Ann.*, vol. 121 (1949), pp. 107 ss.

CHAPTER IV

Dualities

Dualities make their appearance in classical projective geometry under the name of correlation. They are one-to-one and monotone decreasing correspondences between the subspaces of equal or different linear manifolds which in particular interchange points and hyperplanes. In the present chapter we characterize those linear manifolds which are duals of each other and find algebraical representations of them. The algebraical mechanism needed for this purpose is contained in the soc. semibilinear forms which generalize such concepts as bilinear forms, scalar products, Hermitean forms etc. We study in somewhat more detail the polarities. These are involutorial auto-dualities, generalizing the classical concept of the pole-polar-relation with respect to a conic; and their groups generalize those soc. classical groups which are known under names like unitary, symplectic and orthogonal groups; and what we attempt to do here is to provide the basis for a broad study of these groups.

IV.1. Existence of Dualities; Semi-bilinear Forms

A *duality* δ of the linear manifold (F,A) upon the linear manifold (G,B) is a correspondence with the following properties:

(*a*) δ maps every subspace T of A upon a uniquely determined subspace T^δ of B.

(*b*) To every subspace Y of B there exists one and only one subspace X of A such that $X^\delta = Y$.

(*c*) $U \leqslant V$ if, and only if, $V^\delta \leqslant U^\delta$.

In other words a duality is a one-to-one and monotone decreasing mapping of the totality of F-subspaces of A upon the totality of G-subspaces of B.

In II.3 we have exhibited dualities between a linear manifold of finite rank and its adjoint space. It should be recalled, however, that the field F operates from the right on the adjoint space whenever it operates from the left upon the original space. In the present context the elements in the fields F and G are supposed to be left-multipliers; and this is very essential, since in the greater part of our discussion we shall assume that $(F,A) = (G,B)$.

It is clear that the inverse of a duality is a duality, that the product of dualities, if defined, is a projectivity, and that the product of a duality and a projectivity, if defined, is a duality. Dualities of (F,A) upon itself are auto-dualities of (F,A); and a linear manifold possessing an auto-duality will be termed self-dual. Similarly we say that the linear manifolds (F,A) and (G,B) are duals of each other, if there exists a duality of (F,A) upon (G,B). It will be the principal objective of this section to characterize the linear manifolds which possess duals and those which are self-dual and to find algebraical representations for the dualities. We note a few simple consequences of the definition of a duality.

(*d*) Dualities interchange sums and intersections.

This is verified by an argument quite similar to the one used in the proof of III.1, (*e*).

(*e*) If δ is a duality of (F,A) upon (G,B), then $0^\delta = B$, $A^\delta = 0$. An immediate consequence of (*d*).

(*f*) Dualities interchange points and hyperplanes.

This is an immediate consequence of the fact that there is no subspace between 0 and a point and that there is no subspace between a hyperplane and the full space.

The dualities of linear manifolds of rank 1 are trivial, as follows from (*e*). Likewise we infer from (*e*) and (*f*) that any one-to-one mapping of the points on one line upon the totality of points of some line may be considered as a duality. Thus the discussion of dualities ceases to be trivial only if we assume that the rank is at least 3.

Existence Theorem: *The linear manifold* (F,A) *possesses a dual if, and only if, its rank* $r(A)$ *is finite.*

The proof of this fundamental theorem will be obtained as a by-product when characterizing linear manifolds that are duals of each other. In order to enunciate this characterization we need the following concept. *An anti-isomorphism of the field F upon the field G* is a one-to-one mapping α of F upon [the whole of] G such that

$$(x + y)^\alpha = x^\alpha + y^\alpha, \ (xy)^\alpha = y^\alpha x^\alpha \text{ for } x,y \text{ in } F.$$

If F happens to be equal to G, then α is an anti-automorphism of the field F. We note that every isomorphism of a commutative field is at the same time an anti-isomorphism, that the identity is an anti-automorphism of F if, and only if, F is commutative. An example of an anti-automorphism of a non-commutative field is provided by mapping the real quaternion $x_0 + x_1 i + x_2 j + x_3 k$ upon its "conjugate" $x_0 - x_1 i - x_2 j - x_3 k$. [See III.4 for further discussion of anti-automorphisms.]

Duality Theorem: *The linear manifolds (F,A) and (G,B) of rank no less than 3 are duals of each other if, and only if,*

(*a*) $r(A) = r(B)$ *is finite and*

(*b*) *there exists an anti-isomorphism of the field F upon G.*

If we specialize this theorem by letting $(F,A) = (G,B)$, then we obtain the

Self-Duality Theorem: *The linear manifold (F,A) of rank not less than 3 is self-dual if, and only if, $r(A)$ is finite and F possesses an anti-automorphism.*

When proving these theorems we shall derive various facts of independent interest.

Lemma 1: *If δ is a duality of the linear manifold (F,A) upon the linear manifold (G,B), and if X is a subspace of A, then $r(B/X^{\delta}) = r[L(X)]$.*

Note that $L(X)$ is the adjoint space of the linear manifold (F,X) which we introduced and discussed in II.3.

Proof: Consider a basis K of $L(X)$. If k belongs to K, then k is a linear form, not 0, over X; and $S(k)$ [the totality of elements x in X such that $xk = 0$] is a hyperplane in X [see II.3, Remark 2]. We show first.

(1.1) If $k_0, \cdots, k_n$ are finitely many distinct elements in K, then $S(k_1) \cap \cdots \cap S(k_n) \not\leqq S(k_0)$.

Since K is a basis of $L(X)$, the subspaces $\sum_{i=1}^{n} k_iF = U$ and $\sum_{i=0}^{n} k_iF = V$ of $L(X)$ have ranks n and $n+1$ respectively. These ranks are finite; and thus it follows from II.3, Propisition 3, (*a*) that

$$n = r(U) = r[X/S(U)] < n + 1 = r(V) = r[X/S(V)].$$

From $U < V$ we deduce now that $S(V) < S(U)$. One verifies easily that

$$S(V) = S(k_0) \cap S(k_1) \cap \cdots \cap S(k_n) = S(k_0) \cap S(U) < S(U);$$

and this shows the impossibility of $S(U) \leqq S(k_0)$, as we intended to prove.

(1.2) $$\bigcap_{k \in K} S(k) = 0.$$

The cross cut under consideration is easily seen to be the same as

$$S\Big(\sum_{k \in K} kF\Big) = S[L(X)] = S[E(0)] = 0 \text{ by II.3, Proposition 2.}$$

For every k in K we let now $k^* = S(k)^{\delta}$. Since $S(k)$ is a hyperplane in X, k^*/X^{δ} is a point in B/X^{δ}. It follows from (1.2) and property (*d*) of dualities that B/X^{δ} is the sum of the points k^*/X^{δ}; and similarly if follows from (1.1) that the points k^*/X^δ form an independent set of points in B/X^δ. [Remember that a set of points is independent, if every finite subset is independent; and that a finite set of points is independent, if none of them is contained in the sum of the others.] Thus the points k^*/X^δ form a basis of the space B/X^δ. Since the number of points k^*/X^δ is the same as the number of elements in K, and since the former of these numbers has been shown to be the rank of B/X^δ whereas the latter is the rank of $L(X)$, we have verified the desired equality $r(B/X^\delta) = r[L(X)]$.

PROOF *of finiteness of rank*:

Suppose that δ is a duality of (F,A) upon (G,B). Letting $X = A$ in Lemma 1 and remembering that $A^\delta = 0$ we find that $r(B) = r[L(A)]$. It follows from II.3, Theorem 1 that $r(A) \leqslant r[L(A)] = r(B)$. Since the reciprocal of a duality is likewise a duality, we have also

$$r(B) \leqslant r[L(B)] = r(A).$$

Combining these inequalities it follows that $r(A) = r[L(A)]$; and the finiteness of $r(A)$ is a consequence of II.3, Corollary 1.

Corollary 1: *If δ is a duality of (F,A) upon (G,B), then*

$$r(B) = r(X) + r(X^\delta) \textit{ for every subspace } X \textit{ of } A.$$

[Note that this reduces to $r(B) = r(A)$, if we let $X = A$.]

PROOF: From the existence of a duality we have already deduced that (F,A) and (F,B) have finite rank. Now it follows from Lemma 1, the Rank Formulas of II.2 and II.3, Theorem 1 that

$$r(B) - r(X^\delta) = r(B/X^\delta) = r[L(X)] = r(X),$$

as we claimed.

Construction of Canonical Dual Space

The linear manifold (F^*,A^*) to be constructed now will be the dual of (F,A), provided $r(A)$ is finite. We begin by constructing the field F^* and an anti-isomorphism of F upon F^* as follows. The additive group of F^* is identical with the additive group of F; but multiplication of elements in F^* is defined by the rule

$$x * y = yx \text{ [where } yx \text{ is multiplication in } F].$$

The identity mapping is then easily seen to be an anti-isomorphism of F upon F^* [so that in particular F^* is a field]; and we shall denote this anti-isomorphism of F upon F^* by γ.

Now we define a manifold (F^*,A^*) as follows. The additive group of A^*

is identical with the additive group of $L(A)$; but multiplication of the element x in F^* by the element f in A^* is defined by the rule:

$$x * f = fx$$

[where fx is the ordinary product of the linear form f over A by the element x in the field F]. If we denote by γ the identity mapping of $L(A)$ [upon A^*], then γ is an isomorphism of the additive group of $L(A)$ upon the additive group of A^* which satisfies the rule:

$$(fx)^{\gamma} = x^{\gamma} * f^{\gamma} \text{ for } f \text{ in } L(A) \text{ and } x \text{ in } F.$$

It is clear that this identity mapping γ of $L(A)$ constitutes what might be called an anti-isomorphism of the linear manifold $(L(A), F)$ [where the elements in F act as right-multipliers] upon the manifold (F^*,A^*) [where the elements in F^* act as left-multipliers]; and that therefore (F^*,A^*) is a linear manifold. We note finally that every subspace X of $(L(A), F)$ is likewise a subspace of (F^*,A^*), since X and X^{γ} consist of the same elements. Now one deduces from II.3, Theorem 2 without difficulty the following fact:

(g) *If $r(A)$ is finite, then a duality of (F,A) upon (F^*,A^*) is obtained by mapping the subspace X of A upon $E(X)$.*

We had shown before that the existence of a duality implies the finiteness of the rank. Combining this with the result just obtained we have completed the proof of the Existence Theorem.

Lemma 2: *If δ is a duality of (F,A) upon (G,B), and if the rank of A is at least 3, then there exists an anti-isomorphism σ' of F upon G and an isomorphism σ'' of the additive group A upon the additive group $L(B)$ such that*

$$(xa)^{\sigma''} = a^{\sigma''}x^{\sigma'} \text{ for } a \text{ in } A \text{ and } x \text{ in } F \text{ and}$$

$$X^{\sigma''} = E(X^{\delta}) \text{ and } X^{\delta} = S(X^{\sigma''}) \text{ for every subspace } X \text{ of } A.$$

[Such pairs (σ',σ'') we are going to call *anti-semi-linear transformations*; see V. 5.]

Proof: Construct the canonical dual space (G^*,B^*) to (G,B) and denote by ρ both the identity mapping of G upon G^* and the identity mapping of $L(B)$ upon B^*. From the existence of a duality of (F,A) upon (G,B) and previous results [Existence Theorem and Corollary 1] we deduce that (F,A) and (G,B) have equal finite rank; and thus it follows [from (g)] that a duality of (G,B) upon (G^*,B^*) is obtained by mapping the subspace Y of B upon $E(Y)$. If X is a subspace of A, then we may map X upon the subspace $E(X^{\delta})$ of B^*; and it is clear that this correspondence is a projectivity [as the product of two dualities]. We deduce from the First Fundamental Theorem of Projective Geometry [III.1] the existence of a semi-linear transformation σ of (F,A) upon (G^*,B^*) such that $E(X^\delta)=X^\sigma$ for every subspace X of A—here we make use of the hypothesis that $3 \leqslant r(A)$. The isomorphism σ of the field F upon the field G^* effects at the same time an anti-isomorphism σ' of F upon G $[\sigma = \sigma'\rho]$ and the isomorphism σ of A upon B^* effects at the same time an isomorphism σ'' of A upon $L(B)$ $[\sigma = \sigma''\rho]$. By II.3, Proposition 2 we have $Z = S[E(Z)]$; and now it is obvious how to complete the proof of the lemma.

PROOF of *Duality Theorem*:

The necessity of the conditions (*a*), (*b*) is an immediate consequence of the Existence Theorem, Corollary 1 and Lemma 2. Assume conversely the validity of conditions (*a*), (*b*). Construct the Canonical Dual Space (G^*,B^*) to (G,B). From the existence of an anti-isomorphism of F upon G we deduce the existence of an isomorphism of F upon G^*, since the fields G and G^* are anti-isomorphic. It is obvious that (G,B) and (G^*,B^*) have the same rank; and thus (F,A) and (G^*,B^*) have the same rank too. From the Projective Structure Theorem [III.1] we deduce the existence of a projectivity σ of the linear manifold (F,A) upon the linear manifold (G^*,B^*). From the finiteness of the rank of B we deduce the existence of a duality of (G,B) upon (G^*,B^*) [see (*g*)]. Since the inverse of a duality is also a duality, we may now obtain a duality of (F,A) upon (G,B). This completes the proof.

REMARK *on the duality principle of projective geometry*:

The Existence Theorem shows that the theory of projective spaces of finite dimension is self-dual in so far as we have constructed to each finite dimensional linear manifold (F,A) its dual (F^*,A^*) which, as is apparent from its construction, is essentially what often is called the space of the hyperplanes in (F,A). Thus if (F,A) has some property, the dual of this property is satisfied by (F^*,A^*); and if it can be shown that some property is satisfied by every (F,A), then the dual of this property is satisfied by every (F^*,A^*) and is consequently universally true. But, in general, it will not be possible to claim that (F,A) will satisfy the dual of a property P, if (F,A) has property P, since (F,A) need not be self-dual even though finite dimensional. To see this one just has to compare the Self-Duality Theorem and to realize that not every field possesses an anti-automorphism.

Note that every permutation of the points on a line comes from an auto-duality of this line. Hence the preceding fundamental theorems cease to be true, if we omit the hypothesis $2 < r(A)$.

From now on we shall be concerned almost exclusively with auto-dualities [= dualities of (F,A) upon itself].

Corollary 2: *If δ is an auto-duality of (F,A), then the following properties of the subspace M of A imply each other.*

(I) $M \cap M^{\delta} = 0$.

(II) $M + M^{\delta} = A$.

(III) $M \oplus M^{\delta} = A$.

This is an almost immediate consequence of the Rank Formulas [II.2] once one observes that $r(A)$ is finite [by the Existence Theorem] and that $r(A) = r(M) + r(M^{\delta})$ [by Corollary 1].

We note that subspaces with Property (I) will be termed *non-isotropic* with respect to δ.

Lemma 2 when applied to auto-dualities leads us to an algebraical representation of these. For a convenient expression we need the concept of *semi-bilinear form over* (F,A). This is a pair consisting of an anti-automorphism α of the field F and a function $f(x,y)$ with the following properties:

(F.1) $f(x, y)$ is, for every x,y in A, a uniquely determined number in F.

(F.2) $f(a + b,c) = f(a,c) + f(b,c)$, $f(a,b + c) = f(a,b) + f(a,c)$ for a,b,c in A.

(F.3) $f(tx,y) = tf(x,y), f(x,ty) = f(x,y)t^{\alpha}$ for x,y in A, t in F.

Since the anti-automorphism α figures so prominently in the definition we shall refer to f often as to an *α-form over A*. If in particular $\alpha = 1$ [which is possible only if F is commutative], then f will be called a bilinear form.

EXAMPLES: **1.** Let F be the field of real numbers and A the space of n-tuplets $x = (x_1, \cdots, x_n)$ of real numbers x_i. Then (F,A) is the $(n-1)$-dimensional real projective space. The ordinary scalar product

$$xy = \sum_{i=1}^{n} x_i y_i$$

is a bilinear form over A.

2. Let F be the field of complex numbers and A the space of n-tuplets $x = (x_1, \cdots, x_n)$ of complex numbers x_i. Denote by α the automorphism of F which maps every number c in F upon the conjugate complex number $\bar{c}$.

The Hermitean form $x\bar{y} = \sum_{i=1}^{n} x_i\bar{y}_i$ is an α-form over A.

3. Let F be the field of real quaternions; and denote by α the anti-automorphism of F which maps the real quaternion $x = x_0 + ix_1 + jx_2 + kx_3$ upon the conjugate quaternion $\bar{x} = x_0 - ix_1 - jx_2 - kx_3$ [where the x_i are real etc]. Denote by A the space of n-tuplets $q = (q_1, \cdots, q_n)$ of real quaternions q_i [in F]. The Hamiltonian form $x\bar{y} = \sum_{i=1}^{n} x_i\bar{y}_i$ is an α-form over A.

MATRIX REPRESENTATION OF SEMI-BILINEAR FORMS:

Suppose that $f(x,y)$ is an α-form over the linear manifold (F,A) of finite rank n, and that $b_1, \cdots, b_n$ is a basis of A. Let $f(b_i,b_j) = a_{ij}$, $x = \sum_{i=1}^{n} x_i b_i$. Then

$$f(x,y) = \sum_{i,j} x_i a_{ij} y_j^{\alpha}$$

is easily deduced from the properties (F.1) to (F.3). Using matrix notation this formula may be rewritten as follows:

$$f(x,y) = (x_1, \cdots, x_n)\,(a_{ij}) \begin{pmatrix} y_1 \\ \vdots \\ y_n \end{pmatrix}^{\alpha}.$$

If the elements $c_1, \cdots, c_n$ form another basis of A, then $c_i = \sum_{j=1}^{n} b_{ij} b_j$; and we find that

$$f(c_i,c_j) = \sum_{h,k} b_{ih} a_{hk} b_{jk}^{\alpha};$$

and these equations may likewise be expressed in matrix form. No use will be made of these remarks.

If f is a semi-bilinear form over (F,A), and if K is a subset of A, then the totality of solutions x of the equations $f(x,K) = 0$ [i.e. the totality of elements x in A such that $f(x,k) = 0$ for every k in K] is a subspace of A, as is easily verified; and likewise the totality of solutions of the equations $f(K,y) = 0$ is a subspace of A. We shall say that *the duality* δ *of* (F,A) *upon itself is represented by the semi-bilinear form* $f(x,y)$, if X^{δ} is, for every subspace X of A, exactly the totality of solutions x of the equations $f(x,X) = 0$ or in symbols:

$$X^{\delta} = (x \in A\colon f(x, X) = 0).$$

Proposition 1: *Auto-dualities of linear manifolds of rank not less than* 3 *are represented by semi-bilinear forms.*

HISTORICAL NOTE: The first proof of a theorem of this type appears to be due to G. Birkhoff and J. von Neumann: The Logic of Quantum Mechanics, *Ann. of Math.*, vol. 37 (1936), pp. 823-843; see in particular their Appendix I on pp. 837-843.

PROOF: If δ is an auto-duality of the linear manifold (F,A), then we infer from Lemma 2 the existence of an anti-automorphism α of the field F and of an isomorphism σ of the additive group A upon the additive group $L(A)$ such that

$$(xy)^{\sigma} = y^{\sigma} x^{\alpha} \quad \text{for } x \text{ in } F \text{ and } y \text{ in } A,$$
$$X^{\delta} = S(X^{\sigma}) \quad \text{for every subspace } X \text{ of the } F\text{-space } A.$$

If x and y are elements in A, then xy^σ is a well determined element in F, since y^σ belongs to $L(A)$ and is therefore a linear form over A. Consequently we may define:

$$f(x,y) = xy^\sigma \text{ for } x,y \text{ in } A.$$

One verifies without difficulty that $f(x,y)$ is an α-form over A. If X is a subspace of A, then the totality X^* of solutions x of the equations $f(x,X)=0$ is exactly the totality of solutions x of the equations $xX^\sigma = 0$. Thus it follows from our definition of the S-operation that $X^* = S(X^\sigma) = X^\delta$. Hence we have shown that δ is represented by f, as we intended to prove.

Not every semi-bilinear form represents a duality, witness the trivial example: $f(x,y) = 0$ for every x and y. Thus we need a criterion characterizing those forms which represent dualities [the soc. non-degenerate semi-bilinear forms]. To do this we need the following simple connection between semi-bilinear forms over (F,A) and linear forms over (F,A). If $f(x,y)$ is an α-form over (F,A), then we map the element y in A upon the correspondence y^f defined by the rule:

$$xy^f = f(x,y) \text{ for every } x \text{ in } A.$$

One verifies without difficulty that

y^f is, for every y in A, a linear form over (F,A) [belongs to $L(A)$];

$$(a+b)^f = a^f + b^f,\ (ta)^f = a^f t^\alpha \text{ for } a,b \text{ in } A \text{ and } t \text{ in } F.$$

The mapping of y in A upon y^f in $L(A)$ may consequently be termed *an anti-homomorphism of A into $L(A)$*. Naturally such a mapping is an anti-isomorphism if it is one-to-one; and we remind the reader of the fact used before that the identity mapping is an anti-isomorphism of $(L(A),F)$ upon (F^*,A^*).

Proposition 2: *The following properties of the α-form $f(x,y)$ over (F,A) are equivalent.*

(I) *f represents an auto-duality of (F,A).*

(II) *$r(A)$ is finite; and $f(A,y) = 0$ implies $y = 0$.*

(III) *$A^f = L(A)$.*

(IV) *$r(A)$ is finite; and $f(x,A) = 0$ implies $x = 0$.*

(V) *An anti-isomorphism of A upon $L(A)$ is obtained by mapping a in A upon a^f in $L(A)$.*

PROOF: If f represents a duality, then we deduce from the Existence Theorem that $r(A)$ is finite. If y is an element in A such that $f(A,y) = 0$, then also $f(A,Fy) = 0$; and the duality represented by f maps the subspace Fy of A upon A. But 0 and A are interchanged by auto-dualities so that $Fy=0$ and hence $y=0$. This shows that (II) is a consequence of (I).

Assume next the validity of (II). Then the mapping of a in A upon a^f

in $L(A)$ is an anti-isomorphism, since $a^f = 0$ implies $f(A,a) = 0$ and hence $a = 0$. Thus we have obtained an anti-isomorphism of A upon the subspace A^f of $L(A)$; and this implies in particular [as is easily seen in a variety of ways] $r(A) = r(A^f)$. From the finiteness of $r(A)$ and II.3, Corollary 1 we deduce $r(A) = r[L(A)]$. Hence $L(A)$ and its subspace A^f have the same finite rank; and it follows from the Rank Formulas [II.2] that $L(A) = A^f$. Hence (III) is a consequence of (II).

Assume now the validity of (III). Then $L(A) = A^f$ is an anti-homomorphic image of A so that $r[L(A)] \leqslant r(A)$. It is an immediate consequence of II.3, Theorem 1 [and the set theoretical theorem $\aleph < c^{\aleph}$] that $r(A)$ is finite. If $f(x,A) = 0$ for some x in A, then we have $xL(A) = xA^f = f(x,A) = 0$ so that x belongs to $S[L(A)] = S[E(0)] = 0$ by II.3, Proposition 2. Hence (IV) is a consequence of (III).

Assume next the validity of (IV). Then $S(A^f)$ is the totality of solutions x of the equations $0 = xA^f = f(x,A)$; and hence $S(A^f) = 0$. From the finiteness of $r(A)$ we infer the applicability of II.3, Proposition 3; and now it follows that

$$A^f = E[S(A^f)] = E(0) = L(A).$$

Hence (III) is a consequence of (IV). Denote now by W the totality of solutions w of $f(A,w) = 0$. Then the subspace W of A is the kernel of the anti-homomorphism of A upon $A^f = L(A)$ which we obtain by mapping a in A upon a^f in $L(A)$. Because of the finiteness of all the ranks we deduce now from the Rank Formulas [II.2], the property (III) which we just verified and II.3, Theorem 1 that

$$r(A) - r(W) = r(A^f) = r[L(A)] = r(A) \text{ or } r(W) = 0$$

and this implies $W = 0$ so that (II) too is a consequence of (IV). But the properties (II) and (III) together amount just to the property (V); and thus we have shown that (V) is a consequence of (IV).

Assume finally the validity of (V). Then A and $L(A)$ have the same rank and this implies the finiteness of $r(A)$ by II.3, Corollary 1. Now we may apply II.3, Theorem 3 which asserts that the mapping of the subspace N of $L(A)$ upon the subspace $S(N)$ of A constitutes a duality of $L(A)$ upon A. It follows from (V) that mapping the subspace X of A upon the subspace X^f of $L(A)$ constitutes a projectivity of A upon $L(A)$. Thus we obtain an auto-duality σ of A, if we map the subspace X of A upon $S(X^f)$. By definition $S(X^f)$ is the totality of elements x in A which satisfy the equations

$$0 = xX^f = f(x,X);$$

and thus we have shown that the auto-duality σ is represented by f. Hence (I) is a consequence of (V); and this completes the proof.

REMARK 1: If B is a basis of the linear manifold (F,A), and if α is an anti-automorphism of the field F, then there exists one and only one α-form $f(x,y)$ over (F,A) which satisfies

$$f(b',b'') = \begin{cases} 1 \text{ for } b' = b'' \text{ in } B \\ 0 \text{ for } b' \neq b'' \text{ in } B. \end{cases}$$

It is easy to see that f has the properties:

$$f(A,x) = 0 \text{ implies } x = 0; \text{ and } f(a,A) = 0 \text{ implies } a = 0,$$

though the rank of A may be finite or infinite. Thus the finiteness requirement in (II) and (IV) is indispensable for the validity of Proposition 2.

Suppose now that the α-form $f(x,y)$ represents a duality of (F,A). Consider an element $d \neq 0$ in F; and let $g(x,y) = f(x,y)d$. It follows from (F.3) that

$$g(x,ty) = f(x,ty)d = f(x,y)t^{\alpha}d = g(x,y)(d^{-1}t^{\alpha}d)$$

for x,y in A and t in F. If we let $t^{\beta} = d^{-1}t^{\alpha}d$, then β is an anti-automorphism of F [namely α followed by an inner automorphism]; and it is easily verified that g is a β-form over (F,A) which represents the same duality as f. If, for instance, w is an element in A, but not in $(Fw)^f$, then $f(w,w) \neq 0$ and we may let $d^{-1} = f(w,w)$ so that the new form g satisfies $g(w,w) = 1$. This sort of normalization we shall use quite often in the future.

The preceding remarks show that an auto-duality does not determine uniquely the forms representing it. However, the preceding construction exhausts the possibilities, as may be seen from the following fact.

Proposition 3: *If the semi-bilinear forms f and g over (F,A) represent the same auto-duality of A, and if $2 \leqslant r(A)$, then there exists a number $d \neq 0$ in F such that*

$$g(x,y) = f(x,y)d \text{ for every } x,y \text{ in } A.$$

PROOF: If follows from Proposition 2 that mapping a in A upon a^f in $L(A)$ constitutes an anti-isomorphism of A upon $L(A)$ and that the mapping of a in A upon a^g in $L(A)$ constitutes likewise an anti-isomorphism of A upon $L(A)$. Consequently there exists a semi-linear transformation σ of $L(A)$ upon itself such that $(a^f)^{\sigma} = a^g$ for every a in A. If X is a subspace of $L(A)$, then there exists one and only one subspace T of A such that $X = T^f$. The dualities defined by f and g map T upon $S(T^f)$ and $S(T^g)$ respectively; and since f and g represent the same dualities, we have $S(T^f) = S(T^g)$. The existence of dualities implies the finiteness of $r(A)$; and now it follows from II.3, Proposition 3 that

$$T^f = E[S(T^f)] = E[S(T^g)] = T^g.$$

Hence we find that

$$X^\sigma = T^{f\sigma} = T^g = T^f = X.$$

But it follows from $2 \leqslant r(A)$ and III.1, Proposition 3 that "multiplications" are the only semi-linear transformations which leave invariant every subspace. Hence there exists a number $d \neq 0$ in F such that $z^\sigma = zd$ for every z in $L(A)$. Consequently

$$g(x,y) = xy^g = xy^{f\sigma} = xy^f d = f(x,y)d,$$

as we intended to prove.

IV.2. Null Systems

Null systems are a rather special class of dualities. But the results obtained in this section will be used in the next section.

If δ is an auto-duality of the linear manifold (F,A), then δ is said to be *a null system on the subspace W of A* and *W is an N-subspace of A* with respect to δ, if

(N) $\qquad P \leqslant P^\delta$ *for every point P on W.*

A null system then is a duality of (F,A) which is a null system on A.

Lemma 1: *Suppose that the α-form f represents the duality δ of (F,A).*

(*a*) *δ is a null system on the subspace W if, and only if, $f(x,x) = 0$ for every x in W.*

(*b*) *If δ is a null system on W, then $f(x,y) = -f(y,x)$ for x,y in W.*

(*c*) *If δ is a null system on W, and if $f(W,W) \neq 0$, then $\alpha = 1$ so that F is commutative.*

Proof: δ is represented by f. Hence the element a in A belongs to X^δ, for X a subspace of A, if, and only if, $f(a,X) = 0$. Consequently we have $Fx \leqslant (Fx)^\delta$ if, and only if, $f(x,x) = 0$; and (*a*) is an immediate consequence of these remarks.

If the N-subspace W contains the elements x and y, then W contains $x + y$; and it follows from (*a*) that

$$0 = f(x + y,x + y) = f(x,x) + f(y,x) + f(x,y) + f(y,y) = f(x,y) + f(y,x),$$

proving (*b*). If furthermore $f(W,W) \neq 0$, then there exist elements v,w in W such that $t = f(v,w) \neq 0$. For x in F we deduce now from (*b*)

$$xt = xf(v,w) = f(xv,w) = -f(w,xv) = -f(w,v)x^\alpha = f(v,w)x^\alpha = tx^\alpha.$$

Hence $x^\alpha = t^{-1}xt$ for every x in F. Thus the anti-automorphism α is the inner automorphism induced by t. But an anti-automorphism is an automorphism if, and only if, F is commutative; and the only inner automorphism of a commutative field is the identity. Hence $\alpha = 1$ and F is commutative, as we claimed.

Lemma 2: *If the duality δ is represented by the semi-bilinear form f, and if W is an N-subspace such that $f(W,W) \neq 0$, then*

(a) W contains elements u,v such that $f(u,v) = 1$; and

(b) if P,Q are points on W such that P is not on Q^δ, then $P + Q = L$ is a line satisfying $L \cap L^\delta = 0$.

PROOF: From $f(W,W) \neq 0$ we deduce the existence of elements u',v in W such that $t = f(u',v) \neq 0$. Let $u = t^{-1}u'$. Then $f(u,v) = 1$. If $P = Fp$ and $Q = Fq$ are on W, then $f(p,p) = f(q,q) = 0$ [Lemma 1, (a)]. If P is not on Q^δ, then $P \neq Q$, since W is an N-space and $f(p,q) \neq 0$. It follows from Lemma 1, (b) that $f(q,p) = -f(p,q) \neq 0$. If w belongs to $(P + Q) \cap L^\delta$, then $w = hp + kq$ and $0 = f(w,L)$ so that

$$0 = f(w,p) = kf(q,p) \text{ or } k = 0,$$
$$0 = f(w,q) = hf(p,q) \text{ or } h = 0.$$

Hence $w = 0$ and therefore $(P + Q) \cap L^\delta = 0$.

REMARK 1: If the duality δ is represented by the semi-bilinear form f, then the following properties of the subspace W of A are equivalent: $W \not\leq W^\delta$; $W \cap W^\delta < W$; $f(W,W) \neq 0$. This is true in particular, if $W \neq 0$ is non-isotropic [$W \cap W^\delta = 0$].

Proposition 1: *Suppose that the duality δ is represented by the semi-bilinear form f. Then the subspace $W \neq 0$ is a non-isotropic N-subspace if, and only if, $\alpha = 1$, $r(W) = 2k$ and there exists a basis $b_1, \cdots, b_{2k}$ of W such that $f(b_{2i-1}, b_{2i}) = -f(b_{2i}, b_{2i-1}) = 1$ and $f(b_i,b_j) = 0$ otherwise.*

PROOF: Assume first that $\alpha = 1$ [so that F is commutative] and that there exists a basis $b_1, \cdots, b_{2k}$ of W with the desired properties. Then

$$f\Big(\sum_{i=1}^{2k} x_i b_i, \sum_{i=1}^{2k} y_i b_i\Big) = \sum_{i=1}^{k} [x_{2i-1}y_{2i} - x_{2i}y_{2i-1}]; \tag{1.*}$$

and a simple computation shows that $f(w,w) = 0$ for w in W and $0 = W \cap W^\delta$ so that W is a non-isotropic N-subspace of A.

If conversely $W \neq 0$ is a non-isotropic N-subspace, then $f(W,W) \neq 0$; and it follows from Lemma 1, (c) that $\alpha = 1$ and F is commutative. We infer from Lemma 2, (a) the existence of elements b_1, b_2 in W such that $f(b_1,b_2) = 1$; and this implies $f(b_2,b_1) = -1$ by Lemma 1, (b). It follows from Lemma 2, (b) that $L = Fb_1 + Fb_2$ is a line in W such that $L \cap L^\delta = 0$. Hence $A = L \oplus L^\delta$ [IV.1, Corollary 2]; and thus it follows from $L \leqslant W$ that $W = L \oplus (W \cap L^\delta)$. Let $V = W \cap L^\delta$. Then it follows from the Rank Formulas [II.2] that $r(W) = r(L) + r(V) = 2 + r(V)$. Since V is a subspace of the N-subspace W, V itself has Property (N). Since dualities interchange sum and cross cut, we have

$$V \cap V^{\delta} = W \cap L^{\delta} \cap V^{\delta} = W \cap (L + V)^{\delta} = W \cap W^{\delta} = 0,$$

since W is non-isotropic. Thus V is a non-isotropic N-subspace of lower rank than W such that $f(L,V) = f(V,L) = 0$ [by Lemma 1, (b)] and $W = L \oplus V$. Now the proof may be finished by an obvious inductive argument, since the existence of a duality implies finiteness of rank.

Theorem: *The F-space A of rank not less than* 3 *possesses a null system if, and only if, F is commutative and A has finite even rank. This null system is essentially uniquely determined.*

Proof: Assume that the duality δ of (F,A) is a null system. Then δ may be represented by a semi-bilinear form $f(x,y)$ [IV.1, Proposition 1]. Since A has Property (N) and $f(A,A) \neq 0$, it follows from Lemma 1 that F is commutative and f is a skew-symmetric bilinear form. From $A^{\delta} = 0$ it follows that A is a non-isotropic N-subspace; and we deduce from Proposition 1 that A has finite even rank. This proves the necessity of our conditions; and it follows from the representation (1.*) of the skew symmetric bilinear form f that f and consequently δ is essentially uniquely determined. If conversely our conditions are satisfied, then one may define a skew symmetric bilinear form over (F,A) by means of (1.*); and one verifies that this form represents a null system.

Remark 2: If (F,A) happens to be a line, then the duality which maps every point upon itself is a null system so that in this case our conditions would not be necessary. A similar remark applies if (F,A) is a point.

Proposition 2: *If the duality δ is represented by the semi-bilinear form f over (F,A), and if W is an N-subspace of A, then the following properties of the subspace M of W are equivalent:*

(i) $W = M \oplus (W \cap W^{\delta})$.

(ii) $r(W) = r(M) + r(W \cap W^{\delta})$ *and* $M \cap M^{\delta} = 0$.

(iii) *M is a maximal subspace of W with the property*: $M \cap M^{\delta} = 0$.

Proof: Assume first the validity of (i). Then we deduce

$$r(W) = r(M) + r(W \cap W^{\delta})$$

from II.2, Rank Formulas which are applicable, since the existence of a duality implies the finiteness of rank. Next we deduce from Lemma 1, (b) that $f(x,y) = - f(y,x)$ for x,y in W. Hence it follows from $M \leqq W$ that $W^{\delta} \leqq M^{\delta}$ and that therefore

$$f(M \cap M^{\delta}, W \cap W^{\delta}) = - f(W \cap W^{\delta}, M \cap M^{\delta}) \leqq f(W^{\delta}, M) = 0$$

and consequently

$$f(M \cap M^{\delta}, W) \leqq f(M \cap M^{\delta}, M) + f(M \cap M^{\delta}, W \cap W^{\delta}) \leqq f(M^{\delta}, M) = 0.$$

Thus $M \cap M^{\delta} \leqslant W^{\delta}$ and this implies that $M \cap M^{\delta} \leqslant M \cap W \cap W^{\delta} = 0$. Hence (II) is a consequence of (I).

Suppose next that M meets requirement (II) and that the subspace N satisfies $M \leqslant N \leqslant W$ and $N \cap N^{\delta} = 0$. Then $W^{\delta} \leqslant N^{\delta}$ and consequently $N \cap (W \cap W^{\delta}) = N \cap W^{\delta} \leqslant N \cap N^{\delta} = 0$. Now we deduce from (II) and the Rank Formulas of II.2 that

$$\begin{aligned} r(W) &= r(M) + r(W \cap W^{\delta}) \\ &\leqslant r(N) + r(W \cap W^{\delta}) = r(N + [W \cap W^{\delta}]) \leqslant r(W). \end{aligned}$$

Hence $r(N) = r(M)$ and consequently $N = M$, proving the maximality of M. Thus (III) is a consequence of (II).

Assume finally the validity of (III). From $M \cap M^{\delta} = 0$ we infer $A = M \oplus M^{\delta}$ [IV.1, Corollary 2]; and from $M \leqslant W$ we deduce therefore $W = M \oplus (W \cap M^{\delta})$. From $M \leqslant W$ we infer $W^{\delta} \leqslant M^{\delta}$ so that $W \cap W^{\delta} \leqslant W \cap M^{\delta}$. From the Complementation Principle [II.1] we deduce now the existence of a subspace T such that $W \cap M^{\delta} = T \oplus (W \cap W^{\delta})$; and consequently we have $W = N \oplus (W \cap W^{\delta})$ with $N = M \oplus T$. From the fact, already verified, that (II) is a consequence of (I), it follows now that $N \cap N^{\delta} = 0$; and from $M \leqslant N$ and the maximality of M we infer $M = N$. Hence $W = M \oplus (W \cap W^{\delta})$ so that (I) is a consequence of (III).

Remark 3: There exists always a subspace M which meets requirement (I). Thus Proposition 2 establishes a decomposition of the N-subspace W into a non-isotropic component M and the uniquely determined component $W \cap W^{\delta}$. This latter clearly satisfies

$$f(W \cap W^{\delta}, W \cap W^{\delta}) \leqslant f(W^{\delta}, W) = 0$$

so that $W \cap W^{\delta} \leqslant (W \cap W^{\delta})^{\delta}$. Such a subspace is called strictly isotropic. See in this context IV.4, Lemma 2 below.

IV.3. **Representation of Polarities**

An auto-duality σ of the linear manifold (F,A) [of rank not less than 2 as we shall always assume] is called a *polarity*, if $\sigma^2 = 1$. Such a polarity σ pairs the subspaces of A, since $X = Y^{\sigma}$ if, and only if, $Y = X^{\sigma}$. If in particular such a pair has the form P,H where $P = H^{\sigma}$ is a point and $H = P^{\sigma}$ is [because of IV.1, Corollary 2] a hyperplane, then P is the *pole* of the hyperplane H and H is the *polar* of the point P. The reader will realize now that our concept of polarity is a direct generalization of the "polarities with respect to a conic" which interchange points and lines. The basic symmetries of a polarity σ are expressed in the formulas:

$$X \cap X^\sigma = (X + X^\sigma)^\sigma,\ X + X^\sigma = (X \cap X^\sigma)^\sigma$$

for every subspace X of A.

Lemma 1: *Suppose that the semi-bilinear form $f(x,y)$ represents the duality σ. Then σ is a polarity if, and only if,*

(S) $f(x,y) = 0$ *implies* $f(y,x) = 0$.

PROOF: If σ is a polarity and $f(x,y) = 0$, then $Fx \leqslant (Fy)^\sigma$ and consequently $Fy \leqslant (Fx)^\sigma$ so that $f(y,x) = 0$, since σ is represented by f. This shows the necessity of (S).

Assume conversely the validity of (S). If X is a subspace of A, then $0 = f(X^\sigma,X)$, since σ is represented by f. It follows from (S) that $0 = f(X,X^\sigma)$ and consequently $X \leqslant X^{\sigma^2}$. It follows from IV.1, Corollary 1 that $r(A) = r(X) + r(X^\sigma) = r(X^\sigma) + r(X^{\sigma^2})$. Hence $r(X) = r(X^{\sigma^2})$; and now $X = X^{\sigma^2}$ is a consequence of $X \leqslant X^{\sigma^2}$ and the Rank Formulas [II.2] which we may apply because of the finiteness of rank [IV.1, Existence Theorem]. Hence $\sigma^2 = 1$ and σ is a polarity.

Corollary 1: *Null systems are polarities.*

This is an immediate consequence of Lemma 1 and IV.2, Lemma 1, (*b*).

Proposition 1: *If the α-form f represents a polarity σ, and if $f(w,w) = 1$ for some w in A, then $\alpha^2 = 1$ and $f(x,y)^\alpha = f(y,x)$ for every x,y in A.*

PROOF: Let $H = (Fw)^\sigma$. Then H is a hyperplane, since Fw is a point. From $0 = f(H,w)$ we infer that w does not belong to H. Hence $H \cap Fw = 0$; and this implies $A = Fw \oplus H$ by IV.1, Corollary 2. Next we show:

(1.1) If u,v are elements in H and s,t are numbers in F such that $st^\alpha = f(u,v)$, then $ts^\alpha = f(v,u)$.

Since u,v belong to $H = (Fw)^\sigma$, we have $0 = f(u,w) = f(v,w)$; and we infer from Lemma 1, (S) that $0 = f(w,u) = f(w,v)$. If $st^\alpha = f(u,v)$, then we have

$$f(sw + u, tw - v) = sf(w,w)t^\alpha + f(u,w)t^\alpha - sf(w,v) - f(u,v) = 0;$$

and it follows from Lemma 1, (S) that

$$0 = f(tw - v, sw + u) = ts^\alpha - f(v,u),$$

as we intended to show.

(1.2) If u,v are elements in H, then $f(u,v)^\alpha = f(v,u)$.

If we let in (1.1) in particular $t = 1$ and $s = f(u,v)$, then it follows that $f(v,u) = ts^\alpha = f(u,v)^\alpha$, as we wanted to show.

(1.3) $\alpha^2 = 1$.

To prove this we note first that $H \neq 0$ and that therefore [by IV.1, Proposition 2]

$$0 \neq f(H,A) = f(H,Fw \oplus H) = f(H,H),$$

since $0 = f(H,Fw)$ as $H = (Fw)^{\sigma}$. Consequently we may find elements h,k in H such that $f(h,k) = 1$. If $t \neq 0$ is an element in F, then $t^{\alpha} \neq 0$ and there exists one and only one element s in F such that $st^{\alpha} = 1$. Thus $st^{\alpha} = f(h,k)$; and it follows from (1.1) that $ts^{\alpha} = f(k,h)$. It follows from (1.2) that $1 = f(h,k)^{\alpha} = f(k,h)$. Hence $ts^{\alpha} = 1$. But

$$1 = (st^{\alpha})^{\alpha} = t^{\alpha^2}s^{\alpha},$$

since α is an anti-automorphism of F. Hence $t^{\alpha^2} = (s^{\alpha})^{-1} = t$ or $\alpha^2 = 1$.

If x,y are elements in A, then it follows from $A = Fw \oplus H$ that $x = x'w + x''$, $y = y'w + y''$ with x'',y'' in H. Remembering that

$$0 = f(w,x'') = f(x'',w) = f(w,y'') = f(y'', w)$$

we find now that as a consequence of (1.2) and (1.3)

$$f(x,y)^{\alpha} = [x'y'^{\alpha} + f(x'',y'')]^{\alpha} = y'x'^{\alpha} + f(y'',x'') = f(x,y);$$

and this completes the proof of our proposition.

If the α-form $f(x,y)$ satisfies the condition $f(x,y)^{\alpha} = f(y,x)$, then we may say that f is *α-symmetrical* or shorter that f is symmetrical. We note that the three examples of semi-bilinear forms given in IV.1 were all of this type. We note furthermore that it is customary to term anti-automorphisms *involutorial*, if their square equals 1 [as in Proposition 1].

Theorem 1: *Suppose that σ is an auto-duality of the linear manifold (F,A) and that $3 \leqslant r(A)$. Then σ is a polarity if, and only if, either σ is a null system [so that F is commutative and σ is represented by a skew symmetrical bilinear form] or else σ may be represented by a symmetrical α-form with involutorial α.*

PROOF: Because of IV.2, Theorem we may assume that σ is not a null system; and because of IV.1, Proposition 1 we may represent σ by means of an α-form $f(x,y)$. Since σ is not a null system, there exists w in A such that $f(w,w) \neq 0$; and we may assume without loss in generality that $f(w,w) = 1$ [normalization of IV.1, Proposition 3]. If σ is a polarity, then it follows from Proposition 1 that α is an involution and that f is symmetrical. If conversely f is symmetrical, then condition (S), Lemma 1 is satisfied so that σ is a polarity.

Theorem 2: *The linear manifold (F,A) of rank not less than 3 possesses a polarity if, and only if, $r(A)$ is finite and F possesses an involutorial anti-automorphism.*

PROOF: The necessity of our conditions may be inferred from IV.1, Existence Theorem and the preceding Theorem 1 [note that $\alpha = 1$ is an involution]. If conversely $r(A) = n$ is finite and α is an involutorial anti-automorphism of F, then let $b_1, \cdots, b_n$ be some basis of A and denote by f the uniquely determined α-form which satisfies

$$f(b_i, b_j) = \begin{cases} 1 \text{ for } i = j \\ 0 \text{ for } i \neq j. \end{cases}$$

Since $(st^{\alpha})^{\alpha} = ts^{\alpha}$ for s,t in F, f is symmetrical. One deduces from Theorem 1 and IV.1, Proposition 2 that f represents a polarity σ; and this completes the proof.

For a detailed study of involutorial anti-automorphisms the reader might consult A. A. Albert: Structure of Algebras [1].

If f is an α-symmetrical α-form such that $f(A,A) \neq 0$, then it is possible to find elements u,v in A such that $f(u,v) = 1$. From the symmetry of f it follows that $f(v,u) = 1$ and that

$$s^{\alpha^2} = f(u,sv)^{\alpha} = f(sv,u) = sf(v,u) = s \text{ for every } s \text{ in } F.$$

Hence $\alpha^2 = 1$ is a consequence of the symmetry of f. This remark makes it possible to omit the hypothesis that α be involutorial whenever we are assured of the α-symmetry of a non-trivial form f.

IV.4. Isotropic and Non-isotropic Subspaces of a Polarity; Index and Nullity

The subspace X of the linear manifold (F,A) [of rank not less than 2] has been called non-isotropic with respect to a given polarity σ of A, if $X \cap X^{\sigma} = 0$. It is a consequence of IV.1, Corollary 2 that this is equivalent to requiring $A = X + X^{\sigma}$ or $A = X \oplus X^{\sigma}$. From $X = (X^{\sigma})^{\sigma}$ it follows furthermore that with X also X^{σ} is non-isotropic and thus every non-isotropic subspace leads to a decomposition into a direct sum of non-isotropic components [see IV.6 below for a further investigation of this point of view].

Lemma 1: *If σ is a polarity of the linear manifold (F,A), and if M is a non-isotropic subspace of A, then a polarity σ' is defined in the linear manifold (F,M) by the rule*

$$U^{\sigma'} = M \cap U^{\sigma} \textit{ for every subspace } U \textit{ of } M.$$

This polarity σ' we shall call *the polarity induced in M by σ*. This phrase shall always be understood to imply that M is non-isotropic, since σ' otherwise could not be a polarity.

Proof: It is clear that σ' is a single-valued mapping of subspaces of M upon subspaces of M, and that $U \leqslant V \leqslant M$ implies $V^{\sigma'} \leqslant U^{\sigma'}$ [$\leqslant M$].

From $U \leqslant M$, $\sigma^2 = 1$, $M \cap M^\sigma = 0$ and Dedekind's Law we deduce finally

$$U^{\sigma'^2} = M \cap (M \cap U^\sigma)^\sigma = M \cap (M^\sigma + U) = U + (M \cap M^\sigma) = U$$

so that $\sigma'^2 = 1$; and now it is clear that σ' is a polarity.

Naturally we are going to call a subspace X *isotropic*, if it is not non-isotropic; and this is equivalent to either of the following conditions : $X \cap X^\sigma \neq 0$, $X + X^\sigma \neq A$. It is clear that X^σ is isotropic if, and only if, X is isotropic.

The subspace $X \neq 0$ will certainly be isotropic, if $X \leqslant X^\sigma$; and such a subspace we shall call *strictly isotropic*. It will prove convenient to include among the strictly isotropic subspace also the null space. A point, for instance, is either non-isotropic or strictly isotropic. That every subspace is the direct sum of a non-isotropic and a strictly isotropic subspace, will be a consequence of the following fact.

Lemma 2: *If X and Y are subspaces of A such that $X = Y \oplus (X \cap X^\sigma)$, then $X \cap X^\sigma$ is strictly isotropic and Y is a maximal non-isotropic subspace of X.*

PROOF: $X \cap X^\sigma$ is strictly isotropic, since

$$X \cap X^\sigma \leqslant X + X^\sigma = (X \cap X^\sigma)^\sigma.$$

From $Y \leqslant X$ we deduce $X^\sigma \leqslant Y^\sigma$; and now it follows from Dedekind's Law that

$$X^\sigma = [Y + (X \cap X^\sigma)]^\sigma = Y^\sigma \cap (X^\sigma + X) = X^\sigma + (Y^\sigma \cap X)$$

or $Y^\sigma \cap X \leqslant X^\sigma$. Hence

$$Y \cap Y^\sigma = X \cap Y \cap Y^\sigma = Y \cap X \cap Y^\sigma \cap X^\sigma = Y^\sigma \cap (Y \cap X \cap X^\sigma) = 0,$$

since X is the direct sum of Y and $X \cap X^\sigma$. Thus Y is non-isotropic. Assume finally that Z is a non-isotropic subspace between Y and X. Then $X^\sigma \leqslant Z^\sigma$ so that

$$X \cap X^\sigma \cap Z = X^\sigma \cap Z \leqslant Z^\sigma \cap Z = 0$$

and consequently

$$Z = Y \oplus (Z \cap X \cap X^\sigma) = Y,$$

as we claimed.

The reader should compare this result with IV.2, Proposition 2 and IV.2, Remark 3.

Corollary 1: *A line with Property* (N) [*of* IV.2] *is either non-isotropic or strictly isotropic.*

PROOF: If follows from Lemma 2 that the line $L = K \oplus (L \cap L^\sigma)$ where K is non-isotropic and $L \cap L^\sigma$ is strictly isotropic. If L is neither non-isotropic nor strictly isotropic, then K and $L \cap L^\sigma$ are points on L. From $K \cap K^\sigma = 0$ it follows that the point K is not on its polar K^σ; and consequently L does not have Property (N).

If P is a point on the strictly isotropic subspace U of A, then

$$P \leqslant U \leqslant U^\sigma \leqslant P^\sigma;$$

and this implies that *every strictly isotropic subspace has Property* (N). The converse cannot be true in general, since every subspace has Property (N), if σ happens to be a null system. But this is "almost" the only exception, as may be seen from the following facts.

Proposition 1: *The linear manifold* (F,A) *possesses a polarity* σ *which is not a null system, though there exists an* N*-subspace of* σ *which is not strictly isotropic if, and only if,* F *is a commutative field of characteristic* 2 *and* (F,A) *has finite rank not less than* 3. *Any such polarity* σ *may be represented by a symmetrical bilinear form.*

PROOF: Assume first that σ is a polarity of (F,A), that σ is not a null system and that W is an N-subspace of A which is not strictly isotropic. There exists a subspace V of W such that $W = V \oplus (W \cap W^\sigma)$; and it follows from Lemma 2 that V is non-isotropic whereas $W \cap W^\sigma$ is strictly isotropic. Since W is not strictly isotropic, we have $0 < V$; and since W is an N-subspace, V is a non-isotropic N-subspace of A. If V were a point, then V would be on V^σ, since V has Property (N); and thus it follows that $1 < r(V)$. Furthermore it is impossible that $V = A$, since then A would have Property (N) so that σ would be a null system. Hence $3 \leqslant r(A)$; and the finiteness of $r(A)$ follows from IV.1, Existence Theorem. Now it follows from IV.3, Theorem 1 that the polarity σ, not a null system, may be represented by a symmetrical α-semi-bilinear form. Since $V \neq 0$ is a non-isotropic N-subspace of A, it follows from IV.2, Proposition 1 that V has even rank, that $\alpha = 1$ and that therefore F is commutative. Since V is non-isotropic, we have $f(V,V) \neq 0$. Hence there exist elements u,v in V such that $f(u,v) \neq 0$. Now it follows from IV.2, Lemma 1, (b) and the symmetry of f that

$$-f(u,v) = f(v,u) = f(u,v) \text{ and hence } 1 = -1.$$

Thus F is a commutative field of characteristic 2; and f is a symmetrical bilinear form.

Assume next that F is a commutative field of characteristic 2, and that (F,A) has finite rank n, not less than 3. Denote by $b_1, \cdots, b_n$ a basis of A; and consider the form:

$$(*) \qquad f\Big(\sum_{i=1}^n x_i b_i, \sum_{i=1}^n y_i b_i\Big) = \sum_{i=1}^n x_i y_i.$$

It is clear that this is a symmetrical bilinear form which defines a polarity σ [IV.1, Proposition 2 and IV.3, Lemma 1]. From the properties of characteristic 2 one deduces that $(x + y)^2 = x^2 + y^2$ for x,y in F, and that therefore

$$(**) \qquad f\Big(\sum_{i=1}^{n} x_i b_i, \sum_{i=1}^{n} x_i b_i\Big) = \Big[\sum_{i=1}^{n} x_i\Big]^2.$$

Denote now by W the totality of elements $\sum_{i=1}^{n} x_i b_i$ such that $\sum_{i=1}^{n} x_i = 0$. It is easily verified that W is a subspace of A, that W is a hyperplane in A [II.3, Remark 2] and that W has Property (N) by (**). From $1 + 1 = 0$ and $3 \leqslant n$ we deduce the existence of the elements $b_1 + b_2$ and $b_2 + b_3$ in W which satisfy $f(b_1 + b_2, b_2 + b_3) = 1$. Hence $f(W,W) \neq 0$ so that W is not strictly isotropic. From $f(b_i,b_i) = 1$ we infer that σ is not a null system and this completes the proof.

Corollary 2: *Suppose that F is a field of characteristic different from 2 and that the polarity σ of the linear manifold (F,A) is not a null system. Then the subspace U of A is an N-subspace if, and only if, U is strictly isotropic.*

PROOF: We have pointed out before that strictly isotropic subspaces are N-subspaces. That conversely N-subspaces are strictly isotropic, may be inferred from Proposition 1, since σ is not a null system and F has not characteristic 2.

It is clear that the sum of two N-subspaces of A will, in general, not be an N-subspace of A. Thus the following criterion will be of some interest; and it will be needed in various applications.

Lemma 3: *If the polarity σ is represented by the semi-bilinear form f, if U and V are N-subspaces, and if $V \leqslant U + U^\sigma$, then $U + V$ is an N-subspace.*

PROOF: If x belongs to $U + V$, then $x = u + v$ with u in U and v in V. Since $V \leqslant U + U^\sigma$, we have $v = u' + w$ with u' in U and w in U^σ. Since U and V have Property (N), we have

$$0 = f(v,v) = f(u,u) = f(u',u') = f(u + u',u + u');$$

and since σ is a polarity and w belongs to U^σ, we deduce from IV.3, Lemma 1 that

$$0 = f(w,u) = f(u,w) = f(w,u') = f(u',w).$$

Hence

$$\begin{aligned} 0 = f(v,v) &= f(u' + w,u' + w) \\ &= f(u',u') + f(u',w) + f(w,u') + f(w,w) = f(w,w), \end{aligned}$$

$$f(x,x) = f(u + v,u + v) = f(u + u' + w,u + u' + w) = 0$$

so that Fx is on $(Fx)^\sigma$ and $U + V$ has Property (N).

If σ is a polarity of the linear manifold (F,A), then we shall denote by $N(\sigma)$ *the sum of all the non-isotropic N-subspaces of A*; and we define the *nullity* of σ as $n(\sigma) = r[N(\sigma)/(N(\sigma) \cap N(\sigma)^\sigma)]$.

Proposition 2: *If σ is a polarity of the linear manifold (F,A), then $N(\sigma)$ is an N-subspace of A; and every maximal non-isotropic N-subspace M of A has the properties*:

$$N(\sigma) = M \oplus [N(\sigma) \cap N(\sigma)^\sigma],\ r(M) = n(\sigma).$$

Proof: If σ is a null system, then A is the only maximal non-isotropic N-subspace of A; and our contentions are trivially true. Likewise everything is obvious, if 0 is the only non-isotropic N-subspace of A. Thus we assume that σ is not a null system and that there exists a non-isotropic N-subspace $W \neq 0$. We have $1 < r(W) < r(A)$, since W cannot be a point. Hence $3 \leqslant r(A)$ and σ may be represented by a semi-bilinear form [IV. 1, Proposition 1]. Now it follows from the finiteness of $r(A)$ [IV.1, Existence Theorem] and Lemma 3 that $N(\sigma)$ is an N-subspace as the sum of a finite number of non-isotropic N-subspaces, if we just remember that $A = X \oplus X^\sigma$ for every non-isotropic subspace X of A. Every maximal non-isotropic N-subspace of A is contained in $N(\sigma)$; and it follows from IV.2, Proposition 2 that $N(\sigma) = M \oplus [N(\sigma) \cap N(\sigma)^\sigma]$. This implies $n(\sigma) = r(M)$.

Remark 1: If U is a non-isotropic N-subspace, and if V is an N-subspace, then $A = U \oplus U^\sigma$; and it follows from Lemma 3 that $U + V$ is an N-subspace. Hence U is contained in V, if V is a maximal N-subspace of A; and this implies that $N(\sigma)$ is part of every maximal N-subspace of A.

Corollary 3: *The nullity $n(\sigma)$ is always even*; *and $n(\sigma) = 0$ or $r(A)$ if the characteristic of F is not* 2.

This one deduces from IV.2, Proposition 1, and from Propositions 1 and 2.

If one wants to obtain a particularly simple representation of a polarity, then one has to consider a specially selected basis of A. With this in mind we term the points $P_1, \cdots, P_k$ *an O-set of points* [with respect to the polarity σ], if

(0) $$P_i \leqslant P_j^\sigma \text{ if, and only if, } i \neq j.$$

If, for instance, the polarity σ is represented by the semi-bilinear form f, then the points $Fp_1, \cdots, Fp_k$ form an O-set if, and only if,

$$f(p_i,p_j) \begin{cases} \neq 0 \text{ for } i = j \\ = 0 \text{ for } i \neq j. \end{cases}$$

[The reader should note the connections between this concept and the concept of orthogonality; see also below IV.5.]

Lemma 4: *Assume that σ is a polarity of the linear manifold (F,A).*

(*a*) *If the points $P_1,\cdots,P_k$ form an O-set, then they are independent and $\sum_{i=1}^{k} P_i$ is non-isotropic.*

(*b*) *The O-set $P_1,\cdots, P_k$ is maximal if, and only if, $[\sum_{i=1}^{k} P_i]^\sigma$ is an N-subspace.*

(*c*) *If M is a maximal non-isotropic N-subspace of A, then there exists an O-set $Q_1,\cdots,Q_h$ such that $M = [\sum_{i=1}^{h} Q_i]^\sigma$.*

PROOF: If the points $P_1,\cdots, P_k$ form an O-set, then we have $P_i \cap P_i^\sigma = 0$, $P \leqslant P_j^\sigma$ for $i \neq j$. Consequently

$$P_i \cap \sum_{i \neq j} P_j = P_i \cap P_i^\sigma \cap \sum_{i \neq j} P_j = 0;$$

and this proves the independence of the points in an O-set. Next we deduce from Dedekind's Law that

$$\left[\sum_{j=1}^{i} P_j\right] \cap P_i^\sigma = \left[\sum_{j=1}^{i-1} P_j\right] + (P_i \cap P_i^\sigma) = \sum_{j=1}^{i-1} P_j;$$

and now one verifies inductively that

$$\left[\sum_{j=1}^{k} P_j\right] \cap \left[\sum_{j=1}^{k} P_j\right]^\sigma = \left[\sum_{j=1}^{k} P\ \right] \cap P_k^\sigma \cap \cdots \cap P_1^\sigma = \cdots$$

$$= \left[\sum_{j=1}^{i} P_j\right] \cap P_i^\sigma \cap \cdots \cap P_1^\sigma = \cdots = 0.$$

Hence an O-set spans a non-isotropic subspace; and this completes the proof of (*a*).

Suppose that the points $P_1,\cdots,P_k$ form an O-set. Consider some point Q different from all the points P_i. Then the points $Q,P_1,\cdots, P_k$ form an O-set if, and only if, $Q \leqslant [\sum_{i=1}^{k} P_i]^\sigma$ and $Q \neq Q^\sigma$. Hence $P_1, \cdots, P_k$ is not maximal if, and only if, $[\sum_{i=1}^{k} P_i]^\sigma$ does not have Property (N); and this proves (*b*).

Consider now some maximal non-isotropic N-subspace M of A. Then $A = M \oplus M^\sigma$. The property (c) is certainly true if $M^\sigma = 0$; and thus we assume that $M^\sigma \neq 0$. Since M is non-isotropic, so is M^σ; and now we deduce from Lemma 1 that σ induces in M^σ a polarity σ', defined by the rule:

$$U^{\sigma'} = M^\sigma \cap U^\sigma \;[= (M + U)^\sigma] \text{ for every subspace } U \text{ of } M^\sigma.$$

Since the empty set is an O-set, since [by (a)] every O-set is an independent set of points, and since the existence of a polarity implies the finiteness of rank [IV.1, Existence Theorem], there exists a maximal O-set $Q_1, \cdots, Q_k$ of points in M^σ with respect to the induced polarity σ'. It is clear that these points form likewise an O-set with respect to σ. We deduce from (a) and (b) that $V = \left[\sum_{i=1}^{k} Q_i\right]^{\sigma'}$ is a non-isotropic N-subspace with respect to σ' and consequently with respect to σ. Since $V \leqslant M^\sigma$, we have $M \cap V = 0$. We want to prove that $M + V$ has Property (N). This is certainly true if M or V is 0. If neither of them is 0, then both have rank not less than 2, since points with Property (N) are isotropic. Consequently $M \oplus V$ and A have rank not less than 4; and it is possible to represent σ by a semi-bilinear form [IV.1, Proposition 1]. Because of $M + M^\sigma = A$ we may apply Lemma 3 and see that $M + V$ has Property (N) in any case. Letting $W = \sum_{i=1}^{k} Q_i$ we find that $W^{\sigma'} = V$ so that W is non-isotropic with respect to σ' [as is V] and hence with respect to σ. We have furthermore $W \leqslant M^\sigma$ and therefore $M \leqslant W^\sigma$. Now it follows from Dedekind's Law that

$$\begin{aligned}
(V + M) \cap (V + M)^\sigma &= (M + W^{\sigma'}) \cap (M + W^{\sigma'})^\sigma \\
&= [M + (W^\sigma \cap M^\sigma)] \cap [M + (W^\sigma \cap M^\sigma)]^\sigma \\
&= [M + (W^\sigma \cap M^\sigma)] \cap [M^\sigma \cap (W + M)] \\
&= [(M \cap M^\sigma) + (W^\sigma \cap M^\sigma)] \cap (W + M) = W^\sigma \cap M^\sigma \cap (W + M) \\
&= W^\sigma \cap [W + (M^\sigma \cap M)] = W^\sigma \cap W = 0.
\end{aligned}$$

Thus we have shown that $M \oplus V$ is a non-isotropic N-subspace. From the maximality of M we deduce now that $V = 0$. Hence

$$\sum_{i=1}^{k} Q_i = W = V^{\sigma'} = 0^{\sigma'} = M^\sigma \text{ or } M = \sum_{i=1}^{k} Q_i,$$

as we intended to show.

Proposition 3: *If the polarity σ is not a null system, and if the characteristic of the field F is not 2, then the points of every maximal O-set form a basis of A.*

Proof: It follows from Corollary 2 that $n(\sigma) = 0$ so that O is the only non-isotropic N-subspace of A. Our contention is now an immediate consequence of Lemma 4, (*a*), (*b*).

Remark 2: The existence of maximal O-sets is a consequence of Lemma 4, (*a*) and the finiteness of rank, as has already been pointed out in the course of the proof of Lemma 4.

Remark 3: If the hypotheses of Proposition 3 are not satisfied, then either σ is a null system and the vacuous set is the only O-set; or else σ is not a null system, but F has characteristic 2. In the latter case there arises the particularly interesting possibility that two maximal O-sets need not contain the same number of points ["non-invariance of O-rank"], as may be seen from the following general considerations: As in the proof of Proposition 1 we consider the linear manifold (F,A) of rank n not less than 3 over the commutative field F of characteristic 2, a basis $b_1, \cdots, b_n$ of A and the symmetrical bilinear form $f(x,y)$ which satisfies

$$f(b_i, b_j) = \begin{cases} 0 \text{ for } i \neq j \\ 1 \text{ for } i = j. \end{cases}$$

The points $Fb_1, \cdots, Fb_n$ form an O-set of points with respect to the polarity σ defined by f. Next we consider the hyperplane W consisting of all the elements $\sum_{i=1}^{n} x_i b_i$ with $\sum_{i=1}^{n} x_i = 0$. This hyperplane has Property (N), but is not strictly isotropic. Hence $W = M \oplus (W \cap W^\sigma)$ where $M \neq 0$ is a maximal non-isotropic N-subspace. From Lemma 4, (*c*) we deduce the existence of an O-set $Q_1, \cdots, Q_k$ such that $M^\sigma = \sum_{i=1}^{k} Q_i$. It follows that $k < n$, since $r(A) = r(M) + r(M^\sigma)$. We deduce from Lemma 4, (*a*), (*b*) that the Q_i form a maximal O-set too.

Proposition 4: *Any two maximal N-subspaces of (F,A) have the same rank.*

Proof: This is trivially true, if σ is a null system, since then A is the one and only one maximal N-subspace. Thus we assume that σ is not a null system. If A happens to be a line, then either 0 is the only N-subspace of A or else the maximal N-subspaces of A are points; and this shows the validity of our claim for lines. Hence we assume now that $3 \leqslant r(A)$. Then σ is represented by semi-bilinear forms [IV.1, Proposition 1] so that we may apply Lemma 3. Suppose now that W is a maximal N-subspace of A

and that T is some N-subspace of A. Then $T' = T \cap (W + W^\sigma)$ is an N-subspace of $W + W^\sigma$; and thus it follows from Lemma 3 that $W + T'$ is an N-subspace. But W is a maximal N-subspace of A so that $T' \leqslant W$. Hence we have shown:

(4.O) If W is a maximal N-subspace of A, and if T is some N-subspace of A, then $T \cap (W + W^\sigma) \leqslant W$ or $T \cap W = T \cap (W + W^\sigma)$.

[The reader ought to verify that this lemma is also true, if σ is a null system and if $r(A) = 2$.]

Suppose now that U and V are maximal N-subspaces of A. Then it follows from (4.O) that

$$(a) \qquad U \cap (V + V^\sigma) = U \cap V = V \cap (U + U^\sigma).$$

Consider furthermore some maximal non-isotropic N-subspace M of A. We deduce from Proposition 2 that

$$(b) \qquad r(M) = n(\sigma);$$

and from Lemma 3 or Remark 1 that M is part of U and V; and now it follows from IV.2, Proposition 2 that

$$(c) \qquad U = M \oplus (U \cap U^\sigma),\ V = M \oplus (V \cap V^\sigma).$$

Substituting (c) in (a) we infer from Dedekind's Law that

$$(d) \begin{cases} M + [U \cap U^\sigma \cap (V + V^\sigma)] = (V + V^\sigma) \cap [M + (U \cap U^\sigma)] \\ \quad = (V + V^\sigma) \cap U = (U + U^\sigma) \cap V = M + [V \cap V^\sigma \cap (U + U^\sigma)]. \end{cases}$$

Since M is non-isotropic, we have $A = M \oplus M^\sigma$ [IV.1, Corollary 2]. From $M \leqslant U, V$ it follows that $U^\sigma, V^\sigma \leqslant M^\sigma$. If we intersect equation (d) now with M^σ, it follows consequently that

$$(e') \qquad U \cap U^\sigma \cap (V + V^\sigma) = V \cap V^\sigma \cap (U + U^\sigma);$$

and if we apply σ on this equation, then we find

$$(e'') \qquad U + U^\sigma + (V \cap V^\sigma) = V + V^\sigma + (U \cap U^\sigma).$$

Denote by j the rank of the intersection (e') and by s the rank of the sum (e''). Then it follows from the Rank Formulas [II.2] that

$$(f) \quad r(U \cap U^\sigma) + r(V + V^\sigma) = j + s = r(V \cap V^\sigma) + r(U + U^\sigma).$$

From IV.1, Corollary 1 we deduce that

$$(g) \quad r(A) = r(U \cap U^\sigma) + r(U + U^\sigma) = r(V \cap V^\sigma) + r(V + V^\sigma),$$

since $(X \cap X^\sigma)^\sigma = X + X^\sigma$. Substituting (g) in (f) it follows that

$$r(U \cap U^\sigma) - r(V \cap V^\sigma) = r(V \cap V^\sigma) - r(U \cap U^\sigma)$$

or $r(U \cap U^\sigma) = r(V \cap V^\sigma)$. Now it follows from (*b*), (*c*) and the Rank Formulas that

$$r(U) = n(\sigma) + r(U \cap U^\sigma) = n(\sigma) + r(V \cap V^\sigma) = r(V),$$

as we intended to prove.

The common rank of all the maximal N-subspaces with respect to σ may be called $j(\sigma)$. Since every maximal non-isotropic N-subspace of A is part of every maximal N-subspace, we have $n(\sigma) \leqslant j(\sigma)$; and the difference $i(\sigma) = j(\sigma) - n(\sigma)$ is called *the index of* σ. Now we restate the content of the result (*c*) obtained in the course of the preceding proof as follows:

Corollary 4: *If* V *is a maximal* N*-subspace and* M *a maximal non-isotropic* N*-subspace* [*with respect to the polarity* σ *of* (F,A)], *then*

$$V = M \oplus (V \cap V^\sigma), r(M) = n(\sigma), r(V \cap V^\sigma) = i(\sigma), r(V) = n(\sigma) + i(\sigma).$$

Thus $i(\sigma)$ is the rank of the maximal strictly isotropic subspaces.

The σ-Decompositions of a Space

We are now ready to obtain a decomposition of the space with respect to some given polarity which puts into evidence some of the salient features of this polarity and which leads at the same time to a particularly simple representation of the representing semi-bilinear forms.

We assume now that σ is a polarity of the linear manifold (F,A), that σ is not a null system and that $3 \leqslant r(A)$ so that σ may be represented by semi-bilinear forms [IV.1, Proposition 1]. We note furthermore that the rank of A is finite [IV.1, Existence Theorem]. Thus we shall always be assured of the existence of maximal subspaces with a given property.

Denote by W some maximal N-subspace of A and by M some maximal non-isotropic N-subspace of A. It follows from Remark 1 that $M \leqslant W$; and it follows from Corollary 3, Corollary 4 and IV.2, Proposition 2 that

(4.1) $W = (W \cap W^\sigma) \oplus M, \qquad r(M) = n(\sigma), \qquad r(W \cap W^\sigma) = i(\sigma).$

There exist subspaces J such that $W^\sigma = (W \cap W^\sigma) \oplus J$. We prove:

(4.2) J does not contain N-subspaces $\neq 0$.

Suppose that V is an N-subspace of J. Then $V \leqslant J \leqslant W^\sigma \leqslant W + W^\sigma$; and it follows from Lemma 3 that $V + W$ is an N-subspace. But W is a maximal N-subspace of A. Hence $V \leqslant W \cap J = W \cap W^\sigma \cap J = 0$, as we claimed.

(4.3) J is non-isotropic.

This follows from (4.2), since $J \cap J^\sigma$ is strictly isotropic [by Lemma 2].

(4.4) $J \cap M = 0$ and $J \oplus M$ is non-isotropic.

$J \cap M$ is an N-subspace as a subspace of the N-subspace M. But $J \cap M$ is contained in J; and we infer $J \cap M = 0$ from (4.2). Next we deduce from Dedekind's Law and $J \leqslant W^\sigma$, $M \leqslant W$ that $J \leqslant M^\sigma$ and

$$\begin{aligned}(J + M) \cap (J + M)^\sigma &= (J + M) \cap J^\sigma \cap M^\sigma \\ &= [J + (M \cap M^\sigma)] \cap J^\sigma = J \cap J^\sigma = 0\end{aligned}$$

since M and J are non-isotropic. Hence $M + J$ is non-isotropic too.

(4.5) Every maximal O-set of points on J is a basis of J.

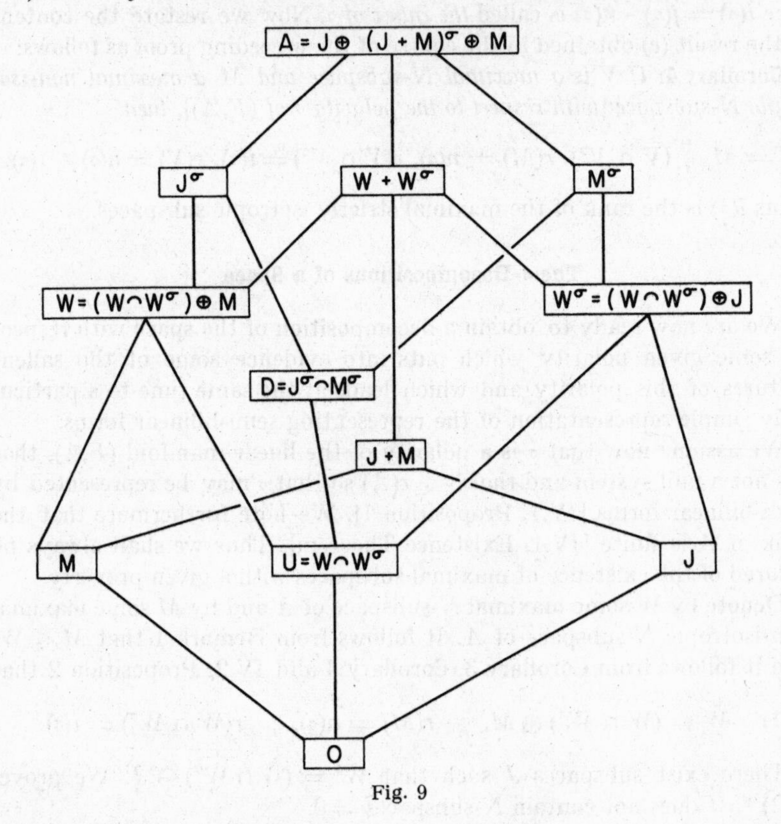

Fig. 9

Since J is non-isotropic, σ induces a polarity σ' in J [Lemma 1]. Now (4.5) is a consequence of (4.2) and Lemma 4.

(4.6) $\quad A = J \oplus M \oplus (J + M)^\sigma, W \cap W^\sigma = (W + W^\sigma) \cap J^\sigma \cap M^\sigma.$

Clearly $W + W^\sigma = M + (W \cap W^\sigma) + J$ by (4.1) and the choice of J. Hence $W \cap W^\sigma = (W + W^\sigma)^\sigma = M^\sigma \cap (W + W^\sigma) \cap J^\sigma$. Next we note that $J \leqslant W^\sigma$ and hence $M \leqslant W \leqslant J^\sigma$. Consequently by Dedekinds' Law

$$M + (J + M)^\sigma = M + (J^\sigma \cap M^\sigma) = J^\sigma \cap (M + M^\sigma) = J^\sigma \cap A = J^\sigma,$$

since $A = M \oplus M^\sigma$ as M is non-isotropic. From $M \cap M^\sigma = 0$ we infer now that $M \cap (J + M)^\sigma = 0$; and thus we see that $J^\sigma = M \oplus (J + M)^\sigma$. Since J is non-isotropic, we find finally that

$$A = J \oplus J^\sigma = J \oplus M \oplus (J + M)^\sigma;$$

and this completes the proof of (4.6).

(4.7) $W \cap W^\sigma \leqslant J^\sigma \cap M^\sigma$.

This follows, for instance, from $J + M \leqslant W + W^\sigma$ or from (4.6).

(4.8) $W \cap W^\sigma$ is strictly isotropic, $r(W \cap W^\sigma) = i(\sigma)$ by Lemma 2, Corollary 4.

The structure of J is essentially elucidated by (4.2), (4.5), the structure of M by IV.2, Proposition 1 and the structure of $W \cap W^\sigma$ is in a way clear by (4.8). What we still need is an elucidation of the structure of $J^\sigma \cap M^\sigma$ with respect to its subspace $W \cap W^\sigma$.

(4.9) $W \cap W^\sigma$ is a maximal N-subspace of $J^\sigma \cap M^\sigma$.

Consider an N-subspace V such that $W \cap W^\sigma \leqslant V \leqslant J^\sigma \cap M^\sigma$. Then

$$J + M \leqslant \left\{ \begin{matrix} V^\sigma \\ W \end{matrix} \right\} \leqslant W + W^\sigma = J + M + (W \cap W^\sigma) \leqslant V^\sigma + V;$$

and it follows from Lemma 3 that $V + W$ is an N-subspace. Since W is a maximal N-subspace, this implies $V \leqslant W$. Hence

$$V \leqslant W \cap M^\sigma \cap J^\sigma \leqslant (W + W^\sigma) \cap M^\sigma \cap J^\sigma = W \cap W^\sigma$$

by (4.6) so that $V = W \cap W^\sigma$.

Since $J + M$ is non-isotropic by (4.4), $J^\sigma \cap M^\sigma = (J + M)^\sigma$ is non-isotropic too. It follows from Lemma 1 that σ induces in $D = J^\sigma \cap M^\sigma$ a polarity δ. We let $U = W \cap W^\sigma$ and prove:

(4.10) $U = U^\delta$, $r(D) = 2r(U)$.

We deduce from (4.6) that

$$\begin{aligned} U^\delta = (W \cap W^\sigma)^\delta &= (W \cap W^\sigma)^\sigma \cap D \\ &= (W^\sigma + W) \cap J^\sigma \cap M^\sigma = W \cap W^\sigma = U; \end{aligned}$$

and $r(D) = r(U) + r(U^\delta) = 2r(U)$ is now a consequence of IV.1, Corollary 1.

(4.11) There exists a non-isotropic subspace V such that $D = U \oplus V$.

We note first that a subspace of D is non-isotropic with respect to σ if, and only if, it is non-isotropic with respect to δ. There exist subspaces X of D which are non-isotropic and satisfy $U \cap X = 0$. Thus we may infer from the finiteness of rank the existence of a maximal non-isotropic subspace V of D such that $U \cap V = 0$. Then $D = V \oplus V^\delta$, since V is non-isotropic [IV.1, Corollary 2]; and we deduce from (4.10) that

$$D = 0^\delta = U + V^\delta.$$

Suppose now that V^δ is not part of $U + V$. Then there exist points on V^δ which do not belong to $U + V$. If one of them, say P, were not an N-point, then $V < V + P$ so that $V + P$ would either be isotropic or else $U \cap (V + P) \neq 0$. The first alternative is impossible, since

$$\begin{aligned}(V + P) \cap (V + P)^\delta &= (V + P) \cap V^\delta \cap P^\delta \\ &= [P + (V \cap V^\delta)] \cap P^\delta = P \cap P^\delta = 0\end{aligned}$$

by Dedekind's Law, since V and P are non-isotropic; and the second alternative is impossible, since

$$\begin{aligned}U \cap (V + P) &= U \cap (U + V) \cap (V + P) \\ &= U \cap [V + ([U + V] \cap P)] = U \cap V = 0;\end{aligned}$$

and thus we have shown that every point on V^δ, but not on $U + V$ is an N-point. Again let P be a point on V^δ, though not on $U + V$. We note that σ as a polarity, not a null system, on the linear manifold (F,A) of rank greater than 2 may be represented by a symmetrical α-form f [IV.1, Proposition 1 and IV.3, Theorem 1]; and it is clear that the polarity δ which is induced by σ in D is also represented by f. Now consider an element $t \neq 0$ in F, an element w in $(U + V) \cap V^\delta$ and a [fixed] element p such that $P = Fp$. Then $F(w + tp)$ is a point in V^δ, but not in $(U + V) \cap V^\delta$; and consequently it is an N-point. Hence we have $0 = f(p,p)$ and

$$\begin{aligned}0 &= f(w + tp, w + tp) = f(w,w) + f(w,tp) + f(tp,w) + tf(p,p)t^\alpha \\ &= f(w,w) + tf(p,w) + [tf(p,w)]^\alpha.\end{aligned}$$

If $f(p,w) = 0$, then this certainly implies $f(w,w) = 0$; and if $f(p,w) \neq 0$, then we let $t = -f(p,w)^{-1}$ and find that $f(w,w) = 2$. If the field has characteristic 2, then this implies again $f(w,w) = 0$. If the field, however, should have characteristic different from 2, then we let $t = +f(p,w)^{-1}$ and find $f(w,w) = -2$ which would be a contradiction. Consequently we have shown that $f(w,w) = 0$ in either case. But this signifies that every point on $(U + V) \cap V^\delta$ is an N-point. Since we had already shown that every point on V^δ, though not on $(U + V) \cap V^\delta$ is an N-point, we have finally verified that every point on V^δ is an N-point. Since

$$U \leqslant D = V \oplus V^\delta,$$

it follows now from Lemma 3 that $U + V^\delta = D$ is an N-subspace. But it follows from (4.9) that U is a maximal N-subspace of D. Thus it follows that $U = D$, and this implies $V^\delta \leqslant U$. Our hypothesis that $V^\delta \not\leqslant U + V$ has led us to a contradiction which proves that V^δ is part of $U + V$ and that therefore $D = V \oplus V^\delta = U + V$. This completes the proof of (4.11).

REMARK 4: It is worth noting that only the maximality of V among the non-isotropic subspaces satisfying $0 = U \cap V$ has been used in the course of the proof of (4.11).

(4.12) If V is a non-isotropic subspace such that $D = U \oplus V$, if $V_1, \cdots, V_k$ is a maximal O-set of points in V, if $U_i = U \cap (V_i + V^\delta)$ and $W_i = V^\delta \cap (U + V_i)$, then

(*a*) $D = U \oplus V = V \oplus V^\delta = V^\delta \oplus U$,

(*b*) $U_i + V_i = V_i + W_i = W_i + U_i$.

(*c*) The U_i are points forming a basis of U, the V_i are points forming a basis of V and the W_i are points forming a basis of V^δ,

(*d*) The V_i and W_i together form an O-set of points in D.

From $D = U \oplus V$ and (4.10) it follows that $D = U^\delta \oplus V^\delta = U \oplus V^\delta$; and from the non-isotropy of V and IV.1, Corollary 2 we deduce $D = V \oplus V^\delta$. This proves (*a*). To prove (*c*) we let $T = \sum_{i=1}^{k} V_i$. Then $T \leqslant V$ and the points V_i form a basis of T by Lemma 4, (*a*). One deduces from Lemma 4, (*a*), (*b*) that $T^\delta \cap V$ is a non-isotropic N-subspace of V; and it follows from Lemma 3 that $U + (V \cap T^\delta)$ is an N-subspace of D. But U is a maximal N-subspace of D by (4.9) so that $V \cap T^\delta \leqslant U$; and $V \cap T^\delta = 0$ is a consequence of $U \cap V = 0$. Since T is non-isotropic by Lemma 4, (*a*), we have $D = T \oplus T^\delta$ [IV.1, Corollary 2] and $T \leqslant V$ implies consequently $V = T \oplus (V \cap T^\delta) = T = \sum_{i=1}^{k} V_i$. Applying the Isomorphism Laws we deduce successively that, since V is the direct sum of the points V_i, $(U + V)/U = D/U$ is the direct sum of the points $(U + V_i)/U$, that therefore and because of $D = U \oplus V^\delta$ the subspace V^δ is the direct sum of the points $(U + V_i) \cap V^\delta = W_i$, that likewise $D/V^\delta = (V + V^\delta)/V^\delta$ is the direct sum of the points $(V^\delta + V_i)/V^\delta$, and that therefore and because of $D = U \oplus V^\delta$ the subspace U is the direct sum of the points $(V^\delta + V_i) \cap U = U_i$. This completes the proof of (*c*); and we note that as a consequence of (*a*) and (*c*) we have $k = r(U) = r(V) = r(W)$ and $2k = r(D)$.

It follows from Dedekind's Law that

$$V_i + W_i = V_i + [(U + V_i) \cap V^\delta] = (U + V_i) \cap (V_i + V^\delta)$$
$$= V_i + [(V^\delta + V_i) \cap U] = V_i + U_i;$$

and this proves (*b*), since U_i, V_i, W_i are three distinct points as points on the mutually exclusive [complementary] subspaces U, V, V^δ.

Since the points V_i form an *O*-set, we have

$$V_i \cap V_j^\delta = \begin{cases} V_i \text{ for } i \neq j \\ 0 \text{ for } i = j. \end{cases}$$

It follows from the definition of W_i that $W_i \leqslant V^\delta \leqslant V_j^\delta$, and consequently $V_j \leqslant W_i^\delta$. Thus it suffices in order to prove (*d*) to verify that the points W_i form an *O*-set. Using Dedekind's Law we deduce from (4.10) that

$$W_j \cap W_i^\delta = V^\delta \cap (U + V_j) \cap [V + (U \cap V_i^\delta)]$$
$$= V^\delta \cap ([(U + V_j) \cap V] + [U \cap V_i^\delta])$$
$$= V^\delta \cap [V_j + (U \cap V_i^\delta)] \qquad \text{since } U \cap V = 0$$
$$= V^\delta \cap (V^\delta + V_j) \cap [V_j + (U \cap V_i^\delta)]$$
$$= V^\delta \cap [V_j + ([V^\delta + V_j] \cap [U \cap V_i^\delta])]$$
$$= V^\delta \cap [V_j + [(V^\delta + [V_j \cap V_i^\delta]) \cap U]].$$

If $j = i$, then this implies

$$W_i \cap W_i^\delta = V^\delta \cap [V_i + (V^\delta \cap U)] = V^\delta \cap V_i = 0;$$

and if $j \neq i$, then we find likewise that

$$W_j \cap W_i^\delta = V^\delta \cap [V_j + ([V^\delta + V_j] \cap U)]$$
$$= V^\delta \cap (V_j + U) \cap (V^\delta + V_j) = W_j;$$

and this completes the proof of (4.12).

We are finally collecting the facts obtained in such a way as to show their effect on the symmetrical α-form f which represents σ.

1. *The maximal non-isotropic N-subspace M of A.*

Its rank is the *nullity* $n(\sigma)$; and on M the form f is just a skew-symmetrical bilinear form. By a proper choice of the basis of M the form f may be brought on M into the normal form

$$f(x,y) = \sum_{i=1}^{h} [x_{2i-1}y_{2i} - x_{2i}y_{2i-1}]$$

where $2h = n(\sigma)$ [see IV.2, (1.*)].

2. *The maximal strictly isotropic N-subspace of A.*
This has been denoted in the preceding discussion by $U = W \cap W^\sigma$; its rank is the *index* $i(\sigma)$; and on it f is identically 0.

3. *The non-isotropic envelope of the maximal strictly isotropic subspaces.*
This has been denoted in the preceding discussion by $D = J^\sigma \cap M^\sigma$; its rank is $2i(\sigma)$. Chosing for a basis of D the O-set V_i, W_i which appears in (4.12) we may easily derive the following normal form for f on D:

$$f(x,y) = \sum_{i=1}^{k} [x_{2i-1} c_i y_{2i-1}^\alpha - x_{2i} c_i y_{2i}^\alpha]$$

where $k = i(\sigma)$ and $c_i = c_i^\alpha \neq 0$.

4. *The component free of N-subspaces.*
This was denoted by J. Its rank is $n - n(\sigma) - 2i(\sigma)$, as follows from (4.6) and other remarks. As the basis of J we may select any maximal O-set of points of J. If we do this, then f has the following normal form on J:

$$f(x,y) = \sum_{i=1}^{j} x_i d_i y_i^\alpha$$

where $j = n - n(\sigma) - 2i(\sigma)$, $d_i = d_i^\alpha \neq 0$, and where $f(x,x) = 0$ for x in J implies $x = 0$. So in a way f is [positive or negative] definite on J.

The coefficients c_i and d_j appearing in these various forms are not in the least uniquely determined, since one may always change them into $t_i c_i t_i^\alpha$ and $s_j d_j s_j^\alpha$. The characterization of semi-bilinear forms by invariants is a fascinating field of study; the reader's attention should be drawn, for instance, to such investigations as

Ernst Witt: Theorie der quadratischen Formen in beliebigen Körpern, *Journal für die reine und angewandte Mathematik*, vol. 176 (1937), pp. 31-48, and to treatises like Dieudonné [1,2], Hua [1], Weyl [1].
For an investigation of the somewhat special situations arising when the characteristic is 2 see, for instance, Cahit Arf: Untersuchungen über quadratische Formen in Körpern der Charakteristik 2. (Teil I.), *Journal für die reine und angewandte Mathematik*, vol. 183 (1941), pp. 148-167.

APPENDIX I

Sylvester's Theorem of Inertia

If α is an involutorial anti-automorphism of the field F, then we define *a domain of α-positivity of* F as a subset P of F with the following properties:

(*a*) 0 *is not in* P *and* 1 *is in* P.

(*b*) *If x and y are in P, then $x + y$ is in P.*
(*c*) *If x is in P, then $x = x^\alpha$.*
(*d*) *If $0 \neq x = x^\alpha$ is not in P, then $-x$ is in P.*
(*e*) *If x is in P, and if $y \neq 0$ is in F, then yxy^α is in P.*

EXAMPLE: If F is the field of complex numbers [or the field of real quaternions], and if x^α is always the conjugate complex number to x, then the positive real numbers form the one and only one domain of α-positivity.

REMARK 1: If α happens to be the identity, then a domain of 1-positivity has great similarity with what we termed before a domain of positivity [III. Appendix III]. But the rules (*a*) to (*e*) do not necessarily imply that the product of numbers in P is likewise in P.

REMARK 2: If P is a domain of positivity [in the sense of III. Appendix III], then P is also a domain of 1-positivity.

REMARK 3: If P is a domain of positivity such that $P = P^\alpha$, and if Q is the set of all the elements q in P which satisfy $q = q^\alpha$, then Q is a domain of α-positivity.

An immediate consequence of (*a*), (*b*) and (*e*) is the following property of a domain of α-positivity.

(*f*) *If $y_1, \cdots, y_k$ are numbers, not 0, in F, then $\sum_{i=1}^{k} y_i y_i^\alpha$ is in P.*

This implies furthermore that such a sum is not 0; and as a consequence of this fact we see that the characteristic of the field F is 0 [so that in particular $+1 \neq -1$].

REMARK 4: It should be noted that there may exist various distinct domains of α-positivity. If, for instance, η is an automorphism of F such that $\alpha\eta = \eta\alpha$, then P^η is with P a domain of α-positivity.

Suppose now that σ is a polarity of the F-space A of rank at least 3. It is possible that σ is a null system, a case which we are going to exclude. Then it follows from IV.4, Corollary 3 that 0 is the only non-isotropic N-subspace of A, provided the characteristic is not 2; and it follows from IV.3, Theorem 1 that σ may be represented by an α-form $f(x,y)$ such that α is an involutorial anti-automorphism of the field F and such that $f(x,y)^\alpha = f(y,x)$ for every x and y.

We assume now that there is given some definite domain P of α-positivity of F; and the following definitions and considerations will all depend on α, P and f.

If $X = Fx$ is some point in A, then there arise three possibilities.

Case 1: $f(x,x)$ is in P.

If $t \neq 0$ is some number in F, then we have $f(tx,tx) = tf(x,x)t^\alpha$; and it follows from Property (*e*) that $f(tx,tx)$ is in P too. Thus $f(y,y)$ belongs to

P for every y such that $X = Fy$; and in this case we call X *a positive point.*

Case 2: $f(x,x) = 0$.

This is the case if, and only if, P is on P^σ; and X is *an N-point.*

Case 3: $f(x,x) \neq 0$ is not in P.

It follows from the symmetry of the form f that $f(x,x) = f(x,x)^\alpha$; and consequently it follows from Property (*d*) that $-f(x,x)$ belongs to P. As under Case 1 we find now that $-f(y,y)$ belongs to P for every y such that $X = Fy$; and we call X *a negative point.*

It is clear that the N-points depend on σ only. If we substitute for f the form $-f$, then we have not made any essential change; but positive and negative points have been interchanged. It is a consequence of IV.1, Proposition 3 that this is the only change which occurs if we substitute for f any other symmetrical form representing σ. Thus our division into three classes of points depends [apart from the naming] only on σ and P.

If we change from one domain of α-positivity to another, then our division into classes may be destroyed altogether, except that N-points remain N-points.

Sylvester's Theorem: *The number of positive points is the same in any two maximal O-sets of* σ.

PROOF: Suppose that $P_1, \cdots, P_m$ and $Q_1, \cdots, Q_n$ are two maximal O-sets of σ. Since the characteristic of F is 0, it follows from IV.4, Lemma 4 that $m = n = r(A)$ [since σ is not a null system by hypothesis]. None of the points P_i, Q_j is an N-point. We assume that the points $P_1, \cdots, P_h$ are positive, $P_{h+1}, \cdots, P_m$ negative; and likewise that $Q_1, \cdots, Q_k$ are positive, $Q_{k+1} \cdots, Q_m$ negative.

(*a*) *Every point in* $P_1 + \cdots + P_h$ *is positive.*

To prove this consider a point in $\sum_{i=1}^{h} P_i$. It has the form Ft with $t = \sum_{i=1}^{h} t_i$ and t_i in P_i. Some of the t_i may be 0; but not all of them, since $t \neq 0$. If $t_i \neq 0$, then $f(t_i,t_i)$ belongs to P, since P_i is a positive point. Since the points P_i form an O-set, we have $f(t_i,t_j) = 0$ for $i \neq j$. Consequently

$$f(t,t) = \sum_{i=1}^{h} f(t_i,t_i);$$

and it follows from Property (*b*) of domains of α-positivity that $f(t,t)$ belongs to P. Hence Ft is a positive point.

By a completely analogous argument we show that

(*b*) *Every point in* $P_{h+1} + \cdots + P_m$ *is negative.*

What is true for the P_i holds, mutatis mutandis, for the points Q_j; and thus every point in $\sum_{i=1}^{k} Q_i$ is positive whereas every point in $\sum_{i=k+1}^{m} Q_i$ is negative. Combining this with (a) and (b) we find that

$$0=\Big[\sum_{i=1}^{h} P_i\Big] \quad \Big[\sum_{i=k+1}^{m} Q_i\Big]=\Big[\sum_{i=h+1}^{m} P_i\Big] \cdot \Big[\sum_{i=1}^{k} Q_i\Big].$$

But the points in an O-set are independent [IV.4, Lemma 4]. Hence

$$h+(m-k) \leqslant r(A) \text{ and } (m-h)+k \leqslant r(A).$$

But $m=r(A)$; and thus we have shown that $h-k \leqslant 0$ and $k-h \leqslant 0$ or $k=h$, as we intended to show.

Remark 5: If we denote by p the number of positive points in all the maximal O-sets of σ [with respect to a definite symmetrical α-form f and a definite domain P of α-positivity], then $n=r(A)-p$ is the number of negative points in all the maximal O-sets. If we change from f to $-f$, then n becomes the number of positive points and p the number of negative points in all the maximal O-sets. Thus only the [unordered] pair $[p,n]$ has an invariant significance. Since $p+n=r(A)$, we introduce now the absolute value $|p-n|$ of the difference of these two numbers as *the index of inertia of* σ [with respect to the domain P of α-positivity]. It is clear that the invariant pair $[p,n]$ is completely determined once we know the two invariants: rank and index of inertia.

Application 1: Assume that F is the field of all [ordinary] real numbers. Then the identity is the only automorphism [and anti-automorphism] of F so that every semi-bilinear form is actually a bilinear form. This field F admits of only one algebraical ordering, since the set P of all squares, not 0, of real numbers is the only domain of positivity.

If the bilinear form $f(x,y)$ defines the polarity σ, as before, and if Fx is a point in A, then we have $f(tx,tx)=t^2f(x,x)$ for every real number t. It follows that every positive point has the form Fu with $f(u,u)=1$, that every negative point has the form Fv with $f(v,v)=-1$. Hence we find in the usual manner that

$$f(x,y)=\sum_{i=1}^{p} x_i y_i - \sum_{i=p+1}^{r(A)} x_i y_i,$$

where p is the number of positive points in maximal O-sets.

Application 2: Assume that F is the field of real quaternions and R its subfield of real numbers. Then R is the center of F; and every anti-automorphism of F leaves invariant every number in R. If α is an involutorial anti-automorphism of F, such that R consists exactly of those numbers in F which are left invariant by α, then a discussion quite similar to the one under Application 1 leads now to the following normal form of f:

$$f(x,y) = \sum_{i=1}^{p} x_i y_i^{\alpha} - \sum_{i=p+1}^{r(A)} x_i y_i^{\alpha}$$

where p is again the number of positive points in maximal O-sets.

REMARK 6: In the preceding examples the index of inertia is actually equal to the index $i(\sigma)$ introduced in section IV.4. In general one can only prove that $i(\sigma)$ does not exceed the index of inertia, but it is fairly easy to construct examples where the index of inertia actually exceeds the index $i(\sigma)$.

APPENDIX II

Projective Relations between Lines Induced by Polarities

If σ is a polarity of the linear manifold (F,A), and if L is a line, then A/L^{σ} has rank 2 [since the only proper subspaces of L are points, the only subspaces properly between L^{σ} and A are hyperplanes; and this is equivalent to saying that $r(A/L^{\sigma}) = 2$]. If L' is some line satisfying $L' \cap L^{\sigma} = 0$, then it follows from $r(A/L^{\sigma}) = 2$ that $A = L' \oplus L^{\sigma}$. But this is equivalent to $A = L'^{\sigma} \oplus L$ so that we have shown:

The following properties of the lines L and L' are equivalent:

$$L' \cap L^{\sigma} = 0, A = L' \oplus L^{\sigma}, A = L \oplus L'^{\sigma}, 0 = L \cap L'^{\sigma}.$$

Consider now a pair of lines L and L' meeting these equivalent requirements. If P is a point on L, then $L' \cap P^{\sigma} = P'$ is a point, since the hyperplane P^{σ} contains L^{σ} and satisfies

$$P^{\sigma} = L^{\sigma} \oplus (P^{\sigma} \cap L') = L^{\sigma} \oplus P'.$$

Furthermore we deduce from Dedekind's Law that

$$L \cap P'^{\sigma} = L \cap (L' \cap P^{\sigma})^{\sigma} = L \cap (L'^{\sigma} + P) = P + (L \cap L'^{\sigma}) = P.$$

Thus we see that mapping the point P on L onto the point $P' = L' \cap P^{\sigma}$ on L' and mapping the point Q on L' onto the point $L \cap Q^{\sigma}$ on L are reciprocal mappings; and this shows in particular that these two mappings are both one-to-one and exhaustive. It will be convenient to term the mapping of the point $P \leqslant L$ onto the point $P^{\sigma} \cap L'$ *the σ-mapping of the line L upon the line L'*; and we note that we can speak of this σ-mapping if, and only if, $A = L' \oplus L^\sigma$. Our objective in the present appendix is an investigation of these mappings.

Lemma 1: *Assume that the polarity σ is represented by the symmetrical α-form $f(x,y)$, that $L = Fp + Fq$ is a line in the linear manifold (F,A), and that the lines L and L' satisfy $A = L' \oplus L^\sigma$. Then there exist elements p',q' such that*

$$L' = Fp' + Fq',\ Fq' = L' \cap (Fq)^\sigma,\ F(p' + x^\alpha q') = L' \cap [F(p + xq)]^\sigma$$

for every number x in F.

Proof: We let $P' = L' \cap P^\sigma$ for every point P on L. Then mapping P onto P' is the σ-mapping of L upon L'; and this mapping is one-to-one and exhaustive, since $A = L' \oplus L^\sigma$, as we have pointed out before. The points $(Fp)'$, $(Fq)'$ and $[F(p + q)]'$ are consequently three distinct points on L'; and we may as usual show the existence of elements p'', q'' such that $(Fp)' = Fp''$, $(Fq)' = Fq''$, $[F(p + q)]' = F(p'' + q'')$. From $Fp'' \leqslant (Fp)^\sigma$ we deduce $f(p,p'') = f(p'',p) = 0$, since σ is represented by f. If $f(p'',q)$ were 0 too, then it would follow that $f(p'',L) = 0$ so that Fp'' would be on L^σ. But $L' \cap L^\sigma = 0$ by hypothesis and Fp'' is a point on L'. Hence $t = f(p'',q) \neq 0$; and we may let $p' = t^{-1}p''$, $q' = t^{-1}q''$. Then we have naturally

$$(Fp)' = Fp',\ (Fq)' = Fq',\ [F(p + q)]' = F(p' + q'),\ 1 = f(p',q)$$

so that in particular $L' = Fp' + Fq'$. As before we note that $f(p,p') = f(p',p) = 0$, $f(q,q') = f(q',q) = 0$; and we find likewise

$$0 = f(p + q, p' + q') = f(p,p') + f(p,q') + f(q,p') + f(q,q') = f(p,q') + 1$$

or $f(p,q') = f(q',p) = -1$, since $f(x,y) = f(y,x)^\alpha$ because of the assumed symmetry of the form f. Now we verify that

$$\begin{aligned} f(p + xq, p' + x^\alpha q') &= f(p,p') + f(p,q')x + xf(q,p') + xf(q,q')x \\ &= -x + x = 0, \end{aligned}$$

since α is involutorial. Hence $F(p' + x^\alpha q')$ is on $[F(p + xq)]^\sigma$ as well as on L'; and now our Lemma is an immediate consequence of the fact that the σ-mapping of L upon L' is a one-to-one and exhaustive mapping of the points on L onto the points on L'.

Proposition 1: *Suppose that the linear manifold (F,A) has rank not less than 3 and that the polarity σ of (F,A) is not a null system.*

(*a*) *The σ-mappings of lines are induced by auto-projectivities of (F,A) if, and only if, F is commutative.*

(*b*) *The σ-mapping of lines are induced by collineations of (F,A) if, and only if, σ may be represented by bilinear forms.*

Proof: It is a consequence of IV.3, Theorem 1 that σ may be represented by a symmetrical α-form f where α is an involutorial anti-automorphism of the field F. If L is a line, then there exists a subspace T of A such that $A = L \oplus T$. Naturally $L' = T^\sigma$ is a line such that $A = L \oplus L'^\sigma$; and thus we have shown that there exist σ-mappings of all lines.

It follows from Lemma 1 and III.4, Lemma 3 that the σ-mappings of lines are induced by auto-projectivities if, and only if, α is an automorphism of F; and that the σ-mappings of lines are induced by collineations if, and only if, α is an inner automorphism of A. But the anti-automorphism α of F is an automorphism if, and only if, F is commutative; and α is inner if, and only if, $\alpha = 1$; and this completes the proof of our proposition.

REMARK 1: If the lines L and L' satisfy $A = L \oplus L'^\sigma$, then it follows from Lemma 1 and the definition of cross ratios [in III.4] that

$$\begin{bmatrix} L' \cap P^\sigma & L' \cap Q^\sigma \\ L' \cap S^\sigma & L' \cap R^\sigma \end{bmatrix} = \begin{bmatrix} P & Q \\ S & R \end{bmatrix}^\alpha$$

holds for any four distinct points P,Q,R,S on L, provided σ is represented by a symmetrical α-form. Thus apart from auto-projectivities also the polarities provide a means of constructing mappings of the points on a line which conserve equality of cross ratios.

Proposition 2: *Assume that L is a line and σ a polarity of the linear manifold (F,A).*

(*a*) *If $L \cap L^\sigma$ is a point, then $L \cap L^\sigma$ is the one and only one N-point on L.*

(*b*) *If σ is represented by a semi-bilinear form, if the characteristic of F is not 2, and if L carries one and only one N-point, then $L \cap L^\sigma$ is a point.*

PROOF: If $P = L \cap L^\sigma$ is a point, then $0 < L^\sigma$ and $L < A$ so that $3 \leqslant r(A)$. Hence σ may be represented by a semi-bilinear form [IV.1, Proposition 1]. From $P \leqslant L \leqslant L + L^\sigma = (L^\sigma \cap L)^\sigma = P^\sigma$ we infer that P is an N-point on L. Assume that Q is a second N-point on L. Then $L = P + Q$, $Q \leqslant L \leqslant P^\sigma$; and it follows from IV.4, Lemma 3 that L is an N-line. Consequently we deduce from IV.4, Corollary 1 that $L \cap L^\sigma = 0$ or L, contradicting our hypotheses that $L \cap L^\sigma$ is a point. Thus P is the only N-point on L. Hence (*a*) is true.

Assume now that σ may be represented by some semi-bilinear form, that the characteristic of F is not 2, and that L carries one and only one N-point P. The line L carries some point $P' \neq P$. Then P' is on L, but not on P'^σ; and this implies in particular that L is not part of the hyperplane P'^σ. Hence $P'' = L \cap P'^\sigma$ is a point which is different from P', since otherwise P' would be an N-point. Either $P'' = P$ or else there exist, as usual, elements p',p'' such that $P' = Fp'$, $P = F(p' + p'')$, $P'' = Fp''$. There exists an α-form $f(x,y)$ which represents σ and which satisfies

$1 = f(p',p')$, since P' is not an N-point; and it follows from IV.3, Proposition 1 that $f(x,y)$ is α-symmetrical and $\alpha^2 = 1$. Since P'' is on P'^{σ} we have $0 = f(p'',p') = f(p',p'')$; and since P is an N-point, we have

$$0 = f(p' + p'', p' + p'') = f(p',p') + f(p'',p'') = 1 + f(p'',p'')$$

or $f(p'',p'') = -1$. Since the characteristic of F is not 2, the point $F(p' - p'')$ is different from P; and this point is clearly an N-point, since $0 = f(p' - p'', p' - p'')$. Thus we have been led to a contradiction by assuming that $P \neq P''$. Hence $P = P'' = L \cap P'^{\sigma}$. Consequently

$$P \leqslant L \cap P^{\sigma} \cap P'^{\sigma} = L \cap (P + P')^{\sigma} = L \cap L^{\sigma},$$

since P is an N-point on L. It is impossible that $L = L \cap L^{\sigma}$, since then every point on L would be an N-point. Since L is a line, this implies $P = L \cap L^{\sigma}$, completing the proof.

REMARK 2: If A is a plane and the characteristic of F is different from 2, then the point L^{σ} is the pole of the line L; and Proposition 2 asserts that the pole of the line L is on L if, and only if, L carries one and only one N-point.

REMARK 3: If $A = L$ is a line, then $L^{\sigma} = 0$, and every permutation of the points on L which possesses one and only one fixed point provides an example of a polarity for which (*b*) is not valid. Thus the hypothesis that σ be represented by a semi-bilinear form is indispensable. We note that this hypothesis is satisfied whenever $3 \leqslant r(A)$ as follows from IV. 1, Proposition 1.

REMARK 4: If the characteristic of the field F is 2, then (*b*) ceases to be valid, as follows from the following often used example. Denote by F any commutative field of characteristic 2 and by (F,A) the plane over F. If we represent the elements in A, as we may, by triplets $x = (x_1, x_2, x_3)$ of numbers x_i in F, then we may consider the bilinear form

$$f(x,y) = \sum_{i=1}^{3} x_i y_i$$

[= scalar product]. The point $F(x_1,x_2,x_3)$ is an N-point if, and only if, it is on the line $\sum_{i=1}^{3} x_i = 0$, since $f(x,x) = \sum_{i=1}^{3} x_i^2 = \left[\sum_{i=1}^{3} x_i\right]^2$ as the characteristic of F is 2. This N-line L is met by every other line in the plane so that every line not L carries one and only one N-point. But the line $x_1 = 0$ has the pole $F(1,0,0)$ which is not on its polar so that this line carries one and only one N-point, though it does not carry its pole.

Lemma 2: *The involutorial anti-automorphism α of the field F is the identity-automorphism if, and only if,*

(*a*) $rr^\alpha = 1$ *implies* $r = \pm 1$ *and*

(*b*) $xx^\alpha = x^\alpha x$ *for every x in F whenever the characteristic of F is two.*

REMARK: It seems to be an open question whether condition (*b*) is a consequence of condition (*a*).

PROOF: The necessity of our conditions is fairly obvious. Assume now the validity of condition (*a*). Then we prove first that

(*a*.1) $xx^\alpha = x^\alpha x$ *implies* $x^\alpha = \pm x$.

Since this is certainly true, if $x = 0$, we may assume that $x \neq 0$; and then we may form $r = x^{-1}x^\alpha$. Since x and x^α commute, we find that

$$rr^\alpha = x^{-1}x^\alpha x x^{-\alpha} = 1.$$

It follows from (*a*) that $r = \pm 1$; and this is clearly equivalent to $x^\alpha = \pm x$.

If the field F has characteristic 2, then $\alpha = 1$ is an immediate consequence of (*a*.1) and (*b*) and therefore of (*a*) and (*b*).

Assume next that the characteristic of F is different from 2; and assume the validity of (*a*) and (*a*.1). If t is any element in F, then the element $s = (t^\alpha - t) - 1$ certainly commutes with the element

$$s^\alpha = t - t^\alpha - 1 = -(t^\alpha - t) - 1 = -s - 2.$$

Hence it follows from (*a*.1) that

$$t - t^\alpha - 1 = s^\alpha = \pm s = \pm (t^\alpha - t - 1).$$

The minus sign is impossible in this equation, since this would imply $-1 = +1$ contradicting our assumption that the characteristic of F is not two. Hence we have $t - t^\alpha - 1 = t^\alpha - t - 1$ or $2t^\alpha = 2t$; and this implies $t^\alpha = t$, since we assumed the characteristic of F to be different from 2. Thus we have shown that $\alpha = 1$; and this completes the proof.

Proposition 3: *Suppose that the polarity σ of the linear manifold (F,A) possesses N-points and that $3 \leqslant r(A)$. Then σ may be represented by ordinary bilinear forms if, and only if,*

(*a*) *lines containing more than two N-points are N-lines and*

(*b*) *F is commutative.*

If the characteristic of F is different from 2, then condition (*b*) *is a consequence of condition* (*a*).

PROOF: Since $3 \leqslant r(A)$, it is possible to represent σ by a semi-bilinear form [IV.1, Proposition 1]; and it should be noted that this will be the only application of the hypothesis $3 \leqslant r(A)$. It follows from IV.1, Proposition 3 that every form representing σ is an ordinary bilinear form, if one of the representing forms has this property.

Assume now that it is possible to represent σ by ordinary bilinear forms. Then the identity automorphism is an anti-automorphism of F so that F is a commutative field. Suppose now that the line L carries the three distinct N-points P,Q,R. Then there exist elements p,q such that $P = Fp$, $Q = Fq$, $R = F(p + q)$. If f is some ordinary bilinear form representing σ, then we have

$$0 = f(p,p) = f(q,q) = f(p,q) + f(q,p);$$

and it follows from the commutativity of F and the bilinearity of f that

$$\begin{aligned} f(xp + yq,xp + yq) &= xf(p,p)x + xf(p,q)y + yf(q,p)x + yf(q,q)y \\ &= xy[f(p,q) + f(q,p)] = 0. \end{aligned}$$

Hence L is an N-line. This completes the proof of the necessity of conditions (*a*) and (*b*). The reader ought to investigate the connection between condition (*a*), Proposition 1 and the fact that a linear transformation of a line with three fixed points is the identity.

We assume now the validity of (*a*). There exists at least one N-point P. We distinguish two cases.

Case 1: Every point, not on P^σ, is an N-point.

If Q is a point on P^σ, but $Q \neq P$, then $Q^\sigma \neq P^\sigma$ so that the hyperplane Q^σ carries at least one point Q' not on P^σ. Then the line $L = Q + Q'$ is not on P^σ and meets therefore P^σ in the point Q. If $Q = Fq$, $Q' = Fq'$ and $R = F(q + q')$, then Q' and R are certainly not on P^σ and are therefore N-points. If f is any semi-bilinear form representing σ, then we have consequently

$$0 = f(q,q') = f(q',q) = f(q',q') \text{ and}$$
$$0 = f(q + q',q + q') = f(q,q)$$

so that Q itself is an N-point. Thus we have shown that every point in A is an N-point. Hence σ is a null system; and it follows from IV.2, Proposition 1 that f is an ordinary bilinear form.

Case 2: There exists a point W which is neither an N-point nor on P^σ.

Let $W = Fw$. Then it is possible to select an α-form f which represents σ and satisfies $f(w,w) = 1$; and we may infer from IV.3, Proposition 1 that α is involutorial and f symmetrical. We have $f(w,P) \neq 0$, since W is not on P^σ; and consequently we may select an element p such that $P = Fp$ and $1 = f(w,p)$. Because of the symmetry of f we have $f(p,w) = 1$; and $f(p,p) = 0$, since P is an N-point. Every point on the line $L = P + W$ with the exception of W has the form $F(p + xw)$; and the correspondence between the numbers x in F and the points different from W on L is a one-to-one correspondence. The point $F(p + xw)$ is an N-point if, and only if,

$$0 = f(p + xw,p + xw) = x + x^\alpha + xx^\alpha = (x + 1)(x + 1)^\alpha - 1.$$

The number of N-points on L is consequently exactly the number of solutions x of the equation $(x+1)(x+1)^{\alpha}=1$, if we remember that W is not an N-point. As this point W is on L, the line L is not an N-line; and it follows from condition (a) that L carries at most two N-points. Consequently we have shown that:

(a^*) *The equation* $(x+1)(x+1)^{\alpha}=1$ *has at most two solutions* x *in* F.

If we let $y=x+1$, then we see that (a^*) is equivalent with the assertion that the equation $yy^{\alpha}=1$ has at most two solutions y in F. This equation has certainly the solutions $y=\pm 1$. If it possessed a solution $z\neq\pm 1$, then $z^{-1}\neq z$ would be a further solution which is different from ± 1; and thus we would have at least three solutions of $yy^{\alpha}=1$ [at least four if $+1\neq -1$]. Since this is impossible, we have shown that:

(a^{**}) $jj^{\alpha}=1$ *implies* $j=\pm 1$.

Thus we have seen that our present condition (a) implies the validity of Lemma 2, (a); and Lemma 2, (b) is certainly a consequence of our present condition (b) which we need apply only in case the characteristic of F is two. Thus we may deduce $\alpha=1$ from Lemma 2; and we have shown again that σ may be represented by ordinary bilinear forms. This completes the proof.

Remark 5: The question whether our condition (b) is a consequence of our condition (a) is equivalent to the corresponding question concerning Lemma 2, since (a^{**}) has been shown to be a consequence of (a) alone. On the other hand it is completely impossible to omit the hypothesis assuring the existence of N-points, as may be seen from easily constructed examples of "positive definite semi-bilinear forms."

Remark 6: Suppose that the polarity σ of the linear manifold (F,A) possesses N-points, that lines carrying more than two N-points are N-lines, that $3\leqslant r(A)$ and that the characteristic of F is two. Then we show:

(c) *Lines carrying at least two N-points are N-lines.*

Suppose namely that P and Q are different N-points and that f is some semi-bilinear form representing σ. As the characteristic of F is two, we may assume without loss in generality that f is symmetrical. If firstly P is on Q^{σ}, then we have

$$0=f(P,P)=f(P,Q)=f(Q,P)=f(Q,Q);$$

and this implies clearly that $0=f(P+Q,P+Q)$ so that $P+Q$ is an N-line. If secondly P is not on Q^{σ}, then $f(P,Q)\neq 0$; and we may select elements p and q such that $P=Fp$, $Q=Fq$ and $1=f(p,q)=f(q,p)$. Since we have $0=f(p,p)=f(q,q)$, it follows that

$$f(p+q,p+q)=f(p,q)+f(q,p)=2=0;$$

and we have shown the existence of a third N-point on $P+Q$. It follows from our hypothesis that $P+Q$ is an N-line; and this completes the proof of (c).

From (c) one deduces without difficulty that:

(d) *The totality of N-points is a subspace.*

Some information on such a subspace may be found in IV.4.

Proposition 4: *If each of the lines L' and L'' carries at least two N-points, though neither of them is an N-line, then L' and L'' carry the same number of N-points.*

Proof: This proposition is vacuously true, if either $r(A) < 3$ or our polarity σ is a null system. Thus we assume that σ is not a null system and that $3 \leqslant r(A)$. Then it follows from IV.3, Theorem 1 that σ may be represented by a symmetrical α-form f with $\alpha^2 = 1$. Consider now a line L which is not an N-line, but carries the two distinct N-points P and Q. Then P is not on Q^σ, since otherwise L would be an N-line [IV.4, Lemma 3]. Let $Q = Fq$. Then $f(q,q) = 0$, but $f(P,q) \neq 0$. Consequently there exists p such that $P = Fp$ and $f(p,q) = 1$. Since f is symmetrical, we have $f(q,p) = 1$; and since P is an N-point, we have $f(p,p) = 0$. To every point X on L different from P and Q, there exists one and only one number $x \neq 0$ in F such that $X = F(p + xq)$. This point is an N-point if, and only if,

$$0 = f(p + xq, p + xq) = x + x^\alpha.$$

If we denote now by C the cardinal number of elements x in F such that $x^\alpha = -x$ [the skew-symmetric elements in F], then we have found that:

The number of N-points on L is exactly $C + 1$.

This last assertion is actually slightly sharper than our proposition.

APPENDIX III

The Theorem of Pascal

Our discussion of the celebrated Theorem of Pascal is rather similar in intent to our previous discussion of the Theorem of Pappus [III, Appendix II]. Whereas this theorem is of universal validity in the classical theory of conics, it will be shown that from our more general point of view it must be considered to be valid only under quite exceptional circumstances. The reader may consult Coxeter [1] or Levi [1] for a discussion of Pascal's Theorem in its classical framework.

It will be convenient to assume throughout this appendix that the linear manifold (F,A) is a plane $[r(A) = 3]$. This will simplify our discussion considerably without constituting an essential loss in generality, since every polarity induces a polarity in every non-isotropic subspace [IV.4, Lemma 3].

Pascal's Theorem concerns itself with certain hexagons. We will need some preliminary discussion of such configurations. We define a *hexagon*

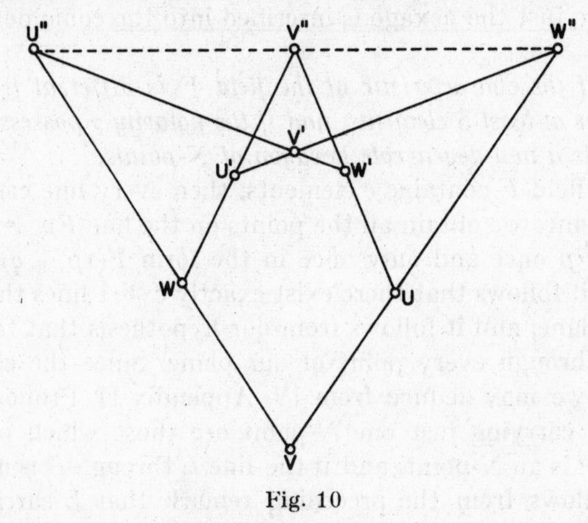

Fig. 10

as a cyclically ordered set of points $[U, V, W, U', V', W']$ in our plane (F, A) subject to the condition that

$$\text{(H)} \quad \begin{cases} U + V \text{ and } U' + V' \text{ are distinct lines meeting in the point } W'', \\ V + W \text{ and } V' + W' \text{ are distinct lines meeting in the point } U'', \\ W + U' \text{ and } W' + U \text{ are distinct lines meeting in the point } V''. \end{cases}$$

This hexagon shall be termed *non-degenerate*, if it consists of six distinct points and if no three consecutive points of the hexagon are collinear [consecutive is to be understood in the sense of the cyclic ordering of the points of the hexagon].

Lemma 1: *If* $[U, V, W, U', V', W']$ *is a non-degenerate hexagon, then its opposite sides meet in three distinct points.*

PROOF: The opposite sides of our hexagon meet in the points U'', V'', W'' [using the notation of (H)]. Assume by way of contradiction that $U'' = V''$. Then this point $U'' = V''$ is on the lines $V + W$, $W + U'$, $V' + W'$ and $W' + U$. Since V, W, U' are consecutive points of the hexagon, they are not collinear; and thus the first two lines meet in W so that $U'' = V'' = W$. Likewise we see that $U'' = V'' = W'$. Hence $W = W'$ contradicting the non-degeneracy of our hexagon.

The polarity σ of the plane will be called *Pascalean*, if it has the following property:

(P) *If the points of the hexagon* $[U, V, W, U', V', W']$ *are N-points of* σ, *then opposite sides of the hexagon meet in collinear points* [so that in the notation of (H) the points U'', V'', W'' are collinear].

The relation of this property to Pascal's Theorem in classical geometry is best seen if one remembers that von Staudt has defined a conic as the totality of N-points of a polarity [in our terminology] so that hexagons of N-points are just the hexagons inscribed into the conic defined by our polarity.

Lemma 2: *If the characteristic of the field F is different from* 2, *if the field F contains at least* 5 *elements, and if the polarity σ possesses N-points, then there exists a non-degenerate hexagon of N-points.*

PROOF: If the field F contains c elements, then every line carries exactly $c + 1$ points, since we obtain all the points on the line $Fp + Fq$ with the exception of Fp once and only once in the form $F(xp + q)$ for x in F. Consequently it follows that there exist exactly $c + 1$ lines through every point of our plane; and it follows from our hypothesis that there exist at least 6 lines through every point of our plane. Since the characteristic of F is not 2, we may deduce from IV. Appendix II, Proposition 2 that the only lines carrying just one N-point are those which pass through their pole. If P is an N-point, and if the line L through P is not the polar P^σ, then it follows from the preceding remark that L carries a second N-point.

There exists by hypothesis some N-point $P(1)$. Suppose that for some $i < 5$ we have already constructed the N-points $P(1), \cdots, P(i)$ in such a way that they all are different and that no three consecutive points $P(j-1)$, $P(j)$, $P(j+1)$ are collinear. Since $i < 5$, there exists a line L through $P(i)$ which is different from the i lines:

$$P(i) + P(1), \cdots, P(i) + P(i-1), \; P(i)^\sigma.$$

This line L carries an N-point $P(i+1)$ different from $P(i)$; and it is clear that $P(i+1)$ is different from all the preceding points and that $P(i-1)$, $P(i)$, $P(i+1)$ are not collinear. Thus it follows by induction that there exist five distinct N-points $P(1), \cdots, P(5)$ such that $P(j-1)$, $P(j)$, $P(j+1)$ are not collinear.

There exists a line Z through $P(5)$ which is different from the 5 lines $P(5)^\sigma, P(5) + P(1), \cdots, P(5) + P(4)$. This line carries an N-point Y which is different from $P(5)$. It is clear that Y is different from all the $P(i)$; and that Y, $P(5)$, $P(4)$ are not collinear. But it could happen that Y is on the line $P(1) + P(2)$. If this should be the case, then the line $P(1) + P(2)$ carries at least 3 N-points; and it follows from IV. Appendix II, Proposition 4 that Z carries at least 3 N-points. Thus Z would carry an N-point X different from Y and $P(5)$; and this shows that an N-point $P(6)$ completing our non-degenerate hexagon can always be found.

Theorem 1: *If the characteristic of the field F is different from 2, and if the polarity σ of the plane (F,A) possesses N-points, then the following properties of σ are equivalent:*

(I) *σ is Pascalean.*

(II) *No line carries more than two N-points.*

(III) *σ may be represented by ordinary [symmetrical] bilinear forms.*

PROOF: We note first that there cannot exist N-lines. For if L were an N-line, then it would follow from IV.4, Corollary 1 that $L \cap L^\sigma$ is either 0 or L. Since L^σ is a point, the second possibility is ruled out. Thus L would be a non-isotropic N-line which is impossible, since the characteristic of F is not 2 [IV.4, Corollary 3]. Now the equivalence of Properties (III) and (II) is an immediate consequence of IV. Appendix II, Proposition 3, again since the characteristic of F is not 2.

If the equivalent properties (II) and (III) are satisfied by σ, then σ may be represented by an ordinary symmetrical bilinear form and F is commutative [IV.3, Theorem 1]; and that σ is consequently Pascalean may now be verified by practically the same arguments as are customarily used in the proof of the classical Theorem of Pascal. We leave the details as an exercise to the reader. Thus (I) is certainly a consequence of (II) and/or (III).

Assume now the validity of (I). If the field F contains fewer than 5 elements, then F contains 3 elements, since its characteristic is not 2. Thus F consists only of the integral multiples of 1 so that its only automorphism or anti-automorphism is the identity. But then every semi-bilinear form is an ordinary bilinear form so that in this case (III) is certainly a consequence of (I). If we assume next that F contains at least 5 elements, then we infer from Lemma 2 and the existence of N-points the existence of a non-degenerate hexagon $[U,V,W,U',V',W']$ of N-points. Opposite sides of this hexagon meet in points U'',V'',W'' [using the notation of (H)]; and it follows from Lemma 1 that these points are distinct. Now we apply condition (I) and see that these points are collinear. Hence $T = U'' + V'' = V'' + W'' = W'' + U''$ is a line. Since opposite sides of our hexagon are distinct, and since consecutive sides of our hexagon are distinct, at most one of the sides of our hexagon can equal T [all six sides of the hexagon are distinct]; and thus we may assume without loss in generality that neither $U + V$ nor $U' + V'$ equals T. Then either of these sides meets T in W''. Since the opposite sides are distinct, we may assume furthermore without loss in generality that U is not on $U' + V'$; and this implies in particular that $U \neq W''$. Assume now by way of contradiction the existence of lines carrying more than two N-points. Since the line $V + W$ carries at least two N-points, it follows from IV. Appendix II, Proposition 4 that $V + W$ carries a third N-point V^*. It is easy to verify that $[U, V^*, W, U', V', W']$ is a hexagon of N-points. From $V^* + W = V + W$ we infer that opposite sides of this hexagon meet in the points $(U + V^*) \cap (U' + V'), U'', V''$. It follows from (I) that these points are collinear; and thus these lines $U + V^*$, $U' + V'$, $U'' + V'' = T$

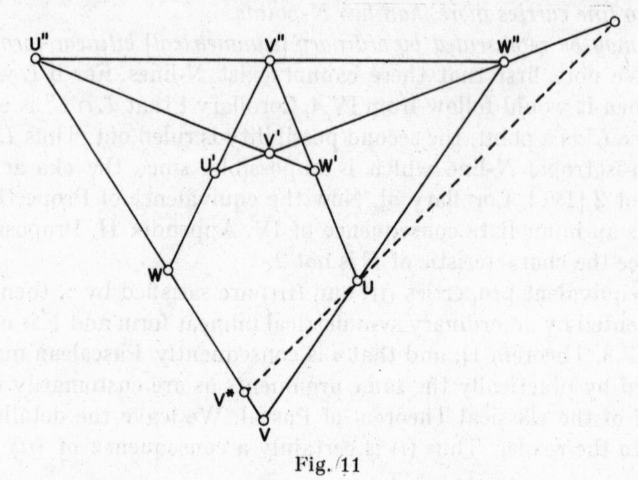

Fig. 11

meet in a point which necessarily is W''. Hence the lines $U + V$ and $U + V^*$ both carry the point W''; and since $U \neq W''$, it follows that $U + V = U + W'' = U + V^*$. Consequently V^* is a common point of the two distinct lines $U + V$ and $V + W$; and this would imply $V = V^*$, an impossibility. This contradiction proves that (I) implies (II); and this completes the proof.

It will be a good exercise for the reader to construct an algebraical proof of our last implication instead of the synthetic proof which we gave.

REMARK 1: The question to which extent we really need the hypothesis that the characteristic of F be different from 2 has as yet not been completely explored. However, if F is a commutative field of characteristic 2, then it is easy to define a symmetrical bilinear form over the plane (F,A) which defines a polarity σ with the property: the N-points are just the points on a certain line [see for further details, for instance, IV. 4, Proof of Proposition 1]. This polarity σ has Property (III), but not Property (II); and it certainly does not possess any hexagons of N-points so that Lemma 2 fails to be true.

Theorem 2: *Suppose that F is a field of characteristic different from 2. Then every polarity of the plane (F,A) is Pascalean if, and only if, there do not exist involutorial anti-automorphisms of F different from the identity.*

PROOF: If there do not exist involutorial anti-automorphisms different from 1, then every polarity of (F,A) may be represented by an ordinary bilinear form [IV.1, Proposition 1]; and it follows from Theorem 1 that every polarity of (F,A) is Pascalean. Suppose conversely that every polarity of the plane (F,A) is Pascalean; and consider some involutorial anti-automorphism α of F. If b_1,b_2,b_3 is a basis of A, then we may define an α-form f over A by the rule:

$$f\left(\sum_{i=1}^{3} x_i b_i, \sum_{i=1}^{3} y_i b_i\right) = x_1 y_1^\alpha + x_2 y_2^\alpha - x_3 y_3^\alpha.$$

It is easily verified that f represents a polarity σ, since f is α-symmetrical [IV.1, Proposition 1 and IV.3, Lemma 1]. This polarity possesses N-points, for instance $F(b_1 - b_3)$. Since σ is Pascalean by hypothesis, it follows from Theorem 1 that $\alpha = 1$; and this completes the proof.

Remark 2: If the field F does not possess any involutorial anti-automorphisms, then the plane (F,A) does not possess any polarities either [IV.3, Theorem 2]. Thus our theorem would be vacuously satisfied in the absence of involutorial anti-automorphisms of F. If we add to the general hypotheses of our theorem the assumption that there exist involutorial anti-automorphisms, then the condition that the identity be the only involutorial anti-automorphism of F implies that F is a commutative field.

Remark 3: It appears justified to say that the Theorem of Pascal holds in the plane (F,A) if all the polarities of (F,A) are Pascalean. Theorem 2 characterizes the planes in which the Theorem of Pascal is valid.

Remark 4: In order to determine the planes in which the Theorem of Pascal is valid, we have to determine the commutative fields F of characteristic different from 2 with the property:

(*) *F does not possess automorphisms of order* 2.

An obvious example of a field with property (*) is the field of all real numbers; and an obvious example of a field which does not have property (*) is the field of complex numbers. To give some further examples of fields with property (*) we have to make use of some elementary facts from Galois Theory; the needed facts may be found in Artin [1], Birkhoff-MacLane [1], van der Waerden [1] etc. Suppose that F is a commutative field and that α is an automorphism of F whose order is 2. Denote by D the totality of elements d in F such that $d = d^\alpha$. Then D is a subfield of F and F has degree 2 over D. Conversely to every subfield S of F such that F has degree 2 over S there belongs such an automorphism of order 2. Thus we see that condition (*) is equivalent to the following condition:

(**) *F does not have degree* 2 *over any of its subfields.*

If, for instance, F is an extension of odd degree of the field of rational numbers [or of some other prime field], then F certainly has this property (**). The following criterion which is of a rather different type may not be without interest, in particular it proves useful in applications.

The commutative field F has Property (**), if it meets the following two requirements:

(*a*) $\sum_{i=1}^{n} x_i^2 = 0$ *implies* $x_1 = \cdots = x_n = 0$.

(*b*) *If* $x \neq 0$ *is an element in* F, *but not the square of an element in* F, *then* $-x$ *is the square of an element in* F.

[Fields with Property (*a*) are termed *formally real*; and fields with Properties (*a*) and (*b*) are termed *euclidean*.]

Proof: Suppose that F has Properties (*a*) and (*b*); and assume by way of contradiction the existence of a subfield D of F such that F has degree 2 over D. Then F may be obtained from D by adjoining an element e such that $e^2 = d$ belongs to D. Since $F \neq D$, $e \neq 0$; and since $(-e)^2 = d$, we may assume without loss in generality that e is the square of an element b in F—here we apply condition (*b*). The element b, as every element in F, has the form $b = b' + b''e$ with b', b'' in D; and now it follows that

$$e = b^2 = (b' + b''e)^2 = b'^2 + b''^2 d + 2b'b''e.$$

This implies $2b'b'' = 1$ and $0 = b'^2 + b''^2 d = b'^2 + (b''e)^2$. We apply (*a*) on the second equation and find that $b' = b''e = 0$. But this is absurd, since $2b'b'' = 1$. Thus we have been led to a contradiction which proves the validity of (**).

IV.5. The Group of a Polarity

Throughout we shall assume that the F-space A has rank not less than 3. Then it is possible to represent every duality by a semi-bilinear form [IV.1, Proposition 1] and every projectivity is induced by a semi-linear transformation [III.1, First Fundamental Theorem of Projective Geometry].

If δ is a duality and π a projectivity of the F-space A, then we term π *a δ-admissible projectivity*, if $\pi\delta = \delta\pi$. This is equivalent to saying that $U^{\pi\delta} = U^{\delta\pi}$ holds for every subspace U of A. It is clear that the totality of δ-admissible projectivities is a group. This group has first right to the name "group of the duality δ." But there are various other groups which will be shown to have similar importance.

Proposition 1: *Suppose that the semi-bilinear form f represents the duality δ. Then the semi-linear transformation σ of A induces a δ-admissible projectivity if, and only if, there exists a number $c \neq 0$ in F such that*

$$f(x^\sigma, y^\sigma) = f(x,y)^\sigma c \text{ for every } x,y \text{ in } A.$$

PROOF: Assume first the validity of our condition. If M is a subspace of A, then x belongs to M^{δ} if, and only if, $f(x,M) = 0$. This is equivalent to saying that $0 = f(x,M)^{\sigma}c = f(x^{\sigma},M^{\sigma})$. Hence x belongs to M^{δ} if, and only if, x^{σ} belongs to $(M^{\sigma})^{\delta}$. Consequently $M^{\delta\sigma} = M^{\sigma\delta}$; and this shows that the projectivity induced by σ is δ-admissible.

Assume conversely that the projectivity induced by σ is δ-admissible. Then we define

$$g(x,y) = f(x^{\sigma},y^{\sigma})^{\sigma^{-1}} \text{ for } x,y \text{ in } A.$$

It is clear that $g(x,y)$ is additive in x and y. If t is a number in F, then

$$g(tx,y) = f([tx]^{\sigma},y^{\sigma})^{\sigma^{-1}} = f(t^{\sigma}x^{\sigma},y^{\sigma})^{\sigma^{-1}} = [t^{\sigma}f(x^{\sigma},y^{\sigma})]^{\sigma^{-1}} = tg(x,y).$$

There exists an anti-automorphism α of the field F such that f is an α-form. We find that

$$g(x,ty) = f(x^{\sigma},[ty]^{\sigma})^{\sigma^{-1}} = f(x^{\sigma},t^{\sigma}y^{\sigma})^{\sigma^{-1}} = [f(x^{\sigma},y^{\sigma})t^{\sigma\alpha}]^{\sigma^{-1}} = g(x,y)t^{\sigma\alpha\sigma^{-1}}$$

We note that $\sigma\alpha\sigma^{-1} = \beta$ is likewise an anti-automorphism of F; and that we have shown that $g(x,y)$ is a β-form.

Suppose that M is a subspace of A. Then x belongs to M^{δ} if, and only if, $f(x,M) = 0$. Likewise x belongs to M^{δ} if, and only if, x^{σ} belongs to $M^{\delta\sigma}$. But σ induces a δ-admissible projectivity; and so $M^{\delta\sigma} = M^{\sigma\delta}$. Thus we have shown that $f(x,M) = 0$ if, and only if, x^{σ} belongs to $M^{\sigma\delta}$; and this latter statement is equivalent to saying that $f(x^{\sigma}, M^{\sigma}) = 0$. Since this is true if, and only if, $0 = f(x^{\sigma},M^{\sigma})^{\sigma^{-1}} = g(x,M)$, we have finally shown that

$$f(x,M) = 0 \text{ if, and only if, } g(x,M) = 0;$$

and thus it follows that the semi-bilinear forms $f(x,y)$ and $g(x,y)$ represent the same duality δ. Now we infer from IV.1, Proposition 3 the existence of a number $z \neq 0$ in F such that $g(x,y) = f(x,y)z$. Hence we have

$$f(x,y)z = g(x,y) = f(x^{\sigma},y^{\sigma})^{\sigma^{-1}} \text{ or } f(x^{\sigma},y^{\sigma}) = f(x,y)^{\sigma}z^{\sigma};$$

and this shows the necessity of our condition.

We recall that σ is a linear transformation if it effects an automorphism of the additive group of A and if $(xy)^{\sigma} = xy^{\sigma}$ holds for x in F and y in A. If furthermore $\sigma^2 = 1$, then we shall call σ just an *involution* [= involutorial linear transformation].

Lemma 1: *Suppose that σ is an involution of the F-space A. Denote by $J^{+}(\sigma)$ the set of all elements x in A such that $x^{\sigma} = x$; and by $J^{-}(\sigma)$ the set of all elements x in A such that $x^{\sigma} = -x$. Then $J^{+}(\sigma)$ and $J^{-}(\sigma)$ are subspaces of A; and $A = J^{+}(\sigma) \oplus J^{-}(\sigma)$ provided the characteristic of F is not 2.*

The proof is very simple, once one remarks that, for every x in A, $x + x^\sigma$ is in $J^+(\sigma)$ and $x - x^\sigma$ is in $J^-(\sigma)$.

Lemma 2: *Assume that the characteristic of the field F is not 2, that the semi-bilinear form $f(x,y)$ represents the polarity π, and that σ is an involution.*

(*a*) *σ induces a π-admissible projectivity if, and only if, $f(x,y) = ef(x^\sigma,y^\sigma)$ for every x and y in A where $e = \pm 1$.*

(*b*) *$f(x,y) = f(x^\sigma,y^\sigma)$ for every x and y in A if, and only if,*

$$J^+(\sigma)^\pi = J^-(\sigma) \text{ and } J^-(\sigma)^\pi = J^+(\sigma).$$

(*c*) *$f(x,y) = - f(x^\sigma,y^\sigma)$ for every x and y in A if, and only if,*

$$J^+(\sigma)^\pi = J^+(\sigma) \text{ and } J^-(\sigma)^\pi = J^-(\sigma).$$

Proof: It is a consequence of Proposition 1 that the condition, given under (*a*), is sufficient. If conversely σ induces a π-admissible projectivity, then we infer from Proposition 1 the existence of a number $c \neq 0$ in F such that $f(x^\sigma,y^\sigma) = f(x,y)c$ for every x and y, since σ is a linear transformation. From $\sigma^2 = 1$ we deduce now that

$$f(x,y) = f(x^{\sigma^2},y^{\sigma^2}) = f(x^\sigma,y^\sigma)c = f(x,y)c^2.$$

Since $f(x,y)$ is not always 0, this implies $c^2 = 1$ or $c = \pm 1$, as we claimed.

Suppose now that $J^+(\sigma)^\pi = J^-(\sigma)$ and $J^-(\sigma)^\pi = J^+(\sigma)$. Then we have $f(x,y) = f(y,x) = 0$ whenever x is in $J^+(\sigma)$ and y is in $J^-(\sigma)$. If now u and v are elements in A, then $u = u' + u''$, $v = v' + v''$ with u',v' in $J^+(\sigma)$ and u'',v'' in $J^-(\sigma)$ [by Lemma 1]. Hence

$$f(u^\sigma,v^\sigma) = f(u' - u'',v' - v'') = f(u',v') + f(u'',v'') = f(u,v).$$

Assume conversely that $f(x^\sigma,y^\sigma) = f(x,y)$ for every x and y in A. If x is in $J^+(\sigma)$ and y in $J^-(\sigma)$, then it follows that

$$f(x,y) = f(x^\sigma,y^\sigma) = f(x, -y) = - f(x,y);$$

and this implies $f(x,y) = 0$, since the characteristic of F is not 2. Consequently $J^+(\sigma) \leqslant J^-(\sigma)^\pi$. But it follows from Lemma 1 and IV.1, Corollary 1 that $r[J^+(\sigma)] = r[J^-(\sigma)^\pi]$. Hence $J^+(\sigma) = J^-(\sigma)^\pi$ and consequently $J^-(\sigma) = J^+(\sigma)^\pi$. This proves (*b*).

Assume now that $J^+(\sigma)^\pi = J^+(\sigma)$ and $J^-(\sigma)^\pi = J^-(\sigma)$. If x and y are elements in A, then we have $x = x' + x''$, $y = y' + y''$ with x',y' in $J^+(\sigma)$ and x'',y'' in $J^-(\sigma)$ [by Lemma 1]. Hence $f(x',y') = f(x'',y'') = 0$; and consequently

$$\begin{aligned} f(x^\sigma,y^\sigma) &= f(x' - x'',y' - y'') = - f(x'',y') - f(x',y'') \\ &= -[f(x',y') + f(x',y'') + f(x'',y') + f(x'',y'')] = - f(x,y). \end{aligned}$$

If conversely $f(x^\sigma,y^\sigma) = -f(x,y)$ is satisfied by every x and y in F, and if u,v belong both to $J^+(\sigma)$, then we have

$$-f(u,v) = f(u^\sigma,v^\sigma) = f(u,v);$$

and $f(u,v) = 0$ is a consequence of the fact that the characteristic of F is not 2. Likewise we show that $f(u,v) = 0$ if u and v belong both to $J^-(\sigma)$. Thus we have shown that

$$J^+(\sigma) \leqslant J^+(\sigma)^\pi \text{ and } J^-(\sigma) \leqslant J^-(\sigma)^\pi.$$

Now we deduce from Lemma 1 and IV.1, Corollary 1 that

$$r[J^+(\sigma)] \leqslant r[J^+(\sigma)^\pi] = r[J^-(\sigma)] \leqslant r[J^-(\sigma)^\pi] = r[J^+(\sigma)]$$

so that all these four ranks are equal. This implies together with the above inequalities the desired equations $J^+(\sigma) = J^+(\sigma)^\pi$, $J^-(\sigma) = J^-(\sigma)^\pi$; and this completes the proof of (*c*).

REMARK 1: We note that in the case (*c*) it follows from Lemma 1 and the results obtained in the course of the proof that

$$r[J^+(\sigma)] = r[J^-(\sigma)] = \tfrac{1}{2}r(A),$$

so that $\frac{1}{2}r(A)$ is the index of our polarity [in the notation of IV.4].

If the involution σ induces a π-admissible polarity such that

$$f(x^\sigma,y^\sigma) = f(x,y)$$

for x and y in A where f is some semi-bilinear form representing π, then the same holds clearly for every semi-bilinear form representing π; and we call σ *a π-admissible involution of the first kind*. All the other involutions which induce π-admissible projectivities shall be termed *π-admissible involutions of the second kind*. The group of linear transformations which is generated by the π-admissible involutions of the first kind shall be denoted by $\Gamma(\pi)$. Every element in $\Gamma(\pi)$ is a linear transformation of A, induces a π-admissible projectivity; and if $f(x,y)$ is some semi-bilinear form representing the polarity π, then we have $f(x^\sigma,y^\sigma) = f(x,y)$ for every x,y in A and every σ in $\Gamma(\pi)$.

This last remark shows that $\Gamma(\pi)$ does not contain π-admissible involutions of the second kind, provided, of course, that the characteristic of F is not 2. We note furthermore that as a consequence of Remark 1 the π-admissible involutions of the second kind are fairly rare whereas the π-admissible involutions of the first kind are fairly common as may be seen from the following fact.

Corollary 1: *Assume that the characteristic of the field F is not 2, that the semi-bilinear form $f(x,y)$ represents the polarity π, and that U is a subspace of the space A. Then U is non-isotropic if, and only if, there exists a π-admissible involution σ of the first kind such that $U = J^+(\sigma)$.*

PROOF: The sufficiency is immediately deduced from

$$U \cap U^{\pi} = J^{+}(\sigma) \cap J^{-}(\sigma) = 0$$

which is true, since the characteristic is not 2.

Assume conversely that $U \cap U^{\pi} = 0$. It follows from IV.1, Corollary 2 that $A = U \oplus U^{\pi}$. Consequently there exists one and only one linear transformation σ of A such that $x = x^{\sigma}$ for x in U and $-x = x^{\sigma}$ for x in U^{π}. Clearly σ is an involution such that $U = J^{+}(\sigma)$ and $U^{\pi} = J^{-}(\sigma)$; and one deduces from Lemma 2, (*a*), (*b*) that σ is a π-admissible involution of the first kind.

REMARK 2: Instead of $U = J^{+}(\sigma)$ we could have required equally well $U = J^{-}(\sigma)$.

That polarities are determined by their groups, is the essential content of the following theorem.

Theorem 1: *If the characteristic of the field F is not 2, then the following properties of the polarities π' and π'' of the F-space A [with $2 < r(A)$] are equivalent.*

(I) $\pi' = \pi''$.

(II) *The same projectivities are π'-admissible and π''-admissible.*

(III) *The same collineations are π'-admissible and π''-admissible.*

(IV) $\Gamma(\pi') = \Gamma(\pi'')$.

PROOF: It is clear that (I) implies (II) and that (II) implies (III). Assume now that (III) is true. Denote by Θ the group of linear transformations which induce π'-admissible collineations. Then it follows from (III) that Θ is likewise the group of all linear transformations which induce π''-admissible collineations. It is a consequence of the definition of the group Γ that Θ contains both $\Gamma(\pi')$ and $\Gamma(\pi'')$. We show now

(IV.*a*) If σ is a π'-admissible involution of the first kind, then σ is also a π''-admissible involution of the first kind.

If (IV.*a*) were not true, then σ would be a π''-admissible involution of the second kind, since, by (III), σ induces a π''-admissible collineation. Thus we would have

$$J^{+}(\sigma) = J^{+}(\sigma)^{\pi''} = J^{-}(\sigma)^{\pi'},\ J^{-}(\sigma) = J^{-}(\sigma)^{\pi''} = J^{+}(\sigma)^{\pi'};$$

and we deduce from Remark 1 that

$$k = r[J^{+}(\sigma)] = r[J^{-}(\sigma)] = \tfrac{1}{2}r(A) \text{ and hence } 1 < k;$$

and from Lemma 1 that $A = J^{+}(\sigma) \oplus J^{-}(\sigma)$.

It follows from Corollary 1 that $J^{+}(\sigma)$ is non-isotropic with respect to π'. Since the characteristic is not 2, and since the rank of $J^{+}(\sigma)$ is greater than 1, it follows from IV.4, Corollary 2 that either π' is a null system or else $J^{+}(\sigma)$ does not have Property (N) with respect to π'.

Case 1: $J^+(\sigma)$ does not have Property (N) with respect to π'.

Then there exists a point P on $J^+(\sigma)$ such that $P \cap P^{\pi'} = 0$. This point P is non-isotropic with respect to π'; and we infer from Corollary 1 the existence of a π'-admissible involution σ' such that $P = J^+(\sigma')$, $P^{\pi'} = J^-(\sigma')$. Clearly σ' belongs to Θ and induces consequently a π''-admissible involution. Since $J^+(\sigma)$ is strictly isotropic for π'', the point P has Property (N) with respect to π''; and so σ' cannot be a π''-admissible involution of the first kind [Corollary 1]. Since $r(P) = 1 < \frac{1}{2}r(A)$, it follows from Lemma 2 and Remark 1 that σ' cannot be a π''-admissible involution of the second kind either. Thus we have been led to a contradiction.

Case 2: $J^+(\sigma)$ has Property (N) with respect to π'.

Then it follows from a previous remark [IV.4, Corollary 2] that π' is a null system. We deduce from IV.2, Lemma 1, (*b*), (*c*) that π' may be represented by some [ordinary] skew-symmetric bilinear form $g(x,y)$; and it follows from IV.2, Proposition 1 that the non-isotropic subspace $J^+(\sigma)$ contains a non-isotropic subspace U of rank 2 which possesses a basis a,b such that $g(a,b) = 1 = -g(b,a)$. Denote by τ the uniquely determined linear transformation of A with the following properties:

$$a^\tau = b,\ b^\tau = -a,\ x^\tau = x \text{ for } x \text{ in } U^{\pi'}.$$

Since $A = U \oplus U^{\pi'}$ is a consequence of the non-isotropy of U [IV.1, Corollary 2], we may represent every element x in A in one and only one way in the form: $x = x'a + x''b + x^0$ with x',x'' in F and x^0 in $U^{\pi'}$. Consequently we find that

$$g(x,y) = g(x'a + x''b + x^0, y'a + y''b + y^0) = x'y'' - x''y' + g(x^0,y^0),$$

since $g(U,U^{\pi'}) = g(U^{\pi'},U) = 0$, $g(a,a) = g(b,b) = 0$. Hence

$$\begin{aligned} g(x^\tau,y^\tau) &= g(-x''a + x'b + x^0, -y''a + y'b + y^0) \\ &= -x''y' + x'y'' + g(x^0,y^0) = g(x,y); \end{aligned}$$

and it follows from Proposition 1 that τ induces a π'-admissible collineation. Consequently τ belongs to the group Θ; and this implies that τ induces also a π''-admissible collineation. Now we represent π'' by some semi-bilinear form $h(x,y)$ [IV.1, Proposition 1]. From Proposition 1 we deduce the existence of a number $c \neq 0$ in F such that $h(x^\tau,y^\tau) = h(x,y)c$ for every x and y in A.

Since Fa and Fb are different points, $(Fa)^{\pi''}$ and $(Fb)^{\pi''}$ are different hyperplanes. From $Fa \leqslant J^+(\sigma) = J^+(\sigma)^{\pi''}$ we infer that $J^+(\sigma) \leqslant (Fa)^{\pi''}$; and $J^+(\sigma) \leqslant (Fb)^{\pi''}$ is seen likewise. From $U \leqslant J^+(\sigma)$ we deduce that $J^-(\sigma) = J^+(\sigma)^{\pi'} \leqslant U^{\pi'}$. Since $(Fa)^{\pi''}$ and $(Fb)^{\pi''}$ are different hyperplanes, and since therefore neither of these hyperplanes is part of the other one, there exists an element z in $(Fa)^{\pi''}$ which does not belong to $(Fb)^{\pi''}$. Since $A = J^+(\sigma) \oplus J^-(\sigma)$ [by Lemma 1] $z = w + u$ with w in $J^+(\sigma)$ and u in $J^-(\sigma) \leqslant U^{\pi'}$. Since $J^+(\sigma) \leqslant (Fa)^{\pi''} \cap (Fb)^{\pi''}$, it follows that w is in $(Fa)^{\pi''} \cap (Fb)^{\pi''}$ and that therefore u is in $(Fa)^{\pi''}$, but not in $(Fb)^{\pi''}$. Hence $h(u,a) = h(a,u) = 0$ whereas neither $h(u,b)$ nor $h(b,u)$ is 0 [IV.3, Lemma 1, (S)]. Now one verifies that

$$h(b,u)c = h(b^\tau,u^\tau) = h(-a,u) = 0$$

which is impossible, since neither c nor $h(b,u)$ is 0. Thus we have been led again to a contradiction; and this completes the proof of (IV.a).

For reasons of symmetry we may restate (IV.a) in the form: σ is a π'-admissible involution of the first kind if, and only if, σ is a π''-admissible involution of the first kind. From this fact the equality of the groups $\Gamma(\pi')$ and $\Gamma(\pi'')$ is an obvious inference; and thus we have shown that (III) implies (IV).

Assume now the validity of (IV). Suppose that the subspace U of A is non-isotropic with respect to π'. It follows from Corollary 1 that there exists a π'-admissible involution σ of the first kind such that $U = J^+(\sigma)$. Then σ belongs also to $\Gamma(\pi'')$ [by (IV)]. Since [as has been pointed out before] $\Gamma(\pi'')$ does not contain involutions of the second kind, σ is also a π''-admissible involution of the first kind. Now we deduce from Lemma 2, (a), (b) that $U^{\pi'} = J^-(\sigma) = U^{\pi''}$. A corresponding result holds for subspaces V, non-isotropic with respect to π''; and thus we have shown:

(I.a) The subspace U is non-isotropic with respect to π' if, and only if, U is non-isotropic with respect to π''; if U is non-isotropic with respect to π' and π'', then $U^{\pi'} = U^{\pi''}$.

Next we need the following lemma.

(1.a) *If π is a polarity of the linear manifold A, and if the point P has Property* (N), *then P is the intersection of two non-isotropic lines.*

Proof *of* (1.a): We have $P \leqslant P^\pi$ by hypothesis. Since P^π is a hyperplane, and since $2 < r(A)$, there exist at least two different lines L' and L'' through P which are not contained in P^π. It is clear that $L' \cap P^\pi = P$. Since L' is not part of P^π, P is not on L'^π; and since P is on L', we have $L'^\pi \leqslant P^\pi$. Hence

$$L' \cap L'^\pi = L' \cap P^\pi \cap L'^\pi = P \cap L'^\pi = 0;$$

and that L'' is non-isotropic, is shown likewise. This proves (1.a).

We return to the proof of our theorem; and we are going to use only the validity of (I.a) [which we have shown to be a consequence of (IV)]. If P is a point in A, then we infer from (I.a) that P has Property (N) with respect to π' if, and only if, P has Property (N) with respect to π''. If firstly P does not have Property (N), then $P^{\pi'} = P^{\pi''}$ as a consequence of (I.a). If secondly P has Property (N) [with respect to π' and π''], then it follows from (1.a) that $P = L' \cap L''$ where L' and L'' are lines which are both non-isotropic with respect to π'. It follows from (I.a) that L' and L'' are both non-isotropic with respect to π'', and that $L'^{\pi'} = L'^{\pi''}$ and $L''^{\pi'} = L''^{\pi''}$. Consequently we find that

$$P^{\pi'} = (L' \cap L'')^{\pi'} = L'^{\pi'} + L''^{\pi'} = L'^{\pi''} + L''^{\pi''} = (L' \cap L'')^{\pi''} = P^{\pi''}.$$

Hence we have shown that

$$P^{\pi'} = P^{\pi''} \text{ for every point } P \text{ in } A.$$

But every subspace of A is the sum of a finite number of points in A; and so it follows that

$$S^{\pi'} = \left[\sum_{i=1}^{s} P_i\right]^{\pi'} = P_1^{\pi'} \cap \cdots \cap P_s^{\pi'}$$

$$= P_1^{\pi''} \cap \cdots \cap P_s^{\pi''} = \left[\sum_{i=1}^{s} P_i\right]^{\pi''} = S^{\pi''}$$

or $\pi' = \pi''$. Thus (I) is a consequence of (I.a); and this completes the proof of the theorem.

The following important application of this theorem will be given. If $f(x,y)$ is a semi-bilinear form, then we say of the transformation τ that it *preserves* $f(x,y)$, if $f(x^\tau, y^\tau) = f(x,y)$ for every x and y in A. We note that π-admissible involutions of the first kind preserve the forms representing π whereas those of the second kind do not. We note furthermore that a semi-linear transformation τ preserves the semi-bilinear form $f(x,y)$ which is not identically 0 if, and only if, τ is linear.

Corollary 2: *If the characteristic of the field F is not 2, and if the semi-bilinear forms f and g over the F-space A [with $2 < r(A)$] represent polarities, then f and g are preserved by the same linear transformations if, and only if, there exists a number $c \neq 0$ such that $f(x,y) = g(x,y)c$ for every x and y in A, and this is equivalent to saying that f and g represent the same polarity.*

PROOF: It is obvious that the condition is sufficient. Hence assume that the same linear transformations preserve f and g. Denote by π' and π'' the polarities represented by f and g respectively. It follows from our hypothesis that the same collineations are π'-and π''-admissible. Hence $\pi' = \pi''$ is a consequence of Theorem 1. Thus f and g represent the same polarity; and we deduce from IV.1, Proposition 3 the existence of a number $c \neq 0$ in F such that $f(x,y) = g(x,y)c$.

REMARK 3: An interesting geometrical interpretation of the last result may be obtained as follows. We consider A as the $r(A)$-dimensional vector space over F. The semi-bilinear form $f(x,y)$ has the following meaning. $f(x,x)$ "measures the length of the vector x" [$f(x,x)$ "is the squared length" of x] whereas $f(x,y)$ gives a sort of measurement for the "angle between the vectors x and y." The linear transformations preserving f form then "the group of the geometry defined in the vector space A by the form f." The Corollary 2 asserts in this terminology that the geometry and its group determine each other apart from a possible change in "the unit of measurement." The vectors x and y are orthogonal to each other in the f-geometry if $f(x,y) = 0$. This orthogonality relation between vectors leads to an orthogonality relation between subspaces which is essentially the same as the polarity represented by f. It is a consequence of IV.1, Proposition 3 that this polarity and therefore this orthogonality relation determine the form up to a constant factor. Thus we see the essential identity of the f-geometry, the f-orthogonality relation and the groups determined by either of them. Since $f(x,x)$ may be 0 for every vector x in A, we see that the "squared length" of the vectors does not always determine the f-geometry completely. But if we exclude these null systems, the situation changes, as we shall show now.

We point out first that every polarity π which is not a null system may be represented by a symmetrical α-form $f(x,y)$ where $\alpha^2 = 1$ and $f(x,y) = f(y,x)^\alpha$ for every x and y in A. [IV.3, Theorem 1.]

Proposition 2: *Suppose that F is a field of characteristic different from 2, and that the symmetrical semi-bilinear form $f(x,y)$ [over the F-space A] represents the polarity π. Then the following properties of the subspaces U and V of A are equivalent.*

(I) $U \leqslant V^\pi$.

(II) $f(U,V) = 0$.

(III) $f(u,u) + f(v,v) = f(u + v, u + v)$ *for every u in U and v in V.*

Using the interpretation of $f(x,x)$ as "squared length of the vector x" and of $f(U,V) = 0$ as "U and V are orthogonal," we may feel justified in calling Proposition 2 the *Theorem of Pythagoras*.

PROOF: The equivalence of (I) and (II) is a trivial consequence of the way we defined the term "π is represented by f." That (II) implies (III), is verified by direct computation. Assume finally the validity of (III); and assume [by way of contradiction] the existence of elements u and v in U and V respectively such that $t = f(u,v) \neq 0$. Then $t^{-1}u$ belongs to U too; and it follows from (III) that

$$\begin{aligned} f(t^{-1}u,t^{-1}u) + f(v,v) &= f(t^{-1}u + v, t^{-1}u + v) \\ &= f(t^{-1}u,t^{-1}u) + f(t^{-1}u,v) + f(v,t^{-1}u) + f(v,v). \end{aligned}$$

We remember that f is a symmetric α-form. Thus we find that

$$0 = f(t^{-1}u,v) + f(v,t^{-1}u) = t^{-1}f(u,v) + f(v,u)t^{-\alpha}$$
$$= 1 + f(u,v)^{\alpha}t^{-\alpha} = 1 + [t^{-1}f(u,v)]^{\alpha} = 1 + 1;$$

and this contradicts our assumption that the characteristic of F be different from 2. Consequently $f(u,v) = 0$; and this shows that (II) is a consequence of (III).

Corollary 3: *Assume that the characteristic of the field F be different from 2, and that the symmetrical semi-bilinear forms f' and f'' represent polarities π' and π'' respectively. Then $f' = f''$ if, and only if, $f'(x,x) = f''(x,x)$ for every x in A.*

PROOF: Assume that $f'(x,x) = f''(x,x)$ for every x in A. Then we infer from Proposition 2 the equivalence of the following assertions concerning subspaces U and V of A:

$$U \leqslant V^{\pi'};\ f'(u,u) + f'(v,v) = f'(u + v,u + v) \text{ for every } u \text{ in } U,\ v \text{ in } V;$$
$$f''(u,u) + f''(v,v) = f''(u + v,u + v) \text{ for every } u \text{ in } U,\ v \text{ in } V;\ U \leqslant V^{\pi''}.$$

From the fact that $U \leqslant V^{\pi'}$ if, and only if, $U \leqslant V^{\pi''}$ we deduce that $\pi' = \pi''$; and now it follows from IV.1, Proposition 3 that there exists a number $c \neq 0$ in F such that $f'(x,y) = f''(x,y)c$ for every x and y in A. Since null systems are represented by skew-symmetrical bilinear forms [IV.2, Lemma 1] and since symmetrical semi-bilinear forms cannot be skew-symmetrical as the characteristic is not 2, there exists some w such that $f'(w,w) = f''(w,w) \neq 0$; and this implies $c = 1$ or $f' = f''$.

Corollary 4: *Assume that the characteristic of the field F is different from 2, and that the symmetrical semi-bilinear form f represents the polarity π. If the semi-linear transformation σ satisfies $f(x^{\sigma},x^{\sigma}) = f(x,x)^{\sigma}$ for every x in A, then it satisfies $f(x^{\sigma},y^{\sigma}) = f(x,y)^{\sigma}$ for every x and y in A.*

PROOF: Let $g(x,y) = f(x^{\sigma},y^{\sigma})^{\sigma^{-1}}$. One verifies without difficulty that g is a symmetrical bilinear form too; and that g represents a polarity [namely $\sigma^{-1}\pi\sigma$]. Since $g(x,x) = f(x,x)$ for every x in A, we deduce $g = f$ from Corollary 3 so that $f(x^{\sigma},y^{\sigma}) = f(x,y)^{\sigma}$ for every x and y in A.

REMARK 4: If σ is in particular linear, then this corollary asserts that the linear transformation σ preserves the form f, if it preserves the "squared length" $f(x,x)$.

APPENDIX IV

The Polarities with Transitive Group

Suppose that π is a polarity of the F-space A of rank not less than 3. We say then that the group of π is *transitive*, if there exists to every pair of points P,Q in A a π-admissible projectivity which maps P on Q. This is, of course, a somewhat restricted concept of transitivity, since preference is given to the points; and it is somewhat wide insofar as we admit all the π-admissible projectivities. Both these remarks will be elaborated in the course of the present discussion.

If we assume now that the group of π is transitive, then we are led immediately to the following dichotomy.

A. Every point has Property (N).

B. No point has Property (N).

These are the only possibilities, since π-admissible projectivities map points with the Property (N) upon points with the Property (N). For $P \leqslant P^{\pi}$ and $\sigma\pi = \pi\sigma$ for σ a projectivity imply $P^{\sigma} \leqslant P^{\pi\sigma} = P^{\sigma\pi}$.

If every point has Property (N), then π is just a null system. If no point has Property (N), then it follows from

$$U \cap U^{\pi} \leqslant U + U^{\pi} = (U \cap U^{\pi})^{\pi},$$

that every subspace is non-isotropic.

Theorem 1: *If the null system π is represented by the [skew-symmetrical, bilinear] form $f(x,y)$, and if P,Q are points in A, then there exists a linear transformation σ which preserves f and maps P upon Q.*

Proof: We note first that it is a consequence of IV.2, Lemma 1 that null systems are represented by skew-symmetrical bilinear forms; and that consequently the field F is commutative.

Consider now any point X. Since $X \leqslant X^{\pi}$, and since X^{π} is a hyperplane, there exists a point Y, not on X^{π}; and such a point Y is different from X. One verifies easily that the line $L = X \oplus Y$ is non-isotropic. Suppose that $X = Fx$ and $Y = Fy$. Then $f(x,y) = -f(y,x) \neq 0$; and it is always possible to chose x and y in such a way that $f(x,y) = 1$ [if $f(x,y)$ were not 1, then we may substitute $f(x,y)^{-1}x$ for x]; and this we assume.

Since L is non-isotropic, L^{π} is non-isotropic and $A = L \oplus L^{\pi}$ [IV.1, Corollary 2]. On L^{π} we apply IV.2, Proposition 1 and find a basis x'_i, x''_i for $i = 1, \cdots, k$ of L^{π} such that

$$1 = f(x'_i,x''_i) = -f(x''_i,x'_i) \text{ for } i = 1, \cdots, k;$$
$$0 = f(x'_i,x'_j) = f(x'_i,x''_j) = f(x''_j,x'_i) = f(x''_j,x''_i) \text{ for } i \neq j.$$

We apply this result on P and on Q and find a basis p, p_0, p'_i, p''_i with $Fp = P$ and a basis q, q_0, q'_i, q''_i with $Fq = Q$ which meet the requirements:

$$\begin{aligned} 1 = f(p,p_0) &= -f(p_0,p) = f(p'_i,p''_i) = -f(p''_i,p'_i) \\ = f(q_0,q) &= -f(q_0,q) = f(q'_i,q''_i) = -f(q''_i,q'_i) \end{aligned}$$

and $f(x,y) = 0$, if x and y are any pair of elements in one of these bases not already enumerated.

There exists one and only one linear transformation σ such that $p^\sigma = q$, $p_0^\sigma = q_0$, $p_i'^\sigma = q_i'$, $p_i''^\sigma = q_i''$ for $i = 1, \cdots, k$. One verifies easily that σ preserves f; and this completes the proof.

REMARK 1: The method of proof which we used can easily be modified in such a way as to prove:

If U and V are non-isotropic subspaces of equal rank, then there exists a linear transformation σ which preserves f and maps U upon V.

REMARK 2: It is fairly obvious how to find a non-isotropic line L' and a strictly isotropic line L''; and it is clear that no π-admissible projectivity can map L' upon L''. Thus the group of a null system though transitive on the points is not transitive on the lines.

The group of linear transformations preserving the given skew-symmetric bilinear form $f(x,y)$ is nowadays called *symplectic*; and we have shown in Theorem 1 that the symplectic group is transitive on the points and [Remark 1] on the non-isotropic subspaces of rank $2i$. In the case B [where all subspaces are non-isotropic] the situation does not seem to be as clear cut as in the case of null systems.

Theorem 2: *If the α-form $f(x,y)$ represents a polarity π, and if $f(w,w) = 1$ for some w in A, then the following properties are equivalent.*

(I) *If P and Q are points in A, then there exists a linear transformation which preserves f and maps P upon Q.*

(II) *If $x \neq 0$ is in A, then there exists a number $y = 0$ in F such that $f(x,x) = yy^\alpha$.*

(III) *F, α and f have the following two properties:*

(*a*) *To every number s in F there exists a number $t \neq 0$ in F such that $1 + ss^\alpha = tt^\alpha$.*

(*b*) *There exists a basis $b_1, \cdots, b_n$ of A such that*

$$f(b_i, b_j) = \begin{cases} 1 \text{ for } i = j \\ 0 \text{ for } i \neq j. \end{cases}$$

PROOF: Assume first the validity of condition (I). If $x \neq 0$, then Fx is a point; and there exists a linear transformation σ which preserves f and which maps Fw upon Fx. Consequently there exists a number $v \neq 0$ in F such that $w^\sigma = vx$. But σ preserves f; and thus we have

$$1 = f(w,w) = f(w^\sigma,w^\sigma) = f(vx,vx) = vf(x,x)v^\alpha$$

or $f(x,x) = yy^\alpha$ with $y = v^{-1}$. Hence (II) is a consequence of (I).

Assume next the validity of (II). If Fx is a point, then there exists a number $y \neq 0$ in F such that $f(x,x) = yy^\alpha$. Let $x' = y^{-1}x$. Then $Fx = Fx'$ and $f(x',x') = y^{-1}f(x,x)y^{-\alpha} = 1$; and thus we have shown that every point Z may be represented in the form $Fz = Z$ with $f(z,z) = 1$.

If P is a point, then $f(P,P) \neq 0$ by the previous paragraph. Hence P is not an N-point so that P may be imbedded into a maximal O-set $P = P_1, \cdots, P_n$. It follows from the absence of N-points and from IV.4, Lemma 4 (*a*), (*b*) that the points P_i form a basis of the F-space A. From the result in the preceding paragraph we deduce the existence of elements p_i such that $P_i = Fp_i$ and $f(p_i,p_i) = 1$. Since the points P_i form an O-set, $p_i \leqslant P_j^\pi$ for $i \neq j$ and consequently $f(p_i,p_j) = 0$ for $i \neq j$. Clearly the elements p_i form a basis of A; and thus we have shown that (III.*b*) is a consequence of (II).

If Q is another point, then we may find likewise a basis $q_1, \cdots, q_n$ of A such that $Q = Fq_1$ and $f(q_i,q_j) = \begin{cases} 1 \text{ for } i = j \\ 0 \text{ for } i \neq j \end{cases}$. There exists one and only one linear transformation σ of A such that $p_i^\sigma = q_i$ for $i = 1, \cdots, n$. One verifies by computation that σ preserves f; and it is clear that $P^\sigma = Q$. Thus we have shown that (I) is a consequence of (II).

Assume again the validity of (II). Then we have already verified the existence of a basis $b_1, \ldots, b_n$ of A such that $f(b_i,b_j) = \begin{cases} 1 \text{ for } i = j \\ 0 \text{ for } i \neq j \end{cases}$. If s is any number in F, then $b_1 + sb_2 \neq 0$; and we infer from (II) the existence of a number $t \neq 0$ such that

$$tt^\alpha = f(b_1 + sb_2, b_1 + sb_2) = 1 + ss^\alpha.$$

Thus (III.*a*) too is a consequence of (II).

Assume finally the validity of (III). We prove first:

(P) *If $x_1, \cdots, x_k$ are numbers, not 0, in F, then there exists a number $y \neq 0$ in F such that $\sum_{i=1}^{k} x_i x_i^\alpha = yy^\alpha$.*

This is clearly true for $k = 1$; and thus we may make the inductive hypothesis that (P) holds for less than k summands $[1 < k]$. Consequently there exists an element $z \neq 0$ such that $\sum_{i=1}^{k-1} x_i x_i^\alpha = zz^\alpha$. We infer from (III.*a*) the existence of an element $t \neq 0$ in F such that

$$1 + (z^{-1}x_k)(z^{-1}x_k)^\alpha = tt^\alpha.$$

Hence

$$\sum_{i=1}^{k} x_i x_i^\alpha = zz^\alpha + x_k x_k^\alpha = z[1 + (z^{-1}x_k)(z^{-1}x_k)^\alpha]z^\alpha = ztt^\alpha z^\alpha = yy^\alpha$$

where $y = zt \neq 0$; and this completes the inductive proof of (P).

If $x \neq 0$ is an element in A, then $x = \sum_{i=1}^{n} x_i b_i$ where the x_i are numbers in F, not all 0; and where the b_i form a basis of A which is normalized according to (III.b). Then $f(x,x) = \sum_{i=1}^{n} x_i x_i^\alpha$; and since not all the x_i are 0, we may deduce from (P) the existence of a number $y \neq 0$ such that $\sum_{i=1}^{n} x_i x_i^\alpha = yy^\alpha$; and this proves that (II) is a consequence of (III). This completes the proof.

REMARK 3: We note that the Property (P) of the involutorial anti-automorphism of the field F is a consequence of (III.a) alone. We shall refer to pairs (F,α) with these [equivalent] Properties (III.a) and (P) as to α-*Pythagorean*. An obvious consequence of this fact is that the characteristic of the field F is 0. We note furthermore that in IV. Appendix I, (f) we have shown: If F possesses a domain of α-positivity, then $\sum_{i=1}^{n} x_i x_i^\alpha \neq 0$, if not all the x_i are 0. It is known that this latter property [which we have shown to be a consequence of the Pythagorean character of F,α] is equivalent to the existence of a domain of α-positivity at least if $\alpha = 1$. [See the Artin-Schreier theory of real fields; for instance van der Waerden (1).]

REMARK 4: If we refer the form f to a basis b_i which is normalized according to (III.b) then

$$f(x,y) = \sum_{i=1}^{n} x_i y_i^\alpha$$

so that f is essentially uniquely determined, once α is given.

Corollary 1: *Assume that the α-form $f(x,y)$ represents a polarity π, that $f(w,w) = 1$ for some w, and that F,α,f have the Properties* (I) *to* (III) *of Theorem* 2.

(a) *If $0 < U_1 < \cdots < U_{n-1} < A$ and $0 < V_1 < \cdots < V_{n-1} < A$ are two chains of subspaces, and if $n = r(A)$, then there exists a linear transformation σ which preserves f and satisfies $U_i^\sigma = V_i$ for every i.*

(b) *If U and V are subspaces of equal rank, then there exists a linear transformation which preserves f and which maps U upon V.*

PROOF: Suppose that there are given subspaces U_i such that

$$0 = U_0 < U_1 < \cdots < U_{n-1} < U_n = A.$$

Then it follows from $r(A) = n$ that $r(U_i) = i$. Since isotropy of subspaces would imply the existence of N-points, every U_i is non-isotropic so that $A = U_i \oplus U_i^\pi$ and consequently $U_{i+1} = U_i \oplus [U_{i+1} \cap U_i^\pi]$. It follows from Theorem 2 that the point $U_{i+1} \cap U_i^\pi$ has the form Fu_{i+1} with $f(u_{i+1}, u_{i+1}) = 1$ [for $0 \leqslant i < n$]. The elements $u_1, \cdots, u_n$ then form a basis of A with the following properties:

$$U_i = \sum_{j=1}^{i} Fu_j,\ f(u_i, u_j) = \begin{cases} 1 \text{ for } i = j \\ 0 \text{ for } i \neq j \end{cases}.$$

Next we connect with the chain V_i of subspaces a basis $v_1, \cdots, v_n$ of A which has analoguous properties. There exists one and only one linear transformation σ of A such that $u_i^\sigma = v_i$ for $i = 1, \cdots, n$. It is clear that $U_i^\sigma = V_i$ and that σ preserves f. This proves (*a*); and (*b*) is easily deduced from (*a*) by imbedding U and V as i-th terms [with $i = r(U) = r(V)$] in such chains as considered under (*a*).

Remark 5: Corollary 1, (*b*) ought to be compared with Remark 2. We see then that the conditions (i) to (iii) of Theorem 2 are necessary and sufficient for "complete transitivity."

Remark 6: If $\alpha = 1$, then the group $\Gamma(\pi)$ is called *orthogonal*; and this group is called *unitary*, if F is commutative and $\alpha \neq 1$, provided, of course, that the polarity π has the transitivity properties of Theorem 2 and Corollary 1. If F is not commutative, then no name has as yet been attached to the corresponding groups. The reader should realize that commutativity of F is not a consequence of the existence of a polarity π with the transitivity properties of Theorem 2 and Corollary 1. If, for instance, F is the field of real quaternions, α the involutorial anti-automorphism of F which maps every quaternion onto its conjugate, then one may use Remark 4 for the construction of a polarity with our transitivity properties [see Theorem 2, (iii)].

IV.6. The Non-isotropic Subspaces of a Polarity

A polarity is certainly not characterized by its strictly isotropic subspaces, since it is easy to construct linear manifolds which possess a great number of distinct polarities without strictly isotropic subspaces, not 0. On the other hand a polarity is completely determined by the involutorial permutation it induces in the set of its non-isotropic subspaces—a proof of this fact could be constructed easily on the basis of a lemma (1.*a*) which we proved in the course of the proof of IV.5, Theorem 1—and we shall obtain a proof of this result as a by-product of the investigations of this section. It should be noted, however, that the system of non-isotropic subspaces alone [without the induced involution] does not characterize the polarity, since as we remarked before many different polarities of the same linear manifold may have the property that every subspace is non-isotropic.

We restate our problem as follows. Suppose that (F,A) is a linear manifold of finite rank not less than 3; and consider a set Θ of subspaces of A together with an involutorial permutation τ of the set Θ. The problem is to find necessary and sufficient conditions for the existence of a polarity σ of A such that Θ is the totality of non-isotropic subspaces of σ and such that τ is induced by σ in Θ. Our principal results may now be stated as follows. [Note that we are going to say "Θ-space" instead of "subspace in Θ."]

Uniqueness Theorem: *There exists at most one polarity with Θ as its system of non-isotropic subspaces which induces τ in Θ.*

Existence Theorem: *There exists a polarity with Θ as its system of non-isotropic subspaces and inducing the involutorial permutation τ in Θ if, and only if, the pair Θ,τ satisfies the following conditions.*

I. *0 and A belong to Θ.*

II. $A = S \oplus S^\tau$ *for every S in Θ.*

III. *The Θ-spaces V_i and W_j satisfy $V_1 \cap \cdots \cap V_m \leqslant W_1 + \cdots + W_n$ if, and only if, $W_1^\tau \cap \cdots \cap W_n^\tau \leqslant V_1^\tau + \cdots + V_m^\tau$.*

IV. *If M' and M'' are maximal Θ-spaces contained in the subspace S, then $S \cap M'^\tau = S \cap M''^\tau$.*

If N' and N'' are minimal Θ-spaces containing the subspace S, then $S + N'^\tau = S + N''^\tau$.

V. *If the subspace S of A is contained in some Θ-space, not A, then S is the intersection of all Θ-spaces containing S.*

If the subspace S of A contains some Θ-space, not 0, then S is the sum of all Θ-spaces contained in S.

VI. *If T is a subspace of A such that 0 is the only Θ-space contained in T and A is the only Θ-space containing T, then*

(*a*) $r(A) = 2r(T)$ *and*

(*b*) $M \leqslant T + N$ *if, and only if,* $T \cap N^\tau \leqslant M^\tau$ *for every M,N in Θ.*

To prove the necessity of the conditions I to VI and to prove the validity of the Uniqueness Theorem we investigate first a polarity σ of the linear manifold (F,A) of rank not less than 3. We denote by $\Theta(\sigma)$ the totality of non-isotropic subspaces of A with respect to σ. We recall IV.1, Corollary 2 asserting that membership in $\Theta(\sigma)$ is equivalent to either of the following equivalent conditions:

$$U \cap U^\sigma = 0, \qquad A = U \oplus U^\sigma, \qquad A = U + U^\sigma.$$

This shows the validity of postulates I and II; and it assures us of the fact that U belongs to $\Theta(\sigma)$ if, and only if, U^σ belongs to $\Theta(\sigma)$. Consequently σ induces an involutorial permutation in $\Theta(\sigma)$ which permutation we shall denote by σ too. The validity of III is an immediate consequence of the fact that polarities interchange sums and intersections. The proofs of postulates IV to VI we precede by the proofs of several lemmas which are of interest in themselves.

Lemma 1: *If U and V are subspaces of A such that*

$$0 = U \cap U^\sigma \cap V, \qquad A = U + U^\sigma + V,$$
$$U = (U \cap U^\sigma) + V, \qquad U = (U + U^\sigma) \cap V,$$

then V is a maximal non-isotropic subspace contained in U.   *then V is a minimal non-isotropic subspace containing U.*

Proof: We note that the right-hand statement arises from the left-hand statement by dualization [polarization]; and thus it suffices to prove the assertion on the left. Thus assume that U and V are subspaces such that $U = (U \cap U^\sigma) \oplus V$. Then we deduce from Dedekind's Law successively that [because of $V \leqslant U$ and $U^\sigma \leqslant V^\sigma$]

$$U^\sigma = [(U \cap U^\sigma) + V]^\sigma = (U^\sigma + U) \cap V^\sigma = U^\sigma + (U \cap V^\sigma) \text{ or } U \cap V^\sigma \leqslant U^\sigma,$$
$$V \cap V^\sigma = V \cap U \cap V^\sigma = V \cap U \cap V^\sigma \cap U^\sigma = 0.$$

Thus V is non-isotropic. If W is a non-isotropic subspace of U such that $V \leqslant W$, then

$$W = V \oplus (W \cap U \cap U^\sigma) = V \oplus (W \cap U^\sigma) = V,$$

since $V \leqslant W \leqslant U = V \oplus (U \cap U^\sigma)$, and since therefore

$$W \cap U^\sigma \leqslant W \cap W^\sigma = 0.$$

This completes the proof.

Verification of Postulate IV:

Since the right-hand assertion of IV arises from the left-hand one by dualization, it suffices to verify the left-hand part of IV. Thus assume that S is a subspace of A and that M' and M'' are maximal non-isotropic subspaces of S. Then we infer from $M' \leqslant S$ and $A = M' \oplus M'^\sigma$ that $S = M' \oplus (S \cap M'^\sigma)$; and it follows from $S^\sigma \leqslant M'^\sigma$ that

$$S \cap S^\sigma \leqslant S \cap M'^\sigma.$$

Consequently there exists a subspace H such that

$$S \cap M'^\sigma = (S \cap S^\sigma) \oplus H$$

[II.1, Complementation Principle]. Then

$$S = (S \cap S^\sigma) \oplus H \oplus M';$$

and it follows from Lemma 1 that $H \oplus M'$ is non-isotropic. Now we deduce $M' = H \oplus M'$ from the maximality of M'; and this implies $H = 0$ so that $S \cap M'^{\sigma} = S \cap S^{\sigma}$. Likewise we show that

$$S \cap M''^{\sigma} = S \cap S^{\sigma};$$

and this completes the verification of IV. Note that we have verified also the following fact.

Corollary 1: *Suppose that U and V are subspaces of A.*

If V is a maximal non-isotropic subspace contained in U, then

$$U \cap V^{\sigma} = U \cap U^{\sigma}.$$

If V is a minimal non-isotropic subspace containing U, then

$$U + V^{\sigma} = U + U^{\sigma}.$$

VERIFICATION OF POSTULATE V:

Again it suffices to verify one of two dual parts of this property; and we shall verify the right-hand assertion. Thus assume that the subspace S contains some non-isotropic subspace $\neq 0$. Then S contains a maximal non-isotropic subspace M which is necessarily different from 0. It follows from Corollary 1 that $S \cap S^{\sigma} = S \cap M^{\sigma}$; and now it follows from $M \leqslant S$ and $A = M \oplus M^{\sigma}$ that $S = M \oplus (S \cap M^{\sigma}) = M \oplus (S \cap S^{\sigma})$. From $M \neq 0$ we infer that $M = Fm \oplus N$ with $m \neq 0$. If t is an element in $S \cap S^{\sigma}$, then we infer that

$$S = M \oplus (S \cap S^{\sigma}) = Fm \oplus N \oplus (S \cap S^{\sigma}) = F(m + t) \oplus N \oplus (S \cap S^{\sigma});$$

and it follows from Lemma 1 that $F(m + t) + N$ is non-isotropic too. Thus m and $m + t$ belong to non-isotropic subspaces of S; and this shows that t belongs to the sum of all the non-isotropic subspaces of S. Consequently $S \cap S^{\sigma}$ is part of the sum of all non-isotropic subspaces of S; and now it is clear that S is the sum of all its non-isotropic subspaces.

Lemma 2: *The subspace S satisfies $S = S^{\sigma}$ if, and only if, 0 is the only non-isotropic subspace contained in S and A is the only non-isotropic subspace containing S.*

PROOF: Assume first that $S = S^{\sigma}$ and that U is a non-isotropic subspace of S. Then $U \leqslant S = S^{\sigma} \leqslant U^{\sigma}$ and therefore $0 = U \cap U^{\sigma} = U$; and that A is the only non-isotropic subspace containing S is seen likewise.

Assume conversely that 0 is the only non-isotropic subspace of S and that A is the only non-isotropic subspace containing S. There exists a subspace T of S such that $S = T \oplus (S \cap S^{\sigma})$. It follows from Lemma 1 that T is non-isotropic; and it follows from our hypothesis that $T = 0$ and hence $S = S \cap S^{\sigma}$ or $S \leqslant S^{\sigma}$. Likewise [or by duality] we see that $S^{\sigma} \leqslant S$. Hence $S = S^{\sigma}$, as we claimed.

VERIFICATION OF POSTULATE VI:

Suppose that T is a subspace of A such that 0 is the only non-isotropic subspace of T and such that A is the only non-isotropic subspace which contains T. It follows from Lemma 2 that $T = T^\sigma$; and now it follows from IV.1, Corollary 1 that $r(A) = r(T) + r(T^\sigma) = 2r(T)$. Suppose next that the non-isotropic subspaces M,N satisfy $M \leqslant T + N$. Then it follows from the basic properties of polarities that

$$T \cap N^\sigma = T^\sigma \cap N^\sigma = (T + N)^\sigma \leqslant M^\sigma;$$

and the converse of this statement follows by duality.

This completes the proof of the necessity of the conditions of the Existence Theorem. The proof of the Uniqueness Theorem is now an almost immediate consequence of Property V and Lemma 2; we leave its details to the reader.

Assume now that the set Θ of subspaces of A and the involutorial permutation τ of Θ meet the requirements I to VI. If S is any subspace of A, then we denote by

S_Θ *the totality of* Θ*-spaces contained in* S *and by* S^0 *the intersection of all the* X^τ *for* X *in* S_Θ.	S^Θ *the totality of* Θ*-spaces containing* S *and by* S_0 *the sum of all the* X^τ *for* X *in* S^Θ.

These two operations will lead us ultimately to the construction of the desired polarity. Our arguments will be considerably simplified by the following remarks. The special case $m = n = 1$ of III asserts that the Θ-spaces S and T satisfy $S < T$ if, and only if, $T^\tau < S^\tau$. The involutorial permutation τ of Θ is therefore an "incomplete polarity." Furthermore all the postulates I to VI are stated in a self-dual fashion so that the dual of any proposition derived from them will be automatically true too. If X is in S_Θ and Y in S^Θ, then $X \leqslant S \leqslant Y$. Hence $Y^\tau \leqslant X^\tau$ and now it is easy to see that

$$S_0 \leqslant S^0.$$

We derive now a number of properties of these operations.

(1) If S^Θ does not consist of A alone, then $S_{0\Theta} = S^{\Theta\tau}$; and if S_Θ does not consist of 0 alone, then $S^{0\Theta} = S_\Theta^\tau$.

Proof: If X is in S^Θ, then $S \leqslant X$; and it follows from the definition of S_0 that $X^\tau \leqslant S_0$ and that therefore X^τ belongs to $S_{0\Theta}$. Hence $S^{\Theta\tau} \leqslant S_{0\Theta}$. Since S^Θ does not consist of A alone, we may infer from V that S is the intersection of all the Θ-spaces in S^Θ. Since A has finite rank, there exists a finite number of Θ-spaces in S^Θ whose intersection is S; and there exists a finite number of spaces in $S^{\Theta\tau}$ whose sum is S_0. Consequently we may find finitely many Θ-spaces $X_1, \cdots, X_k$ such that simultaneously

$$S = X_1 \cap \cdots \cap X_k \text{ and } S_0 = X_1^\tau + \cdots + X_k^\tau.$$

Now it follows from III that the Θ-space X belongs to $S_{0\Theta}$ if, and only if, $S \leqslant X^\tau$; and this is equivalent to saying that X^τ belongs to S^Θ. Now the desired equation $S_{0\Theta} = S^{\Theta\tau}$ is a consequence of the involutorial character of τ. This proves the first statement (1) and the second one follows by dualization.

(2) If S_Θ does not consist of 0 only and if S^Θ does not consist of A only, then $S_0 = S^0$.

PROOF: It follows from our hypotheses that S is both the sum of all the Θ-spaces in S_Θ and the intersection of all the Θ-spaces in S^Θ; and using the finiteness of rank of A we show as in the preceding proof the existence of Θ-spaces $X_1, \cdots, X_h$ and $Y_1, \cdots, Y_k$ such that simultaneously

$$S = X_1 + \cdots + X_h = Y_1 \cap \cdots \cap Y_k,$$
$$S_0 = X_1^\tau \cap \cdots \cap X_h^\tau,\ S^0 = Y_1^\tau + \cdots + Y_k^\tau.$$

Now we deduce from III and the involutorial character of τ the equivalence of the following properties of the Θ-spaces X and Y respectively: $X \leqslant S$ and $S_0 \leqslant X^\tau$, $S \leqslant Y$ and $Y^\tau \leqslant S^0$. Using (1) it follows that $S^{0\Theta} = S_\Theta^\tau = S_0^\Theta$. From $S_0 \leqslant S^0$ and V we deduce that both S_0 and S^0 are the intersection of the Θ-spaces in S_Θ^τ, proving the desired equality $S_0 = S^0$.

We are now ready to *define* S^σ *for every subspace* S *of* A *by the following rules*:

(*a*) If S_Θ does not consist of 0 only, then $S^\sigma = S^0$.

(*b*) If S_Θ consists of 0 only and S^Θ consists of A alone, then $S^\sigma = S$.

(*c*) If S^Θ does not consist of A alone, then $S^\sigma = S_0$.

That the operation σ thus defined maps every subspace S of A upon one and only one subspace S^σ of A, is an immediate consequence of (2) which proposition asserts that rules (*a*) and (*c*) give the same result, if they are both applicable.

Suppose that S is in particular a Θ-space. If $S=0$, $S^\tau = A$ and rule (*c*) gives us $S_0 = A$ so that $0^\sigma = A$; and $A^\sigma = 0$ is seen likewise. If on the other hand the Θ-space S is neither 0 nor A, then S belongs to both S^Θ and S_Θ; and one finds that $S_0 = S^0 = S^\tau$ so that we have again $S^\sigma = S^\tau$. Next we prove that

(3) $\sigma^2 = 1$.

We have to prove that $S = S^{\sigma^2}$ for every subspace S of A. This is trivially true, if S is a Θ-space, since on these σ coincides with the involutorial τ; and it is equally trivially true, if 0 is the only space in S_Θ and A is the only space in S^Θ [rule (*b*)]. If 0 is not the only subspace in S_Θ, then $S^\sigma = S^0$ and $S^{0\Theta} = S_\Theta^\tau$. But then $S^{\sigma^2} = S^0{}_0$ is the sum of all the Θ-spaces in $S^{0\Theta\tau} = S_\Theta$, since τ is involutorial; and it follows from V that $S^{\sigma^2} = S$. The dual argument applies, if A is not the only subspace in S^Θ; and this completes the proof of (3).

From (3) we deduce in particular that σ is an involutorial permutation of the system of all the subspaces of A which coincides with τ on Θ.

(4) *σ is a polarity of A.*

To prove this we have only to show that $S \leqslant T$ implies $T^\sigma \leqslant S^\sigma$. To do so we distinguish various cases.

Case 1: 0 is not the only subspace in S_Θ.

From $S \leqslant T$ we deduce that S_Θ is part of T_Θ so that neither of these sets consists of 0 only. Hence $T^\sigma = T^0 \leqslant S^0 = S^\sigma$.

Case 2: A is not the only subspace in T^Θ.

Then the argument dual to the one used in Case 1 leads to the desired inequality $T^\sigma \leqslant S^\sigma$.

Case 3: 0 is the only subspace in S_Θ and A is the only subspace in T^Θ.

Here we have to distinguish various subcases.

Case 3.1: S^Θ does not consist of A alone and T_Θ does not consist of 0 only.

Then we find as usual finitely many Θ-spaces X_i, Y_j such that simultaneously

$$S = X_1 \cap \cdots \cap X_h,\ S^\sigma = X_1^\tau + \cdots + X_h^\tau,$$

$$T = Y_1 + \cdots + Y_k,\ T^\sigma = Y_1^\tau \cap \cdots \cap Y_k^\tau;$$

and from these equations, the inequality $S \leqslant T$, and Postulate III one deduces $T^\sigma \leqslant S^\sigma$.

Case 3.2: S^Θ consists of A alone whereas T_Θ does not consist of 0 only.

Then we deduce from rule (*b*) that $S^\sigma = S$ and from rule (*a*) that $T^\sigma = T^0$. We need an intermediate result.

(+) If $M \neq 0$ is a Θ-space, then $(S + M)^\sigma = S \cap M^\sigma = S \cap M^\tau$.

Clearly $M \neq 0$ belongs to $(S + M)_\Theta$ so that $(S + M)^\sigma = (S + M)^0$. It follows from Postulate VI, (*b*) that X belongs to $(S + M)_\Theta$ if, and only if, $S \cap M^\tau \leqslant X^\tau$. Hence $(S + M)^\sigma$ is the intersection of all the Θ-spaces in $(S \cap M^\tau)^\Theta$. It follows from Postulate V that $S \cap M^\tau$ is just this intersection; and this proves $(S + M)^\sigma = S \cap M^\tau$.

Since T^σ is the intersection of all the X^τ with $X \neq 0$ in T_Θ, it follows from (+) that $S \cap T^\sigma$ is the intersection of all the $S \cap X^\tau = (S + X)^\sigma$ with $X \neq 0$ in T_Θ. But $(S + X)^\sigma$ is the intersection of all the Y^τ with Y in $(S + X)_\Theta$, since $X \neq 0$ belongs to this set. Hence $S \cap T^\sigma$ is the intersection of all the Θ-spaces Z with the property:

There exists a Θ-space X such that $X \leqslant T$ and $Z^\tau \leqslant S + X$.

Since $S \leqslant T$, these conditions are equivalent to $Z^\tau \leqslant T$, since then we may chose $X = Z^\tau$; and thus we have shown that $S \cap T^\sigma$ is the intersection of all the Θ-spaces X^τ with X in T_Θ. Hence $S \cap T^\sigma = T^\sigma$ or $T^\sigma \leqslant S = S^\sigma$, as we wanted to show.

Case 3.3: S^{Θ} does not consist of A alone whereas T_{Θ} consists of 0 only. Then we show $T^{\sigma} = S^{\sigma}$ by an argument dual to the one used in Case 3.2.

Case 3.4: S^{Θ} consists of A alone and T_{Θ} consists of Θ only.

Then it follows from rule (*b*) that $T = T^{\sigma}$ and $S = S^{\sigma}$; and it follows from Postulate VI, (*a*) that $r(S) = \frac{1}{2}r(A) = r(T)$. From the finiteness of rank it follows that $T^{\sigma} = T = S = S^{\sigma}$, since $S \leqslant T$.

Thus we have proved the desired inequality in all possible cases; and this completes the proof of the fact that σ is a polarity.

(5) *S is non-isotropic for the polarity σ if, and only if, S belongs to Θ.*

Since the polarity σ coincides on Θ with τ, it follows from Postulate II that all the Θ-spaces are non-isotropic. Assume conversely that S is non-isotropic with respect to σ. Because of rule (*b*) it is impossible that 0 is the only Θ-subspace of S and at the same time A is the only Θ-space containing S. Because of duality [in particular the self-duality of the concept non-isotropic] we may assume without loss in generality that 0 is not the only Θ-space contained in S. Then $S^{\sigma} = S^{0}$; and it follows from the finiteness of rank and the definition of S^{0} that there exists a finite number of minimal Θ-spaces, say $M_1, \cdots, M_k$ in S_{Θ}^{τ} such that $S^{\sigma} = S^{0} = M_1 \cap \cdots \cap M_k$. But M_i is a minimal Θ-space in S_{Θ}^{σ} if, and only if, $M_i^{\tau} = M_i^{\sigma}$ is a maximal Θ-space contained in S. Now we infer from Postulate IV that

$$S \cap M_1 = \cdots = S \cap M_k = D.$$

Consequently

$$0 = S \cap S^{\sigma} = S \cap M_1 \cap \cdots \cap M_k = (S \cap M_1) \cap \cdots \cap (S \cap M_k) = D.$$

Since σ is a polarity, and since M_i as Θ-space is non-isotropic, we find because of $M_i^{\sigma} \leqslant S$ that $S = M_i^{\sigma} \oplus (S \cap M_i) = M_i^{\sigma}$. Thus S belongs to Θ, since with M_i also $M_i^{\sigma} = M_i^{\tau}$ is a Θ-space. This completes the proof of (5).

Combining all these results we find that σ is a polarity of A such that Θ is the totality of non-isotropic subspaces for σ and such that σ coincides with τ on Θ. Thus we have completed the proof of the Existence Theorem.

Remark 1: It should be noted that all the six conditions have been used somewhere in our discussion. The question whether it might be possible to omit part of these conditions has not been investigated. However, it may be pointed out that IV, (*a*) is indispensable, since the system Θ consisting of 0 and A only and τ, the permutation interchanging them, satisfies all the postulates with the exception of VI, (*a*), since $3 = r(A)$.

REMARK 2: If we want to be assured that the polarity σ is a null system, then we just add to our six conditions the further condition that points and hyperplanes should not appear in Θ.

REMARK 3: If Θ happens to contain every subspace of A, then our six conditions are just an overly clumsy way of expressing that τ is a polarity.

REMARK 4: The preceding results may be used as a beginning in an attempt of characterizing the group of a polarity [IV.5] by internal properties. For suppose that σ is some polarity and that Φ is the group of σ-admissible linear transformations [i.e. of those linear transformations which induce σ-admissible collineations]. It follows from IV.5, Corollary 1 that the pairs of non-isotropic subspaces $[U,U^\sigma]$ are just the pairs of subspaces $[J^+(\tau),J^-(\tau)]$ where τ is an involution of the first kind in Φ. If we denote by Γ the subgroup of Φ which is generated by the involutions of the first kind, then Φ is just the normalizer of Γ within the group of all linear transformations as the reader will easily verify. Thus we have just to require of the group Φ of linear transformations that it possesses a system of involutions with the property that their spaces J^+,J^- satisfy the postulates I to VI and that Φ be the normalizer of this system of involutions in order to be assured that Φ is just the group of all admissible involutions for some polarity. But these remarks should only be considered as initial steps in the attempt of characterizing these groups Φ by internal properties. The real problem begins only here and is most fascinating. The reader may consult now such works as Dieudonné [1,2], Hua [1], van der Waerden [2], Weyl [1]. Furthermore we might mention here

R. Baer: Free Mobility and Orthogonality, *Transactions Amer. Math. Soc.*, vol. 68 (1950), pp. 439-460;

C. E. Rickart: Isomorphic Groups of Linear Transformations II, *Amer. Journal of Math.*, vol. 73 (1951), pp. 697-716;

C. E. Rickart: Isomorphisms of Infinite Dimensional Analogues of the Classical Groups, *Bull. Amer. Math. Soc.*, vol. 57 (1951), pp. 435-448;
and the work on group theoretical foundations of elliptic geometry like

R. Baer: The Group of Motions of a Two-Dimensional Elliptic Geometry, *Comp. Math.*, vol. 9 (1951), pp. 271-288;

A. Schmidt: Ueber die Bewegungsgruppe der ebenen elliptischen Geometrie, *Journal f. reine und angew. Math.*, vol. 186 (1949), pp. 230-240.

CHAPTER V

The Ring of a Linear Manifold

It might be well to reflect first on the general setting in which to view the problems of this chapter. It is a particularly illuminating instance of what we might call *the problem of the derived manifolds,* a problem that has proved interesting in various branches of mathematics—the derived manifolds of normed linear spaces, for example, have been treated systematically by G. W. Mackey: Isomorphisms of Normed Linear Spaces, *Annals of Math.*, vol. 43 (1942), 244-260.

The problem of the derived manifolds may be stated as follows: There is given some basic mathematical structure and from it one has derived some other mathematical structure. The question is to determine to which extent the basic structure is determined by the derived one; and in which way the properties of the basic structure are reflected in properties of the derived structure. A very obvious example for this kind of relationship may be seen in a linear manifold [the basic structure] and the partially ordered set of its subspaces [the derived structure]. It is the content of the First Fundamental Theorem of Projective Geometry [III.1] that, apart from the low dimensional exceptions, the partially ordered set of subspaces completely determines the original linear manifold. Another instance of this sort is provided by a polarity [basic structure] and its group [derived structure]; see IV.5.

The ring of a linear manifold which we are going to study in the present chapter reflects in a particularly complete fashion the properties of the basic linear manifold. The basic tool in our investigation is what we call the "three cornered Galois Theory" which relates in a very close way the right ideals of the ring, the subspaces of the linear manifold and the left ideals of the ring and which makes it possible to translate properties of the linear manifold into ring properties and vice versa. In particular we shall obtain a mutual determination of projectivities and ring-isomorphisms, dualities and ring-anti-isomorphisms.

For a treatment of these questions in a more comprehensive framework the reader might consult R. Baer: Automorphism Rings of Primary Abelian Operator Groups, *Annals of Math.*, vol. 44 (1943), 192-226.

V.I. Definition of the Endomorphism Ring

An *endomorphism* of the linear manifold (F,A) is a single-valued mapping σ of A into itself with the following properties:

$$(a' + a'')\sigma = a'\sigma + a''\sigma,\ (xa)\sigma = x(a\sigma) \text{ for } a,a',a'' \text{ in } A \text{ and } x \text{ in } F.$$

[Note that the image of the element a in A under the mapping σ is the element $a\sigma$ in A.] Thus endomorphisms are just what in the theory of vector spaces goes under the names of regular and singular linear transformation of the F-space A into itself; and the linear transformations that we have discussed before are special cases of endomorphisms. On the other hand a semi-linear transformation of (F,A) upon itself is an endomorphism if, and only if, it is linear.

If σ is an endomorphism of (F,A), then the totality $A\sigma$ of the elements $a\sigma$ for a in A is a subspace of A [as is easily seen]; and the rank of $A\sigma$ is often called *the rank of* σ. The totality $K(\sigma)$ of elements x in A such that $x\sigma = 0$ is also a subspace of A and its rank is often called *the nullity of* σ. We shall often refer to $K(\sigma)$ as to the *kernel* of σ; and the endomorphisms constitute a generalization of the "projection of A upon $A\sigma$ in the direction of the kernel $K(\sigma)$."

(1) $$A\sigma \simeq A/K(\sigma).$$

This important fact may be expressed in the following sharper form:

(1*) *σ induces a linear transformation [in the strict sense] of $A/K(\sigma)$ upon $A\sigma$.*

This statement is quite easily verified once one has realised the equivalence of the following statements:

$$a \equiv b \text{ modulo } K(\sigma),\ a - b \text{ belongs to } K(\sigma),\ a\sigma = b\sigma.$$

From (1) one deduces clearly

(2) $$r[A\sigma] = r[A/K(\sigma)];$$

and this formula contains as a special case

(3) $$r(A) = r[A\sigma] + r[K(\sigma)]$$

which may be expressed in words as follows: the rank of a linear manifold is the sum of rank and nullity of each of its endomorphisms. We note that the formulas (2) and (3) are equivalent in case $r(A)$ is finite; but in the case of infinite $r(A)$ formula (2) is the stronger one.

We denote the totality of endomorphisms of the linear manifold (F,A)

by $\mathrm{P}(F,A)$ [$=\mathrm{P}(A)=\mathrm{P}$]; and we define addition and multiplication of elements σ',σ'' in P by the following rules:

$$a(\sigma' + \sigma'') = a\sigma' + a\sigma'',\ a(\sigma'\sigma'') = (a\sigma')\sigma'' \text{ for } a \text{ in } A.$$

One verifies without difficulty that the sum $\sigma' + \sigma''$ and the product $\sigma'\sigma''$ of the endomorphisms σ',σ'' is again an endomorphism of the linear manifold (F,A). Thus P is closed under these operations. Next one verifies the by now almost obvious fact that P is a ring; and thus we are justified in calling $\mathrm{P}(F,A)$ *the endomorphism ring* [or shorter: the ring] *of the linear manifold* (F,A). [In the older literature this ring is often called the automorphism ring of (F,A).]

Idempotents and Decomposition Endomorphisms

The endomorphism σ is called *idempotent* whenever $\sigma = \sigma^2$. The simplest examples of such indempotents are 0 and 1 defined by $a0=0$ and $a1=a$ for every a in A

If σ is an idempotent, then let $\sigma' = 1 - \sigma$. We verify that

$$\sigma\sigma' = \sigma'\sigma = \sigma - \sigma^2 = 0 \text{ and } \sigma'^2 = (1-\sigma)^2 = 1 - 2\sigma + \sigma^2 = 1 - \sigma = \sigma'.$$

Hence $1-\sigma$ is an idempotent "orthogonal" to the idempotent σ; and the two idempotents σ and $\sigma' = 1-\sigma$ are complementary in so far as $\sigma + \sigma' = 1$. Note that the idempotents ν,ω will be called *orthogonal* whenever they satisfy $\nu\omega = \omega\nu = 0$.

(4) *If σ is an idempotent endomorphism, then*

$$K(\sigma) = A(1-\sigma),\ A\sigma = K(1-\sigma),\ A = A\sigma \oplus K(\sigma).$$

This is an almost immediate consequence of the trivial formula:

$$a = a\sigma + a(1-\sigma) \text{ for } a \text{ in } A$$

together with the orthogonality and idempotency relations which we have just verified. We note in particular that σ leaves invariant every element in $A\sigma$ and that $1-\sigma$ leaves invariant every element in $A(1-\sigma)$. [Thus the idempotent endomorphisms are the strict equivalent of what often is called a projection of A upon $A\sigma$ in the direction of $K(\sigma)$.]

(5) *If $\sigma_1, \cdots, \sigma_n$ are finitely many mutually orthogonal idempotents, and if $1 = \sum_{i=1}^{n} \sigma_i$, then $A = A\sigma_1 \oplus \cdots \oplus A\sigma_n$.*

PROOF: The σ_i satisfy the rules $\sigma_i\sigma_j = \begin{cases} \sigma_i \text{ for } i = j \\ 0 \text{ for } i \neq j \end{cases}$. Since 1 is the sum of the σ_i, we have

(5.1) $\quad a = a\sigma_1 + \cdots + a\sigma_n$ for every a in A.

If in particular $0 = \sum_{i=1}^{n} a_i$ with a_i in $A\sigma_i$, then $a_i = b_i\sigma_i$ so that

$$a_i\sigma_j = b_i\sigma_i\sigma_j = \begin{cases} b_i\sigma_i = a_i \text{ for } i = j \\ b_i 0 = 0 \text{ for } i \neq j \end{cases};$$

and multiplication of our equation by σ_i leads to $0 = 0\sigma_i = a_i$. Hence

(5.2) $\quad 0 = a_1 + \cdots + a_n$ with a_i in $A\sigma_i$ implies $a_1 = \cdots = a_n = 0$.

But the facts (5.1) and (5.2) together just establish our proposition (5).

If conversely $A = A_1 \oplus \cdots \oplus A_m$, then we denote by σ_i the uniquely determined endomorphism of A which leaves invariant every element in A_i and which maps onto 0 all the elements in $\sum_{i \neq j} A_j$. It is quite easy to see that the σ_i are mutually orthogonal idempotents whose sum is 1 and that they satisfy $A_i = A\sigma_i$. They are called *the* [*complementary*] *decomposition endomorphisms belonging to our direct decomposition of* A; and if we combine these remarks with (5), then we see that "direct decompositions into finitely many components" and "systems of finitely many mutually orthogonal idempotents with sum 1" are equivalent concepts in so far as the direct decompositions and the systems of idempotents determine each other in a one-to-one fashion.

If there is given a direct decomposition of A into infinitely many summands A_ν, then it is possible again to define the decomposition endomorphisms σ_ν which leave invariant every element in A_ν and which annihilate the elements in the remaining components. One verifies again that the σ_ν are mutually orthogonal idempotents; but we cannot claim that their sum is 1, since sums of infinitely many summands are not defined in the ring P. There are various ways of circumventing these difficulties. The one is based on the fact that it may be shown that the σ_ν form a maximal system of mutually orthogonal idempotents; another one is based on the fact that to every element a in A there exists only a finite number of σ_ν which do not map a onto 0 which fact makes it possible to define the infinite sum of the σ_ν in a sensible fashion. [For further details the reader may consult, for instance, R. Baer: The Decomposition of Abelian Groups into Direct Summands, *Quarterly Journal of Mathematics*, vol. 6 (1935), pp. 222-232.]

Representation of the Endomorphism Ring by Matrices

This representation may help the reader to connect the concepts which we have introduced just now with concepts he is already familiar with. But there will be little opportunity for us to make any use of this representation.

Suppose (F,A) is a linear manifold of finite rank n. Select some basis B of A consisting of the elements $b_1, \cdots, b_n$; and we consider B as an ordered set of elements. If σ is an endomorphism of A, then

$$b_i\sigma = \sum_{j=1}^{n} b_{ij}b_j \text{ for } i = 1, \cdots, n;$$

and thus we may map the endomorphism σ upon the uniquely determined n by n matrix $\sigma^B = (b_{ij})$ [where i indicates the row and j the column in which we find the element b_{ij}]. It should be noted that our mapping depends essentially on the choice of the basis B and that it will already be changed if we permute the elements in B.

From the formula

$$\text{(R)} \qquad \Big[\sum_{i=1}^{n} x_i b_i\Big]\sigma = \sum_{i=1}^{n} x_i \sum_{j=1}^{n} b_{ij}b_j = \sum_{j=1}^{n}\Big[\sum_{i=1}^{n} x_i b_{ij}\Big]b_j$$

we infer that the endomorphism σ is completely determined by the corresponding matrix σ^B. Thus the mapping $\sigma \to \sigma^B$ is a one-to-one correspondence. Reading the formula (R) from right to left we see how to define to a given matrix (b_{ij}) an endomorphism σ such that $\sigma^B = (b_{ij})$; and thus we see that the mapping $\sigma \to \sigma^B$ is a one-to-one mapping of the ring P upon the totality F_n of n by n matrices with coefficients in F.

In F_n addition and multiplication may be defined in the customary fashion by the rules:

$$(a_{ij}) + (b_{ij}) = (a_{ij} + b_{ij}),\ (a_{ij})(b_{ij}) = \Big(\sum_{k=1}^{n} a_{ik}b_{kj}\Big).$$

Using these definitions the reader will find it easy to verify the formulas:

$$\sigma'^B + \sigma''^B = (\sigma' + \sigma'')^B,\ \sigma'^B\sigma''^B = (\sigma'\sigma'')^B$$

which show that our mapping of P upon F_n is an isomorphism.

If B' and B'' are bases of A, then the mappings $\sigma^{B'}$ and $\sigma^{B''}$ constitute isomorphisms of P upon F_n. The reader should verify that these two isomorphisms differ only by an inner automorphism of F_n which is induced by that matrix in F_n which connects the two bases B' and B''.

If the linear manifold (F,A) has infinite rank, then very similar considerations apply. Only this time we are not going to get all matrices, but only the "row-finite" matrices. Furthermore some care is needed in considering the various possible orderings for a basis which may be of essentially different order type.

V.2. The Three Cornered Galois Theory

Since the endomorphism ring P of the linear manifold (F,A) contains an identity element 1, it is possible to define *right ideals in* P as non-vacuous subsets J of P which satisfy

$$J = J \pm J = JP;$$

and *left ideals in* P are non-vacuous subsets H of P which satisfy

$$H = H \pm H = PH.$$

Thus right as well as left ideals contain sums and differences of their elements [and both are closed under multiplication too]. But right ideals

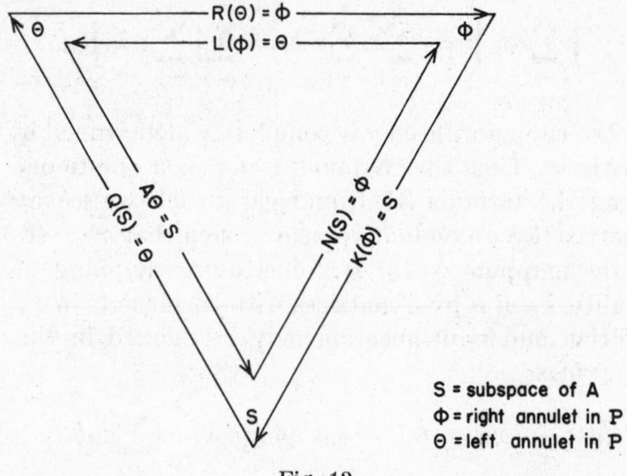

Fig. 12

contain with any element j all its right hand multiples jj' for j' in P and left ideals contain with any element all their left hand multiples.

If S is any subset of the ring P, then the totality $L(S)$ of elements x in P such that $xS = 0$ is clearly a left ideal; and such a left ideal we term a *left annulet*, namely the left annulet determined by S. Likewise we denote by $R(S)$ the totality of elements y in P such that $Sy = 0$. Clearly $R(S)$ is a right ideal; and such a right ideal we term a *right annulet*, namely the right annulet determined by S. [The question under which circumstances all ideals are annulets will be taken up in the next section.]

The three cornered Galois Theory with which we are concerned in this section consists in the construction of a projectivity between the subspaces of A and the left annulets of P and of dualities between the subspaces of A and the right annulets of P and between the right and left annulets.

To obtain these correspondences we need apart from the operations L and R already defined *four further operations*:

If S is a subset of A, then $N(S)$ is the totality of elements σ in P such that $S\sigma = 0$; and $Q(S)$ is the totality of elements σ in P such that $A\sigma \leqslant S$. It is not difficult to see that $N(S)$ is a right ideal; and $Q(S)$ will be a left ideal in P whenever S is closed under addition. The right ideal $N(S)$ might be called the annihilator of S; and $Q(S)$ is some sort of a quotient of S by A — if S happens to be a subspace of A, then $Q(S)$ is really the annihilator of the quotient space A/S.

If J is a subset of P, then we denote by $K(J)$ the totality of elements x in A such that $xJ = 0$ [or $x\sigma = 0$ for every σ in J]. It is quite easy to verify that this kernel $K(J)$ of J is a subspace of A. By AJ we denote the set of all elements aj for a in A and j in J and we note that, in general, AJ need not be a subspace of A. Some difficulties will arise from this possibility; see, however, V.3, Corollary 1 below.

The main content of the "three cornered Galois Theory" may now be stated as follows.

Theorem A: *The correspondences $N(S)$ and $K(\Phi)$ are reciprocal dualities between the subspaces S of A and the right annulets Φ of* P.

Theorem B: *The correspondences $Q(S)$ and $A\Phi$ are reciprocal projectivities between the subspaces S of A and the left annulets Θ of* P.

Theorem C: *The correspondences $L(\Phi)$ and $R(\Theta)$ are reciprocal dualities between the right annulets Φ and the left annulets Θ of* P.

The proof of these three theorems will be effected in a number of steps. In the course of these proofs we shall obtain a number of facts rounding off the theory.

Proposition 1: $K[N(S)] = S = AQ(S)$ *for every subspace S of A.*

PROOF: It follows from the definition of $N(S)$ that $SN(S) = 0$; and thus it follows from the definition of the K-operator that $S \leqslant K[N(S)]$. To prove equality consider an element w in A, but not in S. Then we may form $S \oplus Fw$; and from the Complementation Theorem [II.1] we infer the existence of a subspace W such that $A = S \oplus Fw \oplus W$. There exists one and only one endomorphism σ such that $S\sigma = 0$ and such that $x\sigma = x$ for x in $Fw + W$ [one of the decomposition endomorphisms belonging to our direct decomposition $A = S \oplus (Fw + W)$]. Naturally σ belongs to $N(S)$ and because of $w = w\sigma$ and $w \neq 0$ the element w does not belong to $K[N(S)]$. Hence $S \not< K[N(S)]$ so that $S = K[N(S)]$.

From the definition of the operation Q it follows that $AQ(S) \leqslant S$. Consider now an element s in S. Then Fs is a subspace of S and of A; and from the Complementation Principle we deduce the existence of a subspace T of A such that $A = Fs \oplus T$. There exists one and only one endomorphism τ of A such that $s = s\tau$ and $T\tau = 0$; and this endomorphism has the property: $A\tau = Fs \leqslant S$. Consequently τ belongs to $Q(S)$ and s belongs to $AQ(S)$. Thus we see that $S \leqslant AQ(S) \leqslant S$ or $S = AQ(S)$, as we intended to prove.

Proposition 2: $N(AJ) = R(J)$ *and* $Q[K(J)] = L(J)$ *for every subset* J *of* P.

This is verified by checking the fairly obvious equivalence of the following statements:

$$\sigma \text{ belongs to } R(J); \quad J\sigma = 0; \quad AJ\sigma = 0; \quad \sigma \text{ belongs to } N(AJ)$$

and the equivalence of the following statements:

$$\sigma \text{ belongs to } L(J); \quad \sigma J = 0; \quad A\sigma J = 0; \quad A\sigma \leqslant K(J); \quad \sigma \text{ belongs to } Q[K(J)].$$

Combining Propositions 1 and 2 we find that

$$N(S) = N[AQ(S)] = R[Q(S)]; \; Q(S) = Q[K(N(S))] = L[N(S)]$$

for subspaces S of A; and this we restate as

Proposition 3: $N(S) = R[Q(S)]$ *and* $Q(S) = L[N(S)]$ *for every subspace* S *of* A.

REMARK 1: Proposition 3 is just an improvement on the important and easily verified relation:

$$Q(S)N(S) = 0 \text{ for every subset } S \text{ of } A.$$

Proposition 4: $J = Q(AJ)$ *for every left annulet* J *in* P *and* $H = N[K(H)]$ *for every right annulet* H *in* P.

PROOF: Since J is a left annulet, we have $J = L(T)$ for some subset T of P. Now it follows from Propositions 1 and 2 that

$$J = L(T) = Q[K(T)], AJ = AQ[K(T)] = K(T), Q(AJ) = Q[K(T)] = J.$$

Since H is a right annulet, we have $H = R(M)$ for some subset M of P. The subset AM of A generates a subspace S of A. One verifies easily that $N(AM) = N(S)$; and now one deduces from Propositions 1 and 2 that

$$H = R(M) = N(AM) = N(S), \; K(H) = K[N(S)] = S, \; N[K(H)] = N(S) = H,$$

completing the proof.

PROOF OF THEOREM A: From the Propositions 1, 3, and 4 we deduce that

$$S = K[N(S)], \; N(S) = R[Q(S)] \text{ for every subspace } S \text{ of } A,$$
$$H = N[K(H)] \text{ for every right annulet } H \text{ of } \mathrm{P}.$$

These equations put into evidence that the operations $K(H)$ for H a right annulet in P and $N(S)$ for S a subspace of A are reciprocal correspondences between the system of all subspaces of A and the system of all right annulets in P. This implies in particular that both correspondences are one-to-one and exhaustive.

If S is a subspace of the subspace T of A, then $N(T) \leqslant N(S)$; and if $H \leqslant J$ are subsets of P, then $K(J) \leqslant K(H)$, as follows immediately from the definition of N and K. Consequently K and N are reciprocal dualities between the system of all subspaces of A and the system of all right annulets in P.

Proof of Theorem B: From the Propositions 1, 3, and 4 we deduce that

$$S = AQ(S),\ Q(S) = L[N(S)] \text{ for every subspace } S \text{ of } A,$$
$$J = Q(AJ) \text{ for every left annulet } J \text{ in P}.$$

Now one verifies [similarly to the preceding proof] that the mappings of S onto $Q(S)$ and of J onto AJ constitute reciprocal projectivities between the system of all subspaces of A and the system of all left annulets in P.

Remark 2: This Theorem B shows that the left annulets in P constitute a faithful representation of the projective geometry of all the subspaces of A. To make this representation really useful we would need an internal characterization of the annulets; see in this respect the next section.

Proof of Theorem C: If J is a left annulet, then AJ is a subspace of A [Theorem B]; and it follows from Propositions 2, 3, 4 that

$$L[R(J)] = L[N(AJ)] = Q(AJ) = J;$$

and similarly we have for every right annulet H

$$R[L(H)] = R[Q(K[H])] = N[K(H)] = H;$$

and now it is almost obvious that L and R are reciprocal dualities between the system of all left annulets and the system of all right annulets in P.

Remark 3: It is quite easy to see that the intersection of any number of right annulets is again a right annulet; and an analoguous statement holds for left annulets. The situation is different for sums of annulets. Though we shall be able to prove in the next section that the sum of a finite number of annulets is again an annulet, the corresponding statement for infinitely many annulets will be shown to be generally false.

The connection between the linear manifold and its ring is even closer than the relations expressed in the preceding results seem to indicate. For we shall show next that in a way the linear manifold is contained in its ring. At the same time we want to show that the ring contains also the adjoint space. With this in mind we want to recall a few facts.

If (F,A) is a linear manifold, then its adjoint space shall be denoted by

A^*. We have introduced it in II.3 as the space of all linear forms over A [and in II.3 we denoted this adjoint space by $L(A)$, a notation that we avoid in the present section, since it might lead to confusion with the operator L just introduced]. Since the elements in F are right-multipliers for A^*, we shall also speak of the linear manifold (A^*,F). If a is in A and b is in A^*, then the effect of the linear form b on the element a is denoted by ab; and ab is an element in F. This whole configuration we may indicate therefore by (A^*,F,A). We recall furthermore that $E(T)$ for $T \leqslant A$ consists of all the b in A^* such that $Tb = 0$; and that $S(T)$ for $T \leqslant A^*$ consists of all the a in A such that $aT = 0$.

Proposition 5: *Suppose that* $A = P \oplus H$ *for* P *a point and* H *a hyperplane in the linear manifold* (F,A). *Then there exists an isomorphism* ν *of the field* F *upon the subfield* $N(H) \cap Q(P)$ *of the ring* $\mathrm{P}(F,A)$, *an isomorphism* α *of the additive group* A *upon the subgroup* $N(H)$ *and an isomorphism* α^* *of the additive group* A^* *upon the subgroup* $Q(P)$ [*of the additive group of the ring* P] *with the following properties:*

(*a*) $(bx)^{\alpha^*} = b^{\alpha^*}x^{\nu}$, $(xa)^{\alpha} = x^{\nu}a^{\alpha}$, $(ab)^{\nu} = a^{\alpha}b^{\alpha^*}$ *for* b *in* A^*, x *in* F *and* a *in* A.

(*b*) $S^{\alpha} = Q(S) \cap N(H)$ *and* $E(S)^{\alpha^*} = N(S) \cap Q(P)$ *for every subspace* S *of* A.

Proof: $P = Fp$, since P is a point; and the isomorphisms to be constructed depend on the choice of the element p.

To every element x in F there exists one and only one endomorphism x^{ν} of A with the following properties:

$$px^{\nu} = xp, \; Hx^{\nu} = 0.$$

It is fairly clear that ν effects a one-to-one and additive mapping of F upon $N(H) \cap Q(P)$. This mapping is multiplicative too, since

$$p(xy)^{\nu} = xyp = x(yp) = x(py^{\nu}) = (xp)y^{\nu} = (px^{\nu})y^{\nu} = p(x^{\nu}y^{\nu});$$

and consequently is an isomorphism of the field F upon the subring $N(H) \cap Q(P)$ of P so that $N(H) \cap Q(P)$ is a subfield of P.

To every element a in A there exists one and only one endomorphism a^{α} of A with the following properties:

$$pa^{\alpha} = a, \; Ha^{\alpha} = 0.$$

It is fairly clear that α is a one-to-one and additive mapping of A upon $N(H)$ [since an endomorphism of A which annihilates the hyperplane H is completely determined once we know its value on p and hence on P]. The equation $S^{\alpha} = Q(S) \cap N(H)$ for subspaces S of A is an almost immediate consequence of $A^{\alpha} = N(H)$. If x is in F and a in A, then we find

$$p(xa)^{\alpha} = xa = x(pa^{\alpha}) = (xp)a^{\alpha} = (px^{\nu})a^{\alpha} = p(x^{\nu}a^{\alpha}) \text{ or } (xa)^{\alpha} = x^{\nu}a^{\alpha}.$$

If b is an element in A^*, then b is a linear form over A; and a single-valued mapping b^{α^*} of A into P is defined by the rule:

$$ab^{\alpha^*} = (ab)p \text{ for every } a \text{ in } A.$$

That b^{α^*} is actually an endomorphism in $Q(P)$, is an immediate consequence of $(a' + a'')b = a'b + a''b$ and $(xa)b = x(ab)$. Likewise one sees easily that α^* is a one-to-one and additive mapping of A^* into $Q(P)$. If η is an endomorphism in $Q(P)$, then there exists to every a in A one and only one a' in F such that $a\eta = a'p$; and it is fairly easy to see that the mapping of a upon a' constitutes a linear form b over A such that $b^{\alpha^*} = \eta$. Thus α^* is an isomorphism of the additive group of A^* upon the subgroup $Q(P)$ of the ring P. Next we verify:

$$p(ab)^{\nu} = (ab)p = ab^{\alpha^*} = (pa^{\alpha})b^{\alpha^*} = p(a^{\alpha}b^{\alpha^*}) \text{ or } (ab)^{\nu} = a^{\alpha}b^{\alpha^*};$$

$$a(bx)^{\alpha^*} = [a(bx)]p = (ab)(xp) = (ab)(px^{\nu}) = [(ab)p]x^{\nu} = (ab^{\alpha^*})x^{\nu} = a(b^{\alpha^*}x^{\nu})$$

for every a in A, b in A^* and x in F so that we have also $(bx)^{\alpha^*} = b^{\alpha^*}x^{\nu}$; and we have shown that our isomorphisms meet all the requirements (a). To complete the proof of property (b) we note the equivalence of the following properties of an element b in A^*:

b belongs to $E(S)$; $0 = Sb$; $0 = (Sb)p = Sb^{\alpha^*}$; b^{α^*} belongs to $N(S)$.

Now the validity of $E(S)^{\alpha^*} = N(S) \cap Q(P)$ is an almost immediate consequence of $Q(P) = A^{*\alpha^*}$. This completes the proof.

REMARK 4: Our proposition asserts the following facts:

The pair (ν,α) is a semi-linear transformation of the linear manifold (F,A) upon the linear manifold $[N(H) \cap Q(P),N(H)]$;

The pair (α^*, ν) is a semi-linear transformation of the linear manifold (A^*,F) upon the linear manifold $[Q(P),N(H) \cap Q(P)]$;

and combining these facts with the third equation (a) we may feel justified to say [in an obvious extension of terminology] that

The triplet (α^*,ν,α) is a semi-linear transformation of the linear configuration (A^*,F,A) upon the linear configuration

$$[Q(P),N(H) \cap Q(P),N(H)].$$

V.3. The Finitely Generated Ideals

An internal characterization of the annulets is contained in the following result.

Theorem 1: *An ideal in* P *is an annulet if, and only if, it is finitely generated.*

The proof will be effected in several steps which will give slightly sharper results than those enunciated in the theorem.

Proposition 1: *Every annulet is generated by one of its idempotents.*

PROOF: If J is a right annulet, then we have to prove the existence of an idempotent σ such that $J = \sigma P$; and if J is a left annulet, then we have to show the existence of an idempotent σ such that $J = P\sigma$.

Suppose first that J is a left annulet. Then there exists one and only one subspace S of (F,A) such that $J = Q(S)$ [V.2, Theorem A]. There exists an idempotent σ such that $S = A\sigma$, since every subspace of A is a direct summand [Complementation Principle]. Naturally σ belongs to $Q(S) = J$. From $AP\sigma \leqslant A\sigma = S$ we infer that $P\sigma \leqslant Q(S)$. If conversely τ belongs to $Q(S)$, then $A\tau \leqslant S$. Since σ leaves invariant every element in $S = A\sigma$, it follows that $a\tau = a\tau\sigma$ for every a, as $a\tau$ is in S; and thus we have $\tau = \tau\sigma$ so that τ belongs to $P\sigma$. Hence $P\sigma = Q(S) = J$ as we wanted to show.

Suppose next that J is a right annulet. Then there exists one and only one subspace T of A such that $J = N(T)$ [V.2, Theorem A]. There exists an idempotent endomorphism σ such that $T = K(\sigma)$, since every subspace is a direct summand. Naturally σ belongs to $N(T)$; and since $N(T)$ is a right-ideal we have $\sigma P \leqslant N(T)$. Suppose now that τ is in $N(T)$. If a is an element in A, then we have $a = a\sigma + a(1 - \sigma)$ and $a(1 - \sigma)$ belongs to $K(\sigma) = T$, since σ is idempotent. Hence $a(1 - \sigma)\tau = 0$ so that $a\tau = a\sigma\tau$ or $\tau = \sigma\tau$. Consequently τ belongs to σP or $\sigma P = N(T) = J$; and this completes the proof.

Lemma 1: *To every endomorphism σ of A there exists an endomorphism η of A with the following properties:*

(*a*) *$\sigma\eta$ and $\eta\sigma$ are both idempotent.*

(*b*) $A\sigma = A\eta\sigma,\quad K(\eta) = K(\eta\sigma),\quad K(\sigma\eta) = K(\sigma),\quad A\sigma\eta = A\eta.$

PROOF: Since $A\sigma$ and $K(\sigma)$ are subspaces of A, there exist subspaces S and T such that

$$A = A\sigma \oplus S = T \oplus K(\sigma)$$

[Complementation Principle]. If x and y are elements in T such that $x\sigma = y\sigma$, then $x - y$ belongs to the intersection 0 of T and $K(\sigma)$ so that $x = y$. Consequently σ induces a one-to-one and linear transformation of T upon $T\sigma = T\sigma + K(\sigma)\sigma = A\sigma$. Such a transformation possesses an inverse transformation τ; and we note that τ is a one-to-one and linear transformation of $A\sigma$ upon T which satisfies $x\sigma\tau = x$ for x in T and $y\tau\sigma = y$ for y in $A\sigma$. Since $A = A\sigma \oplus S$, there exists one and only one endomorphism η of A which induces τ in $A\sigma$ and which annihilates S. Thus for every a in T and s in S we have

$$(a\sigma + s)\eta = a\sigma\tau = a;$$

and we note that every element in A has this form $a\sigma + s$, since $T\sigma = A\sigma$. This formula shows already that $K(\eta) = S$ and $A\eta = T$. If we specialize the above formula by letting $s = 0$, then we find that

$$a(\sigma\eta) = a \text{ and } (a\sigma)\eta\sigma = a\sigma \text{ for } a \text{ in } T.$$

The first of these equations implies that $\sigma\eta$ is the idempotent endomorphism such that $A\sigma\eta = T$, $K(\sigma\eta) = K(\sigma)$; and from the second equation one deduces that $\eta\sigma$ is the idempotent endomorphism with $A\eta\sigma = T\sigma = A\sigma$, $K(\eta\sigma) = S$, if we only recall that certainly $S \leqslant K(\eta\sigma)$ and $K(\sigma) \leqslant K(\sigma\eta)$.

Proposition 2: *To every endomorphism σ of the linear manifold (F,A) there exists an endomorphism σ' such that $\sigma = \sigma\sigma'\sigma$.*

Rings with the property expressed in this proposition are often referred to as *regular* rings.

PROOF: According to Lemma 1 there exists an endomorphism σ' such that $\sigma'\sigma$ is idempotent and $A\sigma = A\sigma'\sigma$. It follows that every element in $A\sigma'\sigma = A\sigma$ is left invariant by the idempotent $\sigma'\sigma$. Hence we have $a\sigma = a\sigma\sigma'\sigma$ for every a in A or $\sigma = \sigma\sigma'\sigma$.

Proposition 3: *Sums of finitely many annulets are annulets.*

PROOF: Clearly it suffices to prove that the sum of two annulets is an annulet. Thus assume first that J' and J'' are left annulets. We infer from Proposition 1 the existence of idempotents σ' and σ'' such that $J' = \mathrm{P}\sigma'$ and $J'' = \mathrm{P}\sigma''$. Applying the Complementation Principle twice we find subspaces S and T such that

$$A\sigma'' = (A\sigma' \cap A\sigma'') \oplus S, \qquad A = (A\sigma' + A\sigma'') \oplus T,$$

and consequently

$$A\sigma' + A\sigma'' = A\sigma' \oplus S, \qquad A = A\sigma' \oplus S \oplus T.$$

Naturally there exist idempotent endomorphisms η', η'' with the following properties:

$$A\eta' = A\sigma', K(\eta') = S + T; A\eta'' = S, K(\eta'') = A\sigma' + T.$$

One verifies without difficulty the following relations between these idempotents:

$$\eta'\eta'' = \eta''\eta' = 0, \eta' = \eta'\sigma', \sigma' = \sigma'\eta', \eta'' = \eta''\sigma''.$$

It follows that η' is in J', that η'' is in J''; and that therefore $\eta = \eta' + \eta''$ belongs to $J' + J''$. One verifies next by direct computation that the sum η of two orthogonal idempotents is itself an idempotent; and $\mathrm{P}\eta$ contains certainly $\sigma' = \sigma'\eta' = \sigma'\eta'\eta$. From

$$A\sigma'' \leqslant A\sigma' + A\sigma'' = A\sigma'' \oplus S = A\eta' \oplus A\eta'' = A\eta$$

we deduce that $\sigma'' = \sigma''\eta$, since η is an idempotent. Consequently σ', σ'' both belong to $P\eta$. Thus we have shown that $J' + J'' = P\eta$ where η is idempotent so that $P\eta = L(1 - \eta)$. Thus sums of two [and hence of finitely many] left annulets are left annulets.

Assume next that J' and J'' are right annulets. Then we deduce from Proposition 1 the existence of idempotents σ', σ'' such that $J' = \sigma' P$ and $J'' = \sigma'' P$. We note that $K(\sigma', \sigma'') = K(\sigma') \cap K(\sigma'')$; and we deduce from the Complementation Principle the existence of subspaces S', S'', T such that

$$K(\sigma') = K(\sigma',\sigma'') \oplus S',\ K(\sigma'') = K(\sigma',\sigma'') \oplus S'',\ A = T \oplus [K(\sigma') + K(\sigma'')];$$

and it follows that

$$K(\sigma') + K(\sigma'') = S' \oplus S'' \oplus K(\sigma',\sigma''),\ A = T \oplus S' \oplus S'' \oplus K(\sigma',\sigma'').$$

Clearly there exist idempotent endomorphisms η', η'' such that

$$A\eta' = S'',\ K(\eta') = T + K(\sigma'),\ A\eta'' = S' + T,\ K(\eta'') = K(\sigma'').$$

From $A\eta' \leqslant K(\eta'')$ and $A\eta'' \leqslant K(\eta')$ we deduce that $\eta'\eta'' = \eta''\eta' = 0$ so that the sum $\eta = \eta' + \eta''$ of these orthogonal idempotents is itself an idempotent. From $A = A\eta' \oplus A\eta'' \oplus K(\sigma', \sigma'')$ we deduce now easily that $K(\eta) = K(\sigma', \sigma'')$. We note that for every idempotent σ we have $\sigma P = R(1 - \sigma)$. Thus we may apply V.2, Theorem A and find that $\sigma P = N[K(\sigma P)] = N[K(\sigma)]$. Consequently σP consists of all those endomorphisms $\varkappa$ which satisfy $K(\sigma) \leqslant K(\varkappa)$. From this remark and the inequalities

$$K(\eta) \leqslant K(\sigma') \leqslant K(\eta'),\ K(\eta) \leqslant K(\sigma'') \leqslant K(\eta'')$$

we deduce now that σ' and σ'' belong to ηP, that η' belongs to $J' = \sigma' P$ and that η'' belongs to $J'' = \sigma'' P$. Consequently $\eta = \eta' + \eta''$ belongs to $J' + J'' = \sigma' P + \sigma'' P = \eta P$; and this shows that $J' + J'' = \eta P = R(1 - \eta)$. Thus sums of two [and hence of finitely many] right annulets are right annulets.

Proof of Theorem 1: It is a consequence of Proposition 1 that annulets are generated by one of their idempotents and so a fortiori they are finitely generated. Assume conversely that the left ideal J is finitely generated. Then there exist finitely many elements $j_1, \cdots, j_n$ such that J is the smallest left ideal containing these elements. Denote by J_i the smallest left ideal containing j_i. Then $J = \sum_{i=1}^{n} J_i$. We deduce from Proposition 2 the existence of elements j_i' such that $j_i = j_i j_i' j_i$. Clearly $\sigma_i = j_i' j_i$ is an idempotent belonging to J_i; and the element $j_i = j_i \sigma_i$ belongs to $P\sigma_i$. Hence $J_i = P\sigma_i$. Since σ_i is an idempotent, we have $J_i = P\sigma_i = L(1 - \sigma_i)$ so that J_i is a left annulet. But now it follows from Proposition 3 that J as the sum of finitely many left annulets is itself a left annulet. Similarly we prove that finitely generated right ideals are right annulets.

Corollary 1: *AJ is a subspace of A for every left ideal J in* Ρ.

PROOF: Any two elements in AJ have the form $a'j', a''j''$ with a', a'' in A and j', j'' in J. Denote by H the left ideal generated by j', j''. Clearly $H \leqslant J$; and it follows from Theorem 1 that the finitely generated left ideal H is a left-annulet. We deduce from V.2, Theorem B that AH is a subspace of A. Since $a'j'$ and $a''j''$ are both contained in the subspace AH of A, their sum and difference $a'j' \pm a''j''$ likewise belong to $AH \leqslant AJ$; and this proves that AJ is closed under addition and subtraction. It is clear that $AJ = FAJ$. Hence AJ is a subspace of A.

Corollary 2: (*a*) *If J is a left ideal in* Ρ *such that AJ has finite rank, then $J = Q(AJ)$ is a left annulet.*

(*b*) *If J is a right ideal in* Ρ *such that $A/K(J)$ has finite rank, then $J = N[K(J)]$ is a right annulet.*

PROOF: Assume first that J is a left ideal such that AJ has finite rank—note that AJ is a subspace by Corollary 1. Since AJ is finitely generated, there exist finitely many endomorphisms $\sigma_1, \cdots, \sigma_n$ in J such that $AJ = \sum_{i=1}^{n} A\sigma_i$. Denote by J^* the left ideal generated by $\sigma_1, \cdots, \sigma_n$. Then $AJ = AJ^*$ and J^* is a left annulet, since J^* is finitely generated [Theorem 1]. Now we deduce from V.2, Theorem B that $J^* = Q(AJ^*)$; and this implies clearly

$$J \leqslant Q(AJ) = Q(AJ^*) = J^* \leqslant J$$

so that $J = Q(AJ)$ is a left annulet.

Assume next that J is a right ideal such that $A/K(J)$ has finite rank. Among all the right annulets contained in J there is one, say J^*, such that $K(J^*)/K(J)$ has minimal rank [note that $J^* \leqslant J$ implies always $K(J) \leqslant K(J^*)$]. If $K(J^*)$ were different from $K(J)$, then there would exist an element w in $K(J^*)$, but not in $K(J)$. By definition of $K(J)$ there exists at least one endomorphism σ in J such that $w\sigma \neq 0$. It follows from Theorem 1 that the right ideal σ^* generated by σ is a right annulet; and it follows from Proposition 3 that $J^* + \sigma^*$ is a right annulet. It is clear that $K(J^* + \sigma^*) < K(J^*)$, since w cannot be in $K(J^* + \sigma^*)$; and thus $K(J^* + \sigma^*)/K(J)$ would have smaller rank than $K(J^*)/K(J)$ contradicting our choice of J^*. Consequently $K(J) = K(J^*)$. Since J^* is a right annulet we deduce from V.2, Theorem A that $J^* = N[K(J^*)]$; and this implies clearly [since $K(J)J = 0$] that

$$J \leqslant N[K(J)] = N[K(J^*)] = J^* \leqslant J$$

so that $J = N[K(J)]$ is a right annulet.

Theorem 2: *The following properties of the linear manifold (F,A) are equivalent.*

(I) $r(A)$ *is finite.*

(II) *Every left ideal in* P *is a left annulet.*

(III) *Every right ideal in* P *is a right annulet.*

PROOF: It is an immediate consequence of Corollary 2 that the properties (II) and (III) are consequences of (I). If $r(A)$ is infinite, then we consider the totality Φ of endomorphisms σ of finite rank, i.e. σ belongs to Φ if, and only if, $A\sigma$ has finite rank. The reader will verify without difficulty that Φ is both a right and a left ideal in P, that $A = A\Phi$ and $0 = K(\Phi)$ and that the identity 1 does not belong to Φ. Consequently

$$\begin{aligned} \Phi < \mathrm{P} &= Q(A) = Q(A\Phi) \\ &= N(0) = N[K(\Phi)]; \end{aligned}$$

and it follows from V.2, Theorems A and B that Φ is neither a left nor a right annulet. Hence (I) is a consequence of (II) as well as a consequence of (III)

REMARK 1: It is a consequence of Theorem 1 that every ideal is a sum of annulets, since every ideal is a sum of finitely generated ideals [even of ideals generated by one element]. In case $r(A)$ is infinite, then not every ideal is an annulet [Theorem 2]. Hence sums of infinitely many annulets need not be annulets, showing the impossibility of omitting in Proposition 3 the finiteness condition; see V.2, Remark 3.

REMARK 2: If $r(A)$ is finite, then a combination of Theorem 2 and of V.2, Theorems A and B shows the following facts:

The system of subspaces of A and the system of right ideals in P are duals of each other.

The system of subspaces of A and the system of left ideals in P are projectively equivalent.

V.4. The Isomorphisms of the Endomorphism Ring

If the linear manifold (F,A) has rank 1, then one sees easily that its endomorphism ring P is isomorphic with the field F. Thus we may make the hypothesis in this and the following section that all the linear manifolds under consideration shall have rank not less than 2; and sometimes we shall have to assume that the rank be at least 3. The reader is advised to check which of our theorems cease to be true without these restrictions.

Consider now linear manifolds (F,A) and (G,B) [which may be equal or different]. Each of them leads us to a certain endomorphism ring

$\mathrm{P}(F,A)$ and $\mathrm{P}(G,B)$ respectively; and the question arises to relate the structures of the manifolds and the structure of their rings.

Structure Theorem: *The linear manifolds (F,A) and (G,B) have the same structure if, and only if, their rings $\mathrm{P}(F,A)$ and $\mathrm{P}(G,B)$ have the same structure.*

This signifies that there exists a semi-linear transformation of (F,A) upon (G,B) if, and only if, there exists an isomorphism of $\mathrm{P}(F,A)$ upon $\mathrm{P}(G,B)$ [and ring isomorphisms are naturally one-to-one correspondences which preserve addition and multiplication]. The proof of this Structure Theorem will be obtained as an incidental result in our subsequent discussion in which we want to obtain a closer relation between semi-linear transformations and ring isomorphisms. With this in mind we make first the following simple remark. Suppose that σ is a semi-linear transformation of (F,A) upon (G,B). Then σ^{-1} is a semi-linear transformation of (G,B) upon (F,A). If η is an endomorphism of A, then the element x in B is mapped by σ^{-1} upon the element $x^{\sigma^{-1}}$ in A which latter element is mapped by η upon the element $x^{\sigma^{-1}}\eta$ in A; and the last element is mapped by σ upon the element $(x^{\sigma^{-1}}\eta)^{\sigma}$ in B. The product

$$\eta^{\sigma} = \sigma^{-1}\eta\sigma$$

of these three mappings maps the element x in B upon the element $(x^{\sigma^{-1}}\eta)^{\sigma}$ in B. Now one verifies easily that η^{σ} is an endomorphism of B and that mapping η in $\mathrm{P}(F,A)$ upon η^{σ} in $\mathrm{P}(G,B)$ constitutes an isomorphism of the ring $\mathrm{P}(F,A)$ upon the ring $\mathrm{P}(G,B)$. We shall refer to this ring isomorphism as to *the ring isomorphism induced by the semi-linear transformation* σ. This remark shows among other things the necessity of the condition of the Structure Theorem; and the sufficiency of this condition is clearly contained in the following result.

Theorem 1: *Every isomorphism of $\mathrm{P}(F,A)$ upon $\mathrm{P}(G,B)$ is induced by a semi-linear transformation of (F,A) upon (G,B); and two semi-linear transformations induce the same ring isomorphism if, and only if, they induce the same projectivity of A upon B.*

The proof of this fundamental theorem will be preceded by the proof of some lemmas.

Lemma 1: *The following properties of the semi-linear transformation τ of the linear manifold (F,A) upon itself are equivalent.*

(I) $\tau\sigma = \sigma\tau$ *for every endomorphism σ of (F,A).*

(II) $\tau\sigma = \sigma\tau$ *for every idempotent σ in $\mathrm{P}(F,A)$.*

(III) $S^{\tau} = S$ *for every subspace S of A.*

(IV) *There exists a number $d \neq 0$ in F such that $x^{\tau} = dx$ for every x in A.*

PROOF: It is trivial that (I) implies (II). Suppose that τ satisfies (II). If S is a subspace of A, then there exists an idempotent endomorphism σ of A

such that $S = A\sigma$. If s is an element in S, then $s = s\sigma$; and it follows from (II) that

$$s^{\tau} = (s\sigma)^{\tau} = (s^{\tau})\sigma$$

belongs to $A\sigma = S$. Hence $S^{\tau} \leqslant S$; and from this fact one deduces [in a variety of ways] the validity of (III). That (III) implies (IV), is a consequence of III.1, Proposition 3, since $1 < r(A)$. If finally (IV) is true, then we have for every endomorphism σ

$$(x^{\tau})\sigma = (dx)\sigma = d(x\sigma) = (x\sigma)^{\tau} \text{ or } \tau\sigma = \sigma\tau;$$

and this completes the proof.

Lemma 2: *If the automorphism α of the endomorphism ring $\mathrm{P}(F,A)$ leaves invariant every idempotent, then $\alpha = 1$.*

PROOF: If σ is an endomorphism in P, then the left ideal generated by σ shall be denoted by σ_L; and the right ideal generated by σ we denote by σ_R. It follows from V.3, Theorem 1 and V.3, Proposition 1 that σ_L and σ_R are generated by idempotents. These are left invariant by α; and thus we find that

$$\sigma_L = (\sigma^{\alpha})_L, \; \sigma_R = (\sigma^{\alpha})_R.$$

From these equations we deduce directly that

$$K(\sigma) = K(\sigma_R) = K[(\sigma^{\alpha})_R] = K(\sigma^{\alpha}), \quad A\sigma = A\sigma_L = A(\sigma^{\alpha})_L = A\sigma^{\alpha}.$$

Suppose now that x is an element in A, σ an endomorphism, and that x and $x\sigma$ are independent elements. Then one constructs easily an idempotent $\varkappa$ which leaves invariant $x\sigma$ and annihilates $x - x\sigma$, since these two elements are independent too. Note

$$0 = x\varkappa - x\sigma\varkappa = x\varkappa - x\sigma \text{ or } x\varkappa = x\sigma.$$

Consequently x belongs to $K(\varkappa - \sigma) = K(\varkappa^{\alpha} - \sigma^{\alpha}) = K(\varkappa - \sigma^{\alpha})$ or $x\sigma = x\varkappa = x\sigma^{\alpha}$.

If the element x and its image $x\sigma$ are not independent, but $x \neq 0$, then we infer from $1 < r(A)$ the existence of an element y independent of x. There exists an idempotent ω which annihilates $x - y$ and leaves invariant y so that x and $x\omega = y\omega = y$ are independent. Since x and $x\sigma$ are dependent elements x and $x\sigma + x\omega = x(\sigma + \omega)$ are independent. Applying the result of the preceding paragraph of our proof and remembering that idempotents are left invariant by α we find

$$x\sigma + x\omega = x(\sigma + \omega) = x(\sigma + \omega)^{\alpha} = x\sigma^{\alpha} + x\omega \text{ or } x\sigma = x\sigma^{\alpha}.$$

Hence $\sigma = \sigma^{\alpha}$ for every σ or $\alpha = 1$.

Note that this lemma fails to be true for $1 = r(A)$, since then P is essentially the same as the field F.

Lemma 3: *If σ is a semi-linear transformation of (F,A) upon (G,B), then $\sigma^{-1}N(S)\sigma = N(S^\sigma)$ and $\sigma^{-1}Q(S)\sigma = Q(S^\sigma)$ for every subspace S of A.*

Proof: The following properties of the endomorphism η of (F,A) are equivalent:

η belongs to $N(S)$; $S\eta = 0$; $0 = (S\eta)^\sigma = S^\sigma(\sigma^{-1}\eta\sigma)$; $\sigma^{-1}\eta\sigma$ belongs to $N(S^\sigma)$.

Hence $\sigma^{-1}N(S)\sigma = N(S^\sigma)$.

Likewise the following properties of η are equivalent:

η belongs to $Q(S)$; $A\eta \leqslant S$; $B(\sigma^{-1}\eta\sigma) = (A\eta)^\sigma \leqslant S^\sigma$; $\sigma^{-1}\eta\sigma$ belongs to $Q(S^\sigma)$.

Hence $\sigma^{-1}Q(S)\sigma = Q(S^\sigma)$.

Lemma 4: *If σ is an isomorphism of $\mathrm{P}(F,A)$ upon $\mathrm{P}(G,B)$, then*

(*a*) $K[N(S)^\sigma] = BQ(S)^\sigma$ $[= S^{\sigma^*}]$ *for every subspace S of A and*

(*b*) σ^* *is a projectivity of A upon B.*

Proof: We note first that $L(J)^\sigma = L(J^\sigma)$ for every subset J of $\mathrm{P}(F,A)$. Applying Propositions 1 to 3 of V.2 we find now that

$$K[N(S)^\sigma] = BQ(K[N(S)^\sigma]) = BL[N(S)^\sigma] = B(L[N(S)])^\sigma$$
$$= BQ(S)^\sigma \qquad \text{for every subspace } S \text{ of } A.$$

That mapping S upon this subspace S^{σ^*} constitutes a projectivity of A upon B, follows from V.2, Theorem B, since according to this result σ^* is the product of the projectivity which maps S upon $Q(S)$, the projectivity which maps $Q(S)$ upon $Q(S)^\sigma$ [induced by the isomorphism σ which maps annulets upon annulets] and the projectivity which maps $Q(S)^\sigma$ upon $BQ(S)^\sigma$.

Proof of Theorem 1: The semi-linear transformations σ' and σ'' of (F,A) upon (G,B) induce the same projectivity if, and only if, $S^{\sigma'} = S^{\sigma''}$ for every subspace S of A. If we let $\sigma = \sigma'\sigma''^{-1}$, then σ' and σ'' induce the same projectivity if, and only if, $S = S^\sigma$ for every subspace S of A. It follows from Lemma 1 that this is equivalent with the property: $\sigma\eta = \eta\sigma$ for every η in $\mathrm{P}(F,A)$. But this last condition is equivalent with

$$\eta^{\sigma'} = \sigma'^{-1}\eta\sigma' = \sigma''^{-1}\sigma^{-1}\eta\sigma\sigma'' = \sigma''^{-1}\eta\sigma'' = \eta^{\sigma''} \text{ for every } \eta \text{ in } \mathrm{P}(F,A).$$

Hence σ' and σ'' induce the same projectivity if, and only if, they induce the same ring isomorphism.

Consider now some isomorphism σ of the ring $\mathrm{P}(F,A)$ upon the ring $\mathrm{P}(G,B)$. We deduce from Lemma 4 that mapping the subspace S of A upon the subspace

$$S^{\sigma^*} = K[N(S)^\sigma] = BQ(S)^\sigma$$

constitutes a projectivity of A upon B.

[If the rank of A would be at least 3, then we could apply now the First Fundamental Theorem of Projective Geometry to obtain a semi-linear transformation which meets our requirements; and the reader is advised to work out the details of such a proof. Since we do not want to assume, however, that $2 < r(A)$, we have to use a different method for constructing the desired transformation.]

There exists a direct decomposition $A = P \oplus H$ where P is a point and H is a hyperplane. Let $P^* = P^{\sigma^*}$ and $H^* = H^{\sigma^*}$. Since σ^* is a projectivity, P^* is a point, H^* a hyperplane and $B = P^* \oplus H^*$. From V.2, Proposition 5 and V.2, Remark 4 we deduce the existence of a semi-linear transformation τ of the linear manifold (F,A) upon the linear manifold $[N(H) \cap Q(P), N(H)]$ satisfying $S^\tau = Q(S) \cap N(H)$ for every subspace S of A and the existence of a semi-linear transformation ν of the linear manifold $[N(H^*) \cap Q(P^*), N(H^*)]$ upon the linear manifold (G,B) satisfying $T^{\nu^{-1}} = Q(T) \cap N(H^*)$ for subspaces T of B. The ring isomorphism σ clearly induces a semi-linear transformation of

$$[N(H) \cap Q(P), N(H)] \text{ upon } [N(H^*) \cap Q(P^*), N(H^*)].$$

Consequently $\omega = \tau\sigma\nu$ is a semi-linear transformation of (F,A) upon (G,B).

As a ring isomorphism σ maps cross cuts upon cross cuts and annulets upon annulets. Hence it follows from V.2, Proposition 1 that

$$\begin{aligned} S^{\tau\sigma} &= [Q(S) \cap N(H)]^\sigma = Q(S)^\sigma \cap N(H)^\sigma = Q[BQ(S)^\sigma] \cap N(K[N(H)^\sigma]) \\ &= Q[S^{\sigma^*}] \cap N[H^{\sigma^*}] = Q(S^{\sigma^*}) \cap N(H^*) = S^{\sigma^*\nu^{-1}} \end{aligned}$$

or $S^{\sigma^*} = S^\omega$ for every subspace S of A.

The semi-linear transformation ω induces a ring isomorphism which we denote by ω too. It follows from Lemma 3 that

$$N(S^\omega) = \omega^{-1}N(S)\omega = N(S)^\omega \quad \text{and} \quad Q(S^\omega) = \omega^{-1}Q(S)\omega = Q(S)^\omega;$$

and now we deduce from V.2, Proposition 4 that

$$\begin{aligned} Q(S)^\sigma &= Q[BQ(S)^\sigma] = Q(S^{\sigma^*}) = Q(S^\omega) = Q(S)^\omega, \\ N(S)^\sigma &= N(K[N(S)^\sigma]) = N(S^{\sigma^*}) = N(S^\omega) = N(S)^\omega. \end{aligned}$$

Let $\alpha = \sigma\omega^{-1}$. Then α is an automorphism of the ring $\Lambda(F,A)$ which leaves invariant every $Q(S)$ and $N(S)$. It follows from V.2, Theorems A and B that α leaves invariant every annulet in $\Lambda(F,A)$. If e is some idempotent in $\Lambda(F,A)$, then e^α is likewise an idempotent. Since $\Lambda e = L(1 - e)$ and $e\Lambda = R(1 - e)$ are annulets, it follows that $\Lambda e = \Lambda e^\alpha$ and $e\Lambda = e^\alpha\Lambda$; and this implies $e = ee^\alpha = e^\alpha$. Consequently every idempotent is left invariant by the automorphism α; and we deduce $\alpha = 1$ from Lemma 2. Consequently $\sigma = \omega$; and we have shown that the ring isomorphism σ is induced by the semi-linear transformation ω. This completes the proof of Theorem 1.

Note that we have shown at the same time the following fact.

Corollary 1: *The identity is the only automorphism of* $\mathrm{P}(F,A)$ *which leaves invariant every* $Q(S)$ [*which leaves invariant every* $N(S)$].

Theorem 2: *If* $3 \leqslant r(A)$, *then the group of automorphisms of* $\mathrm{P}(F,A)$ *is essentially the same as the group of projectivities of* A; *and the collineations of* A *correspond under this* [*natural*] *isomorphism to the inner automorphisms of* P.

Proof: If σ is an automorphism of the ring $\mathrm{P}(F,A)$, then we let

$$S^{\sigma^*} = AQ(S)^{\sigma} \text{ for every subspace } S \text{ of } A;$$

and it follows from Lemma 4 that σ^* is, for every automorphism σ of P, an autoprojectivity of A. It is clear that we obtain in this fashion the [natural] homomorphism of the group of automorphisms of P into the group of auto-projectivities of A.

Assume now that $\sigma^* = 1$ for some automorphism σ of P. We infer from Theorem 1 the existence of a semi-linear transformation τ of A which induces the automorphism σ. It follows from Lemma 3 that

$$Q(S)^{\sigma} = Q(S^{\tau}) \text{ for every subspace } S \text{ of } A;$$

and now we deduce from V.2, Proposition 1 that

$$S = S^{\sigma^*} = AQ(S)^{\sigma} = AQ(S^{\tau}) = S^{\tau} \text{ for every subspace } S \text{ of } A.$$

It follows from Lemma 1 that τ induces the identity automorphism in P. Hence $\sigma = 1$ is a consequence of $\sigma^* = 1$ so that the mapping $\sigma \rightarrow \sigma^*$ is an isomorphism.

Assume next that σ^* is a collineation. Then we show as before that there exists a semi-linear transformation τ which induces σ and satisfies

$$S^{\sigma^*} = S^{\tau} \text{ for every subspace } S \text{ of } A.$$

Since σ^* is a collineation, there exists a linear transformation ν, necessarily in P, such that

$$S^{\sigma^*} = S^{\nu} \text{ for every subspace } S \text{ of } A.$$

Since ν and τ induce the same auto-projectivity, it follows from Theorem 1 that they induce the same automorphism of P. Since τ induces σ, so does ν and this shows that σ is the inner automorphism of P induced by ν. The converse of this statement is obvious so that we may say:

(2.1) σ^* *is a collineation if, and only if,* σ *is an inner automorphism of* P.

Assume now that ω is an auto-projectivity of A. Since $2 < r(A)$, we may apply the First Fundamental Theorem of Projective Geometry.

Consequently there exists a semi-linear transformation τ of A which induces ω. This transformation τ induces an automorphism β of P such that $Q(S)^\beta = Q(S^\tau) = Q(S^\omega)$ for every subspace S of A. It follows now as usual that

$$S^{\beta *} = AQ(S)^\beta = AQ(S^\omega) = S^\omega \text{ for every subspace } S \text{ of } A$$

or $\beta^* = \omega$; and this completes the proof of our theorem.

REMARK 1: It should be observed that the hypothesis $2 < r(A)$ has been used only in the last paragraph of the proof. Thus the following proposition is valid under the weaker hypothesis that $2 \leqslant r(A)$:

Mapping the automorphism σ of P upon the auto-projectivity σ^* of A is an isomorphism of the automorphism group of P into the group of auto-projectivities of A; and this isomorphism maps the group of inner automorphisms upon the group of collineations.

Note furthermore that every σ^* is induced by a semi-linear transformation of A. If $r(A) = 2$, then there exist, in general, auto-projectivities which are not so induced; and these latter auto-projectivities can therefore not be represented in the form σ^*.

V.5. The Anti-isomorphisms of the Endomorphism Ring

It will prove useful for our discussion to relate the endomorphism ring of a linear manifold with that of its adjoint space. Thus suppose that (F,A) is a linear manifold and (A^*,F) its adjoint space [using the notation which we introduced towards the end of V.2]. Since the elements in F are right-multipliers for the elements in A^*, we shall denote the effect of the endomorphism η of A^* upon the element x in A^* by ηx [so that the formula $(\eta x)y = \eta(xy)$ holds for x in A^* and y in F] and we shall define the product of endomorphisms by the rule:

$$(\eta\varkappa)x = \eta(\varkappa x) \text{ for } x \text{ in } A^* \text{ and } \eta,\varkappa \text{ endomorphisms of } A^*.$$

Defining addition as usual we obtain in this fashion *the endomorphism ring* $\text{P}(A^*,F)$ of the adjoint space (A^*,F); and we may apply, mutatis mutandis, all the results that we have obtained for endomorphism rings of linear manifolds.

Proposition 1: *There exists one and only one isomorphism* $\eta \rightarrow \eta^*$ *of* $\text{P}(F,A)$ *into* $\text{P}(A^*,F)$ *with the property*:

(*) $\quad a(\eta^* b) = (a\eta)b$ *for* α *in* A, b *in* A^* *and* η *in* $\text{P}(F,A)$.

We shall refer to this isomorphism as to *the natural isomorphism of* $\text{P}(F,A)$ *into* $\text{P}(A^*,F)$.

PROOF: Assume first that η' and η'' are endomorphisms of (A^*,F) such that $a(\eta'b) = a(\eta''b)$ for every a in A and b in A^*. Then $\eta' - \eta''$ is an endomorphism of A^* such that $0 = a[(\eta' - \eta'')b]$ for every a in A and b in A^*. This implies first that $0 = (\eta' - \eta'')b$ for every b in A^* and next that $\eta' - \eta'' = 0$ or $\eta' = \eta''$. Now it is obvious that there exists at most one isomorphism with the desired properties.

If η is an endomorphism of A, b an element in A^*, then b is a linear form over A; and mapping the element a in A upon the element $(a\eta)b$ in F clearly constitutes a linear form over A. As this linear form depends on η and b, we may denote it by η^*b; and a,b,η and η^*b are connected by the formula (*). One verifies without difficulty that mapping b onto η^*b constitutes an endomorphism of A^* which we naturally denote by η^*; and that the mapping of η upon η^* is single-valued and preserves addition and multiplication. If $\eta^* = 0$, then $(a\eta)b = a(\eta^*b) = 0$ for every a in A and b in A^*; and this implies first that $a\eta = 0$ for every a in A and next that $\eta = 0$ [since $xA^* = 0$ implies $x = 0$ by II.3, Proposition 2]. This shows that mapping η onto η^* constitutes the desired isomorphism of $\mathrm{P}(F,A)$ into $\mathrm{P}(A^*,F)$.

Proposition 2: *The following properties of the linear manifold* (F,A) *are equivalent.*

(I) *The natural isomorphism maps* $\mathrm{P}(F,A)$ *upon* $\mathrm{P}(A^*,F)$.

(II) $\mathrm{P}(F,A)$ *and* $\mathrm{P}(A^*,F)$ *are isomorphic.*

(III) $r(A)$ *is finite.*

Proof: It is clear that (I) implies (II). Assume next the existence of an isomorphism σ of $\mathrm{P}(F,A)$ upon $\mathrm{P}(A^*,F)$. Mapping the subspace S of A upon $Q(S)$ constitutes a projectivity of the system of subspaces of A upon the totality of left annulets of $\mathrm{P}(F,A)$ [V.2, Theorem B]. Since σ is an isomorphism, mapping $Q(S)$ upon $Q(S)^\sigma$ constitutes a projectivity of the totality of left annulets of $\mathrm{P}(F,A)$ upon the totality of left annulets of $\mathrm{P}(A^*,F)$. When applying the three cornered Galois Theory of V.2 upon the ring $\mathrm{P}(A^*,F)$ of (A^*,F), we have to interchange right and left. Thus it follows from V.2, Theorem A that mapping $Q(S)^\sigma$ upon $N[Q(S)^\sigma]$ constitutes a duality of the totality of left annulets in $\mathrm{P}(A^*,F)$ upon the totality of subspaces of A^*.Mapping the subspace S of A upon the subspace $N[Q(S)^\sigma]$ of A^* constitutes therefore a duality of A upon A^*. Application of IV.1, Existence Theorem shows the finiteness of $r(A)$; and we have shown that (III) is a consequence of (II).

Assume now the validity of (III). Consider some endomorphism $\varkappa$ of A^*. If a is some definite element in A, then mapping the element b in A^* upon the number $a(\varkappa b)$ in F constitutes a definite linear form a' over A^*, as may be seen by direct verification, since every $\varkappa b$ is a linear form over A. We note that $a,b,\varkappa$ and a' are connected by the formula

$$a'b = a(\varkappa b).$$

It follows now from II.3, Theorem 2 and the assumed finiteness of $r(A)$ that there exists to every linear form over A^* an element in A which induces it. Consequently we can find an element a'' in A such that $a''b = a'b$ for every b in A^*. Thus we have found for every element a in A an element a'' in A such that

$$a''b = a(\varkappa b) \text{ for every } b \text{ in } A^*.$$

One verifies easily that a'' is uniquely determined by a and the preceding equation [since $ub = vb$ for every b in A^* implies $u = v$]; and now it is easy to check that mapping a upon a'' constitutes an endomorphism η of A. This endomorphism η satisfies $(a\eta)b = a(\varkappa b)$ for every a in A and b in A^*; and now we deduce from the definition of the natural isomorphism that $\eta^* = \varkappa$. Thus the natural isomorphism maps $\mathrm{P}(F,A)$ upon the whole of $\mathrm{P}(A^*,F)$. This shows that (I) is a consequence of (III) and completes the proof.

Before taking up the principal objective of this section some genera remarks on the formation of certain concepts are in order. If a certain class T of mappings has among other properties that of preserving a certain multiplication, then the mappings that have all the properties of the T-mappings except that they invert the order of multiplication are customarily called anti-T-mappings. The anti-isomorphisms between fields are a case in point and more generally the anti-isomorphisms between rings which are one-to-one correspondences σ satisfying

$$(x + y)^\sigma = x^\sigma + y^\sigma, \ (xy)^\sigma = y^\sigma x^\sigma.$$

We shall need still another concept of the same type. If (F,A) and (G,B) are linear manifolds, (A^*,F) the adjoint space of (F,A), then *an anti-semi-linear transformation of* (A^*,F) *upon* (G,B) is a pair consisting of an anti-isomorphism σ' of the field F upon the field G and of an isomorphism σ'' of the additive group A^* upon the group B subject to the following requirements:

$$(dx)^{\sigma''} = x^{\sigma'} d^{\sigma''} \text{ for } x \text{ in } F \text{ and } d \text{ in } A^*.$$

The inverse of an anti-semi-linear transformation we shall consider likewise as an anti-semi-linear transformation; and whenever we speak of a certain anti-semi-linear transformation σ, this shall be understood as signifying that the two components of σ are both denoted by σ too.

An example of an anti-semi-linear transformation is provided by the identity mapping of a linear manifold upon its canonical dual space [IV.1, Lemma 2]. Thus every linear manifold possesses anti-semi-linear transformations, showing that the finiteness hypotheses imposed below are indispensable.

Lemma 1: *If (F,A) has finite rank, then there exists to every anti-semi-linear transformation σ of (A^*,F) upon (G,B) one and only one anti-isomorphism σ^* of $P(F,A)$ upon $P(G,B)$ with the following properties.*

(*a*) $(\eta^*d)^\sigma = d^\sigma\eta^{\sigma^*}$ *for η in $P(F,A)$ and d in A^* [where the asterisk indicates the natural isomorphism of Proposition 1].*

(*b*) $N[E(S)^\sigma] = Q(S)^{\sigma^*}$ *and* $Q[E(S)^\sigma] = N(S)^{\sigma^*}$ *for every subspace S of A.*

This anti-isomorphism σ^* will be called the anti-isomorphism *induced* by σ.

Proof: If b is an element in B, then $b^{\sigma^{-1}}$ is an element in A^*, i.e. a linear form over A. If η is an endomorphism of (F,A), then the natural isomorphism maps η upon an endomorphism η^* of (A^*,F) [Proposition 1] so that $\eta^*b^{\sigma^{-1}}$ is a well determined element in A^* for every b in B. We apply σ again and have mapped b upon the element $(\eta^*b^{\sigma^{-1}})^\sigma$ of B. This mapping of B into itself we may denote by η^{σ^*}. The defining equation for this mapping is

$$b\eta^{\sigma^*} = (\eta^*b^{\sigma^{-1}})^\sigma \text{ for every } b \text{ in } B.$$

The mapping of η onto η^{σ^*} is clearly the sequence of the following mappings: first the mapping σ^{-1}, then application of η^* upon the result and finally application of σ. Thus we have

$$\eta^{\sigma^*} = \sigma^{-1}\eta^*\sigma.$$

It follows from Proposition 2 and the finiteness of $r(A)$ that the natural isomorphism is an isomorphism of $P(F,A)$ upon $P(A^*,F)$; and now one sees easily that the mapping of η upon η^{σ^*} constitutes an isomorphism of $P(F,A)$ upon $P(G,B)$ with Property (*a*). To check the validity of (*b*) we just notice the equivalence of the following properties of the endomorphism η of A:

$$\eta \text{ belongs to } Q(S);\ A\eta \leqslant S;\ A[\eta^*E(S)] = (A\eta)E(S) = 0;\ \eta^*E(S) = 0;$$
$$0 = [\eta^*E(S)]^\sigma = E(S)^\sigma\eta^{\sigma^*};\ \eta^{\sigma^*} \text{ belongs to } N[E(S)^\sigma].$$

Hence $N[E(S)^\sigma] = Q(S)^{\sigma^*}$. Likewise we deduce $Q[E(S)^\sigma] = N(S)^{\sigma^*}$ from the following set of equivalences:

$$\eta \text{ belongs to } N(S);\ 0 = S\eta;\ 0 = (S\eta)A^* = S(\eta^*A^*);\ \eta^*A^* \leqslant E(S);$$
$$B\eta^{\sigma^*} = (\eta^*A^*)^\sigma \leqslant E(S)^\sigma;\ \eta^{\sigma^*} \text{ belongs to } Q[E(S)^\sigma].$$

This completes the proof.

Lemma 2: *If (F,A) and (G,B) are linear manifolds, and if σ is an anti-isomorphism of the ring $P(F,A)$ upon $P(G,B)$, then*

(*a*) $K[Q(S)^\sigma] = BN(S)^\sigma\ [= S^{\sigma^*}]$ *for every subspace S of A and*

(*b*) σ^* *is a duality of A upon B.*

This duality σ^* is called the duality *induced* by the anti-isomorphism σ.

PROOF: We note first that σ interchanges right and left annulets so that in particular $R(J)^\sigma = L(J^\sigma)$ for every subset J of $\Gamma(F,A)$. Applying Propositions 1 to 4 of V.2 we find that

$$BN(S)^\sigma = B(R[Q(S)])^\sigma = BL[Q(S)^\sigma] = BQ[K(Q(S)^\sigma)] = K[Q(S)^\sigma]$$

for every subspace S of A. Next we apply V.2, Theorems A and B. Consequently a duality is effected by mapping S upon $N(S)$; mapping $N(S)$ upon $N(S)^\sigma$ constitutes a projectivity of the system of right annulets in $\Gamma(F,A)$ upon the system of left annulets in $\Gamma(G,B)$; and mapping $N(S)^\sigma$ upon $BN(S)^\sigma$ is a projectivity. Thus σ^* is a duality as the product of a duality and two projectivities.

Existence Theorem: *There exists an anti-isomorphism of* $\Gamma(F,A)$ *upon* $\Gamma(G,B)$ *if, and only if,* $r(A)$ *is finite and there exists an anti-semi-linear transformation of* (A^*,F) *upon* (G,B).

The sufficiency of these conditions is contained in Lemma 1. If there exists an anti-isomorphism of $\Gamma(F,A)$ upon $\Gamma(G,B)$, then it follows from Lemma 2, (*b*) that there exists a duality of (F,A) upon (G,B); and this implies the finiteness of $r(A)$ because of IV.1, Existence Theorem. That the existence of a ring-anti-isomorphism implies likewise the existence of anti-semi-linear transformation, will be a consequence of the following sharper proposition.

Theorem 1: *Every anti-isomorphism of the ring* $\Gamma(F,A)$ *upon* $\Gamma(G,B)$ *is induced by an anti-semi-linear transformation of* (A^*,F) *upon* (G,B).

Here we make again the hypothesis $1 < r(A)$; and we say in accordance with Lemma 1, (*a*) that the anti-isomorphism σ of $\Gamma(F,A)$ is induced by the anti-semi-linear transformation τ of (A^*,F), if

$$(\eta^*d)^\tau = d^\tau\eta^\sigma \text{ for } \eta \text{ in } \Gamma(F,A) \text{ and } d \text{ in } A^*$$

where the asterisk indicates again the natural isomorphism of $\Gamma(F,A)$ into $\Gamma(A^*,F)$.

PROOF: Suppose that σ is some anti-isomorphism of the ring $\Gamma(F,A)$ upon $\Gamma(G,B)$. It follows from Lemma 2 that a duality σ^* is defined by mapping the subspace S of A upon the subspace

$$S^{\sigma^*} = BN(S)^\sigma = K[Q(S)^\sigma]$$

of B. The existence of a duality implies finiteness of rank [IV.1, Existence Theorem]. Hence it follows from Proposition 2 that the natural isomorphism $\eta \to \eta^*$ is an isomorphism of $\Gamma(F,A)$ upon the whole ring $\Gamma(A^*,F)$.

There exist direct decompositions $A = P \oplus H$ where P is a point and H a hyperplane. If we let $P^* = P^{\sigma^*}$ and $H^* = H^{\sigma^*}$, then P^* is a hyperplane, H^* a point and $B = P^* \oplus H^*$, since σ^* is a duality. Since σ interchanges right and left annulets, we may deduce from V.2, Theorems A and B and the above equations defining σ^* that

$$N(S)^\sigma = Q(S^{\sigma^*}),\ Q(S)^\sigma = N(S^{\sigma^*}) \text{ for every subspace } S \text{ of } A.$$

Applying V.2, Proposition 5 twice we find a semi-linear transformation σ' of (A^*,F) upon $[Q(P),N(H) \cap Q(P)]$ such that $E(S)^{\sigma'} = N(S) \cap Q(P)$ for every subspace S of A, and a semi-linear transformation σ'' of $[N(P^*) \cap Q(H^*),N(H^*)]$ upon (G,B) which satisfies $S^{\sigma''^{-1}} = Q(S) \cap N(H^*)$ for every subspace S of B. The anti-isomorphism σ induces an anti-semi-linear transformation of $[Q(P), N(H) \cap Q(P)]$ upon

$$[N(H)^\sigma \cap Q(P)^\sigma, Q(P)^\sigma] = [Q(H^*) \cap N(P^*), N(P^*)]$$

which maps $N(S) \cap Q(P)$ upon $N(S)^\sigma \cap Q(P)^\sigma = Q(S^{\sigma^*}) \cap N(P^*)$. The product $\tau = \sigma'\sigma\sigma''$ is consequently an anti-semi-linear transformation of (A^*,F) upon (G,B) such that $E(S)^\tau = S^{\sigma^*}$ for every subspace S of A.

It follows from Lemma 1 and the finiteness of $r(A)$ that τ induces an anti-isomorphism τ^* of $P(F,A)$ upon $P(G,B)$ such that $Q(S)^{\tau^*} = N[E(S)^\tau]$ for every subspace S of A. Consequently

$$Q(S)^{\tau^*\sigma^{-1}} = N[E(S)^\tau]^{\sigma^{-1}} = N(S^{\sigma^*})^{\sigma^{-1}} = Q(S).$$

Since σ and τ^* are anti-isomorphisms, $\tau^*\sigma^{-1}$ is an automorphism of the ring $P(F,A)$ which leaves invariant every $Q(S)$. As in the proof of V.4, Theorem 1 we show now that this implies $\tau^*\sigma^{-1} = 1$ or $\tau^* = \sigma$. Thus we have shown that σ is induced by the anti-semi-linear transformation τ, as we intended to prove.

General Existence Theorem: *The following properties of the linear manifold (F,A) are equivalent.*

(I) *There exists an anti-isomorphism of the ring of (F,A) upon the ring of some linear manifold.*

(II) *There exists a duality of (F,A) upon some linear manifold.*

(III) *$r(A)$ is finite.*

PROOF: If (I) is true, then we infer from Lemma 2 the existence of a duality of A so that (II) is a consequence of (I). If (II) is true, then we deduce the finiteness of $r(A)$ from IV.1, Existence Theorem so that (III) is a consequence of (II). Suppose now that $r(A)$ is finite. In IV.1, Construction of Canonical Dual Space we have shown that the identity map is an anti-semi-linear transformation of (A^*,F) upon the soc. Canonical Dual Space. Because of the finiteness of $r(A)$ we may apply Lemma 1 and see that there exists an anti-isomorphism of the ring of (F,A) upon the ring of the Canonical Dual Space. Thus (I) is a consequence of (III).

Proposition 3: *If (F,A) has finite rank, if σ is an anti-semi-linear transformation of (A^*,F) upon (G,B), if the anti-isomorphism σ' of $P(F,A)$*

upon $P(G,B)$ *is induced by* σ, *and if the duality* σ'' *of* A *upon* B *is induced by* σ', *then*

$$S^{\sigma''} = E(S)^{\sigma} \text{ for every subspace } S \text{ of } A.$$

PROOF: Let $S^{\sigma^*} = E(S)^{\sigma}$ for every subspace S of A. Since the rank n of A is finite, we may apply II.3, Theorem 3. Consequently mapping S upon $E(S)$ constitutes a duality of A upon A^*. Since σ effects a projectivity of the system of subspaces of A^* upon the system of subspaces of B, it follows that σ^* is a duality of A upon B. Hence it follows from IV.1, Corollary 1 that $r(A) = r(B) = n$ and $r(S) + r(S^{\sigma^*}) = n$.

Since σ' is induced by σ, it follows from the definition of σ' [Lemma 1] that

$$(\eta^*d)^{\sigma} = d^{\sigma}\eta^{\sigma'} \text{ for } \eta \text{ in } P(F,A) \text{ and } d \text{ in } A^*$$

where the asterisk indicates the natural isomorphism of $P(F,A)$ upon $P(A^*,F)$ [Propositions 1 and 2] which satisfies

$$(a\eta)d = a(\eta^*d) \text{ for } a \text{ in } A,\ d \text{ in } A^* \text{ and } \eta \text{ in } P(F,A).$$

Since σ'' is induced by σ', it follows from the definition of σ'' [Lemma 2] that

$$S^{\sigma''} = BN(S)^{\sigma'} \text{ for every subspace } S \text{ of } A.$$

Using these formulas we find successively that

$$S[N(S)^*A^*] = [SN(S)]A^* = 0,$$
$$N(S)^*A^* \leqslant E(S),$$
$$S^{\sigma''} = BN(S)^{\sigma'} = A^{*\sigma}N(S)^{\sigma'} = [N(S)^*A^*]^{\sigma} \leqslant E(S)^{\sigma} = S^{\sigma^*}.$$

Since σ'' is a duality, it follows that $r(S) + r(S^{\sigma''}) = n$. Combining this with the result obtained in the first paragraph of this proof we find that $r(S^{\sigma''}) = r(S^{\sigma^*})$. But we have shown that $S^{\sigma''} \leqslant S^{\sigma^*}$; and now it follows from the finiteness of all the ranks that $S^{\sigma''} = S^{\sigma^*}$. This completes the proof.

Uniqueness Theorem: *If the linear manifold* (F,A) *has finite rank, then the following properties of the anti-semi-linear transformations* ν *and* ω *of* (A^*,F) *upon* (G,B) *are equivalent.*

(I) *There exists a number* $g \neq 0$ *in* G *such that* $d^{\nu} = gd^{\omega}$ *for every* d *in* A^*.

(II) ν *and* ω *induce the same anti-isomorphism of* $P(F,A)$ *upon* $P(G,B)$.

(III) ν *and* ω *induce the same duality of* A *upon* B.

Statement (II) has to be interpreted in accordance with Lemma 1; and (III) signifies [in accordance with Proposition 3] that $E(S)^{\nu} = E(S)^{\omega}$ for every subspace S of A.

PROOF: If (I) is true, then the induced anti-isomorphisms ν' and ω' of $P(F,A)$ satisfy

$$d^{\nu}\eta^{\nu'} = (\eta^*d)^{\nu} = g(\eta^*d)^{\omega} = g(d^{\omega}\eta^{\omega'}) = (gd^{\omega})\eta^{\omega'} = d^{\nu}\eta^{\omega'}$$

for every d in A^* and every η in $P(F,A)$. But this implies $\eta^{\nu'} = \eta^{\omega'}$ for every η, since d^{ν} ranges over the whole of B; and thus we have shown that $\nu' = \omega'$. Hence (II) is a consequence of (I).

If ν and ω induce the same anti-isomorphism σ of $P(F,A)$ upon $P(G,B)$, then we deduce from Proposition 3 that

$$E(S)^{\nu} = BN(S)^{\sigma} = E(S)^{\omega} \text{ for every subspace } S \text{ of } A.$$

Hence ν and ω induce the same duality of A upon B and (III) is a consequence of (II).

Assume finally the validity of (III). Then $\nu^{-1}\omega$ is a semi-linear transformation of (G,B); and we have $E(U)^{\nu} = E(U)^{\omega}$ for every subspace U of A. From the finiteness of $r(A)$ and II.3, Theorem 3 we deduce that $T = E[S(T)]$ for every subspace T of A^*. Hence

$$U^{\nu^{-1}\omega} = E[S(U^{\nu^{-1}})]^{\omega} = E[S(U^{\nu^{-1}})]^{\nu} = U \text{ for every subspace } U \text{ of } B.$$

It follows now from III.1, Proposition 3, (*a*) that there exists a number $g \neq 0$ in G such that $b^{\nu^{-1}\omega} = gb$ for every b in B. Letting $d = b^{\nu^{-1}}$ we see that (I) is a consequence of (III), as we wanted to show.

MULTIPLICATIVITY *of the correspondence between isomorphisms and anti-isomorphisms on the one side and the induced projectivities and dualities on the other side.*

Assume first that ν is an anti-isomorphism of $P(F,A)$ upon $P(G,B)$ and that ω is an anti-isomorphism of $P(G,B)$ upon $P(H,C)$. Then $\nu\omega$ is an isomorphism of $P(F,A)$ upon $P(H,C)$. The induced mappings of the systems of subspaces are

$$S^{\nu'} = BN(S)^{\nu} = K[Q(S)^{\nu}], \text{ a duality of } A \text{ upon } B,$$
$$T^{\omega'} = CN(T)^{\omega} = K[Q(T)^{\omega}], \text{ a duality of } B \text{ upon } C,$$
$$S^{(\nu\omega)'} = CQ(S)^{\nu\omega} = K[N(S)]^{\nu\omega}, \text{ a projectivity of } A \text{ upon } C,$$

as follows from Lemma 2 and V.4, Lemma 4. Next we deduce from the results of V.2 that

$$S^{(\nu\omega)'} = C[Q(S)^{\nu}]^{\omega} = C[N(K[Q(S)^{\nu}])]^{\omega} = C[N(S^{\nu'})]^{\omega} = (S^{\nu'})^{\omega'}$$

or $(\nu\omega)' = \nu'\omega'$.

If ν or ω is a projectivity or a duality, then the multiplicativity formula is derived in a quite similar fashion. We leave the details to the reader.

These multiplicativity formulas have various applications.

Corollary 1: *If the anti-isomorphisms* ν *and* ω *of* $P(F,A)$ *upon* $P(G,B)$ *induce the same duality of* A *upon* B, *then* $\nu = \omega$.

For $\nu\omega^{-1}$ is an automorphism of $P(F,A)$ which induces the identity projectivity in A and is therefore [by V.4, Remark 1] equal to 1.

The Extended Groups

The totality of automorphisms and anti-automorphisms of the ring $P(F,A)$ is a group with respect to multiplication; and this group is called *the extended automorphism group of the ring* $P(F,A)$. Likewise we define *the extended projective group of the linear manifold* (F,A) as the totality of auto-projectivities and auto-dualities of (F,A). Using Corollary 1 one proves easily:

Mapping every automorphism and anti-automorphism of the ring $P(F,A)$ upon the induced projectivity and duality of A constitutes an isomorphism of the extended automorphism group of the ring into the extended projective group of the manifold.

In particular: involutorial anti-automorphisms of the ring induce polarities of the manifold.

Theorem 2: *If* $2 < r(A)$, *then every duality of* (F,A) *is induced by an anti-isomorphism of* $P(F,A)$.

Proof: It follows from IV.1, Lemma 2 that every duality δ of (F,A) upon (G,B) is induced by an anti-semi-linear transformation τ of (A^*,F) upon (G,B); and it follows from Proposition 3 that the anti-isomorphism τ' induced by τ induces exactly δ.

Corollary 2: *If* $2 < r(A)$, *then the existence of auto-dualities and the existence of anti-automorphisms are equivalent properties; and the existence of polarities and involutorial anti-automorphisms are likewise equivalent properties.*

The proofs are simple consequences of the preceding results. We leave the details to the reader.

Semi-bilinear Forms and Ring Anti-automorphisms

Assume that (F,A) has finite rank not less than 3. If σ is an auto-duality of A, then σ may be represented by a semi-bilinear form [IV.1, Proposition 1]; and σ is induced by a uniquely determined anti-automorphism of $P(F,A)$ [Theorem 2 and Corollary 1]. We want to establish a relation between the representing semi-bilinear form and the inducing anti-automorphism.

Suppose that the auto-duality σ is represented by the semi-bilinear form $f(x,y)$. If S is a subspace of A, then S^σ is the totality of elements x in A such that $f(x,S) = 0$. Consider next an anti-automorphism ν of the ring $\mathrm{P}(F,A)$. Then ν induces σ if, and only if, $S^\sigma = K[Q(S)^\nu] = AN(S)^\nu$ for every subspace S of A; and this is easily seen to be equivalent to

$$0 = f(AN(S)^\nu, S) \text{ for every subspace } S \text{ of } A$$

where one has to make use of the fact that f represents a duality.

It is a consequence of IV.1, Proposition 2 that the semi-bilinear form f represents an auto-duality if, and only if, an anti-semi-linear transformation of (F,A) upon (A^*,F) is obtained by mapping the element a in A onto the linear form a^f defined by the rule: $xa^f = f(x,a)$ for every x in A [where we have to use the anti-automorphism of F entering into the definition of f as the other component of this anti-semi-linear transformation]. Now it is not difficult to prove on the basis of the results of the present section the following interesting proposition.

Proposition 4: *The anti-automorphism σ of $\mathrm{P}(F,A)$ induces the auto-duality of A which is represented by the semi-bilinear form f if, and only if,*

(*a*) *f represents an auto-duality and*

(*b*) *$f(x,y\eta) = f(x\eta^\sigma,y)$ for every x,y in A and η in $\mathrm{P}(F,A)$.*

We leave the details to the reader.

Remark: Suppose that (F,A) and (G,B) are linear manifolds of equal finite rank n and that α is an anti-isomorphism of the field F upon the field G. It is possible to represent the ring $\mathrm{P}(F,A)$ as the ring of all n by n matrices with coefficients from F; and we may likewise represent in some definite fashion the ring $\mathrm{P}(G,B)$ as the ring of all n by n matrices with coefficients from G. If (a_{ik}) is some matrix in $\mathrm{P}(F,A)$, then we may map this matrix (a_{ik}) in $\mathrm{P}(F,A)$ upon the matrix $(a_{ik})^\sigma = (a_{ik}^\alpha)$ which is obtained by first forming the transposed matrix and then applying the anti-isomorphism α on the coefficients. The reader should verify that σ is an anti-isomorphism of $\mathrm{P}(F,A)$ upon $\mathrm{P}(G,B)$ and that every anti-isomorphism between these rings may be obtained in this fashion.

APPENDIX I

The Two-sided Ideals of the Endomorphism Ring

Suppose that A is an F-space of finite or infinite rank, and that P is the ring of endomorphisms of A. It is our objective to determine all the two-sided ideals of the ring P. We recall that a non-vacuous subset J of P is a two-sided ideal, if $J = J \pm J = J\mathrm{P} = \mathrm{P}J$.

If σ is an endomorphism of the F-space A, then $r[A/K(\sigma)] = r[A\sigma]$ by V.1, (2).

If $\aleph_\nu$ is some [infinite] cardinal number, then we denote by P_ν the totality of endomorphism σ of A such that

$$r[A/K(\sigma)] = r[A\sigma] < \aleph_\nu.$$

(1) P_ν *is a two-sided ideal in* P.

PROOF: If σ' and σ'' belong both to P_ν, then $A(\sigma' \pm \sigma'') \leqslant A\sigma' + A\sigma''$ and $r(A\sigma' + A\sigma'') \leqslant r(A\sigma') + r(A\sigma'') < 2\,\aleph_\nu = \aleph_\nu$ imply that $\sigma' \pm \sigma''$ belongs to P_ν. If σ belongs to P_ν and η to P, then $A\eta\sigma \leqslant A\sigma$ implies $r(A\eta\sigma) \leqslant r(A\sigma) < \aleph_\nu$ so that $\eta\sigma$ belongs to P_ν; and $K(\sigma) \leqslant K(\sigma\eta)$ implies that $r[A/K(\sigma\eta)] \leqslant r[A/K(\sigma)] < \aleph_\nu$ so that $\sigma\eta$ belongs to P_ν. Thus every P_ν is a two-sided ideal in P.

We note that P_0 contains all the σ with finite $r[A/K(\sigma)] = r[A\sigma]$; and this fact implies clearly:

(2) $P = P_0$ *if, and only if,* A *has finite rank.*

It is quite obvious to note that

(3) $0 < P_0 \leqslant \cdots \leqslant P_\nu \leqslant \cdots$

Next we prove.

(4) *If* $\aleph_\nu \leqslant r(A)$, *then* $P_\nu < P_{\nu+1}$; *and if* $r(A) < \aleph_\nu$, *then* $P_\nu = P$.

PROOF: If $\aleph_\nu \leqslant r(A)$, then A possesses a subspace U of exact rank $\aleph_\nu$. There exists an idempotent endomorphism σ which maps A upon U. Then σ belongs to $P_{\nu+1}$, but not to P_ν. If $r(A) < \aleph_\nu$, then the identity endomorphism belongs to the two-sided ideal P_ν which consequently equals P.

(5) *If* J *is a two-sided ideal in* P, *and if* σ *is an element in* J, *then* J *contains every endomorphism* η *such that* $r(A\eta) \leqslant r(A\sigma)$.

PROOF: Since $r(A\eta) \leqslant r(A\sigma)$, there exists a one-to-one mapping of part of a basis of $A\sigma$ upon a basis of $A\eta$; and now it is easy to construct some endomorphism σ'' which maps $A\sigma$ upon $A\eta$. Then $\sigma\sigma''$ belongs to the two-sided ideal J; and we have $A\sigma\sigma'' = A\eta$. There exists by V.3, Lemma 1 an endomorphism σ' such that $\sigma'(\sigma\sigma'')$ is an idempotent endomorphism, satisfying $A\sigma'\sigma\sigma'' = A\sigma\sigma'' = A\eta$. Again it is clear that $\sigma'\sigma\sigma''$ belongs to the two-sided ideal J; and likewise $\eta(\sigma'\sigma\sigma'')$ belongs to J. But $\sigma'\sigma\sigma''$ leaves invariant every element in $A\sigma'\sigma\sigma'' = A\eta$ as an idempotent; and thus it follows that $\eta = \eta\sigma'\sigma\sigma''$ belongs to J, as we intended to prove.

(6) *Every two-sided ideal, not* 0, *of* P *is equal to one of the ideals* P_ν.

PROOF: Suppose that $J \neq 0$ is a two-sided ideal in P. If $J = P$, then $J = P_\nu$ for some ν, as follows from (4). Assume now that $0 < J < P$. It is easily deduced from (5) and $J \neq 0$ that $P_0 \leqslant J$. Since $J \neq P$, there exist some ordinals ν such that J does not contain all the endomorphisms σ satisfying $r(A\sigma) \leqslant \aleph_\nu$ [for instance any ordinal ν with $r(A) < \aleph_\nu$]. Among these ordinals there is a smallest one, τ. Thus J does not contain every endomorphism σ satisfying $r(A\sigma) \leqslant \aleph_\tau$, though it contains every endomorphism σ with the property $r(A\sigma) < \aleph_\tau$. Now we apply (5) to see that $J = \mathrm{P}_\tau$; and this completes the proof.

The results of this appendix ought to be considered as a first step towards a characterization of endomorphism rings by internal properties. In this respect the reader ought to consult some standard work on ring theory like Albert [1], Artin-Nesbitt-Thrall [1], Jacobson [1]; furthermore R. Baer: Automorphism Rings of Primary Abelian Operator Groups, *Annals of Math.*, vol. 44 (1943), pp. 192-227 where a similar, though more comprehensive problem has been solved; and the works on simple rings like

E. Artin: The Influence of J. H. M Wedderburn on the Development of Modern Algebra, *Bull. Amer. Math. Soc.*, vol. 56 (1950), pp. 65-72, in particular pp. 68-70;

E. Artin and G. Whaples: The Theory of Simple Rings, *Amer. Journal of Math.*, vol. 65 (1943), p. 87-107;

N. Jacobson: Structure Theory of Simple Rings without Finiteness Assumptions, *Trans. Amer. Math. Soc.*, vol. 57 (1945), pp. 228-245;

R. E. Johnson : Equivalence Rings, *Duke Math. Journal*, vol. 15 (1948), pp. 787-793.

CHAPTER VI

The Groups of a Linear Manifold

In III.2 we have given a survey of the various groups that one may relate with some given linear manifold (F,A). We recall the group $\Lambda(F,A)$ of its semi-linear transformations, the group $\Gamma(F,A)$ of its linear transformations, the group of auto-projectivities, the group of collineations and the extended group of auto-projectivities that we mentioned in V.5. Considering any one of these groups one may pose the problem whether its structure determines that of the basic linear manifold.

Of these derived groups we are going to investigate Γ and Λ only; and when doing so we shall impose the hypothesis that the characteristic of the basic linear manifold is not 2 and that its rank is not less than 3. Under these hypotheses we are going to prove the equivalence of the following properties of the linear manifolds (F,A) and (G,B).

(I) $\Gamma(F,A) \sim \Gamma(G,B)$;

(II) $\Lambda(F,A) \sim \Lambda(G,B)$;

(III) (F,A) and (G,B) are projectively equivalent or they are duals of each other.

The proof of these equivalences will be a consequence of theorems which permit us to construct in an almost explicit fashion all the isomorphisms, if any, between $\Gamma(F,A)$ and $\Gamma(G,B)$ as well as the isomorphisms between $\Lambda(F,A)$ and $\Lambda(G,B)$. These theorems may be found in VI.5 and VI.8 respectively.

To obtain these results we shall derive in VI.1 to VI.4 certain results whose purpose it is to translate "additive" properties of linear transformations into "multiplicative" ones; here we term a property additive, if it refers to the way the transformation acts on the basic linear manifold, whereas a property is termed multiplicative, if it deals with properties of the abstract group of transformations without reference to its given representation. The results concerning Λ will be derived from those concerning Γ; to do this we shall show in VI.6 that Γ is a characteristic subgroup of Λ. This translation of additive properties into multiplicative ones will provide a means to reconstruct the basic linear manifold within the derived group.

For a discussion of the collineation group within the framework of permutation groups see:

J. Tits: Généralisations des groupes projectifs, *Acad. Roy. Belgique Bull.*, Cl. Sciences, 1949.

For investigations of the isomorphisms and [mainly] automorphisms of the groups of projective geometry the reader ought to consult the classical paper by

O. Schreier and B. L. van der Waerden: Die Automorphismen der projektiven Gruppen, *Abhandlungen aus dem mathematischen Seminar der Hamburgischen Universität*, Bd. 6 (1928), pp. 303-322,

and the more recent treatments by Dieudonné [1,2], Hua [1], Rickart [1] and

Loo-Keng Hua: On the Automorphisms of the Symplectic Group over any Field, *Annals of Math.*, vol. 49 (1948), pp. 739-759;

G. W. Mackey: Isomorphisms of Normed Linear Spaces, *Annals of Math.*, vol. 43 (1942), pp. 244-260;

C. E. Rickart: Isomorphic Groups of Linear Transformations, I, *Amer. Journal of Math.*, vol. 72 (1950), pp. 451-461.

Remark on Notations: In the first part of this chapter we shall investigate the group of linear transformations; and we shall consider this group as the group of units in the endomorphism ring P. Consequently it will prove convenient during this part of our investigation to regard the linear transformations as right-hand multipliers. However, in the last two sections of this chapter we shall investigate the group of semi-linear transformations; and as far as these are concerned the exponential notation is more suggestive. Consequently we shall use multiplicative notation in the first sections and exponential notation in the last two sections.

VI.1. The Center of the Full Linear Group

If (F,A) is a linear manifold, then we denote by $\mathrm{T} = \mathrm{T}(F,A)$ the group of all its linear transformations. *The center* $\mathrm{Z} = \mathrm{Z}(F,A)$ consists of all those linear transformations in T which commute with every linear transformation in T. We shall need a characterization of this characteristic abelian subgroup of T.

Proposition 1: *If* $1 < r(A)$, *and if* σ *is a linear transformation in* T, *then the following properties of* σ *are equivalent.*

(I) σ *belongs to the center* Z *of* T.

(II) σ *leaves invariant every subspace of* A.

(III) *There exists a number* $z \neq 0$ *in the center of* F *such that* $x\sigma = zx$ *for every* x *in* A.

PROOF: Assume first the validity of (I). If the elements x and $x\sigma$ were independent elements in A, then neither of them would be 0; and there would exist a subspace T of A such that $A = Fx \oplus Fx\sigma \oplus T$ [Complementation Principle]. Clearly there exists one and only one linear transformation τ of A such that $y\tau = y$ for y in $Fx + T$ and $x\sigma\tau = x + x\sigma$. But $x\tau\sigma = x\sigma$ so that σ and τ would not commute. This contradicts (I). Hence x and $x\sigma$ are dependent elements for every x in A; and this signifies $Fx = (Fx)\sigma$ for every x in A. Hence (II) is a consequence of (I).

Assume next the validity of (II). Then we infer from III.1, Proposition 3 the existence of a number $z \neq 0$ in F such that $x\sigma = zx$ for every x in A. If g is any number in F, then it follows that

$$(gz)x = g(zx) = g(x\sigma) = (gx)\sigma = z(gx) = (zg)x$$

for every x in A; and this implies $gz = zg$ for every g in F, since x may be chosen different from 0. Hence z belongs to the center of F; and we have shown that (III) is a consequence of (II). That finally (III) implies (I), is obvious.

REMARK 1: If $r(A) = 1$, then the equivalence of (I) and (II) is still true; but the reader will see equally easily that (II) is satisfied by every linear transformation so that (II) and (I) cannot be equivalent unless F is commutative.

Proposition 2: *The center* Z *of* T *is essentially the same as the multiplicative group of the center of* F.

This is a fairly obvious consequence of Proposition 1 together with Remark 1.

REMARK 2: According to Proposition 2 it is clear that the structure of the multiplicative group of the center of F is completely determined by the structure of the linear group T. However, there exist many non-isomorphic commutative fields with isomorphic multiplication group. For instance, it is not too difficult to prove that the multiplication groups of the commutative and algebraically closed fields F' and F'' are isomorphic if, and only if, F' and F'' have the same characteristic and contain the same number of elements. However, it seems to be an open question whether some special field invariant like the characteristic of F is determined by the structure of the multiplicative group of the center of F.

REMARK 3: Because of the close connection between the center of F and the center of T there is no danger of confusion, if we denote, as we shall do, the mapping $x \rightarrow zx$ [for variable x in A and fixed z in the center of F] by the letter z.

VI.2. First and Second Centralizer of an Involution

The reader ought to recall that we denote by $J^+(\sigma)$ the totality of elements x in A such that $x\sigma = x$ and that we denote by $J^-(\sigma)$ the totality of elements x in A such that $x\sigma = -x$. If σ is a linear transformation or more generally an endomorphism, then $J^+(\sigma)$ and $J^-(\sigma)$ are subspaces of the linear manifold (F,A) under consideration. The subspaces $J^+(\sigma)$ and $J^-(\sigma)$ are equal, if the characteristic of the field F is 2; and if the characteristic of F is different from 2, then they have nothing in common except 0. For this reason we shall impose throughout the remainder of this chapter the hypothesis that *the characteristic of F be different from* 2. Though we shall not restate this hypothesis, the reader ought to check which parts of our arguments remain valid without this hypothesis.

If Θ is a subset of the linear group $\mathrm{T}(F,A)$ of our linear manifold, then we denote by $\mathrm{Z}(\Theta)$ the totality of linear transformations ζ which satisfy $\sigma\zeta = \zeta\sigma$ for every σ in Θ. It is clear that $\mathrm{Z}(\Theta)$ is a subgroup of T; and this subgroup is called *the centralizer of* Θ [*in* T]. We shall have to form several times the centralizer of a centralizer; and for this *second centralizer* we shall use the notation: $\mathrm{Z}_2(\Theta) = \mathrm{Z}[\mathrm{Z}(\Theta)]$.

Consider now an involution σ of A. We have pointed out before [IV.4, Lemma 1] that involutions σ are characterized by the property:

$$A = J^+(\sigma) \oplus J^-(\sigma) \text{ if, and only if, } \sigma^2 = 1.$$

The centralizer of an involution is completely determined by the following proposition.

Proposition 1: *The following properties of the involution σ and the linear transformation τ are equivalent.*

(I) $\sigma\tau = \tau\sigma$.

(II) $J^+(\sigma)\tau \leqslant J^+(\sigma)$ *and* $J^-(\sigma)\tau \leqslant J^-(\sigma)$.

(III) $J^+(\sigma)\tau = J^+(\sigma)$ *and* $J^-(\sigma)\tau = J^-(\sigma)$.

PROOF: Assume first the validity of (I). If $x = x\sigma$, then we have $x\tau = x\sigma\tau = x\tau\sigma$. This proves $J^+(\sigma)\tau \leqslant J^+(\sigma)$; and $J^-(\sigma)\tau \leqslant J^-(\sigma)$ is shown likewise. Thus (II) is a consequence of (I). If τ satisfies (I), so does τ^{-1}. Hence it follows from what we have shown already that both τ and τ^{-1} satisfy (II); and consequently we have:

$$J^+(\sigma) = J^+(\sigma)\tau^{-1}\tau \leqslant J^+(\sigma)\tau \leqslant J^+(\sigma) \text{ or } J^+(\sigma)\tau = J^+(\sigma).$$

Likewise we see that $J^-(\sigma) = J^-(\sigma)\tau$; and thus we have shown that (III) is a consequence of (I).

It is obvious that (II) is a consequence of (III). Assume finally the validity of (II). If x is an element in $J^+(\sigma)$, then $x\tau$ belongs to $J^+(\sigma)$ so that we find

$$x\sigma\tau = x\tau = x\tau\sigma;$$

and if x belongs to $J^-(\sigma)$, then we see likewise

$$x\sigma\tau = -x\tau = x\tau\sigma.$$

But $A = J^+(\sigma) \oplus J^-(\sigma)$; and thus the two preceding equations imply $\sigma\tau = \tau\sigma$.

Corollary 1: *If σ is an involution, then $Z(\sigma)$ is essentially the same as the direct product of $T(F,J^+(\sigma))$ and $T(F,J^-(\sigma))$.*

The exact meaning of the terms used will become quite clear in the course of the proof.

PROOF: Denote by $Z^+(\sigma)$ the totality of linear transformations ζ in $Z(\sigma)$ which satisfy $x = x\zeta$ for every x in $J^-(\sigma)$; and denote by $Z^-(\sigma)$ the totality of transformations ζ in $Z(\sigma)$ which satisfy $x = x\zeta$ for every x in $J^+(\sigma)$. It is clear that $Z^+(\sigma)$ and $Z^-(\sigma)$ are subgroups of $Z(\sigma)$ which satisfy:

(1) $$Z^+(\sigma) \cap Z^-(\sigma) = 1.$$

Next we show:

(2) Every element in $Z^+(\sigma)$ commutes with every element in $Z^-(\sigma)$.

For suppose that σ' and σ'' belong to $Z^+(\sigma)$ and $Z^-(\sigma)$ respectively. If x is an element in $J^+(\sigma)$, then $x\sigma'$ belongs also to $J^+(\sigma)$ [by Proposition 1] and consequently we have $x\sigma'\sigma'' = x\sigma' = x\sigma''\sigma'$; and likewise we see that $x\sigma'\sigma'' = x\sigma''\sigma'$ for x in $J^-(\sigma)$. Hence $\sigma'\sigma'' = \sigma''\sigma'$ is a consequence of $A = J^+(\sigma) \oplus J^-(\sigma)$.

(3) $$Z(\sigma) = Z^+(\sigma)Z^-(\sigma).$$

If ζ belongs to $Z(\sigma)$, then it follows from Proposition 1 that $J^+(\sigma)\zeta = J^+(\sigma)$. A linear transformation is therefore induced in $J^+(\sigma)$ by ζ. Consequently there exists one and only one linear transformation ζ' such that

$$x\zeta' = \begin{cases} x\zeta \text{ for } x \text{ in } J^+(\sigma) \\ x \text{ for } x \text{ in } J^-(\sigma) \end{cases}.$$

It is an immediate consequence of Proposition 1 that ζ' belongs to $Z^+(\sigma)$. From the definition of ζ' we deduce next that $\zeta'' = \zeta\zeta'^{-1}$ belongs to $Z^-(\sigma)$. Hence $\zeta = \zeta''\zeta' = \zeta'\zeta''$ [by (2)] belongs to $Z^+(\sigma)Z^-(\sigma)$, proving (3).

The properties (1), (2), (3) are just an elaboration of the fact usually concentrated in the statement:

(4) $$Z(\sigma) = Z^+(\sigma) \otimes Z^-(\sigma)$$

[where $\otimes$ indicates the direct product].

If ζ belongs to $Z^+(\sigma)$, then ζ induces in $J^+(\sigma)$ a definite linear transformation ζ^* [Proposition 1]. One verifies without difficulty that mapping ζ onto ζ^* constitutes an isomorphism of $Z^+(\sigma)$ upon $T(F,J^+(\sigma))$; and in an analoguous fashion we see that $Z^-(\sigma)$ and $T(F,J^-(\sigma))$ are essentially the same. Combining these facts with (4) we obtain Corollary 1.

We are ready now to give a complete determination of the second centralizer of an involution.

Proposition 2: *The linear transformation τ belongs to the second centralizer $Z_2(\sigma)$ of the involution σ if, and only if, there exist numbers h,k [not 0] in the center of the field F such that $x\tau = hx$ for every x in $J^+(\sigma)$ and $x\tau = kx$ for every x in $J^-(\sigma)$.*

PROOF: Assume first that τ satisfies the condition and that ζ belongs to $Z(\sigma)$. If x is in $J^+(\sigma)$, then $x\zeta$ belongs likewise to $J^+(\sigma)$ [Proposition 1]; and we infer from our condition that $x\tau\zeta = hx\zeta = x\zeta\tau$; and if x is in $J^-(\sigma)$, then we see likewise that $x\tau\zeta = kx\zeta = x\zeta\tau$. Now $\zeta\tau = \tau\zeta$ is a consequence of $A = J^+(\sigma) \oplus J^-(\sigma)$; and this shows the sufficiency of our condition.

Assume conversely that τ belongs to $Z_2(\sigma)$. Since σ commutes with itself, σ belongs to $Z(\sigma)$. Hence τ commutes with σ; and it follows from Proposition 1 that $J^+(\sigma)\tau = J^+(\sigma)$ and $J^-(\sigma)\tau = J^-(\sigma)$. It has been pointed out in the proof of Corollary 1 that every linear transformation of $J^+(\sigma)$ is induced by a linear transformation in $Z(\sigma)$. Since τ commutes with every linear transformation in $Z(\sigma)$, it follows that τ induces in $J^+(\sigma)$ a linear transformation which commutes with every linear transformation of $J^+(\sigma)$ [belongs to the center of $T(F,J^+(\sigma))$]. We may apply VI.1, Proposition 1 [and Remark 1 in case $r[J^+(\sigma)] = 1$]. Consequently there exists a number $h \neq 0$ in the center of the field F such that $x\tau = hx$ for every x in $J^+(\sigma)$; and the existence of a number $k \neq 0$ in the center of F such that $x\tau = kx$ for every x in $J^-(\sigma)$ is verified likewise. This completes the proof.

REMARK 1: Denote by $Z = Z(T)$ the center of the group $T(F,A)$ of all linear transformations. If σ is an involution, then denote by $Z_2^+(\sigma)$ the totality of transformations in $Z_2(\sigma)$ which leave invariant every element in $J^-(\sigma)$. One deduces now easily from Proposition 2 and VI.1, Proposition 2 that Z and $Z_2^+(\sigma)$ are both essentially the same as the multiplicative group of the numbers, not 0, in the center of the field F; and that

$$Z_2(\sigma) = Z \otimes Z_2^+(\sigma).$$

Corollary 2: *If σ is an involution, τ a linear transformation in $Z_2(\sigma)$ and x an element in A, then*

$$r\Big(\sum_{i=0}^{\infty} Fx\tau^i\Big) \leq 2 \text{ and } \Big(\sum_{i=0}^{\infty} Fx\tau^i\Big)\tau = \sum_{i=0}^{\infty} Fx\tau^i.$$

PROOF: We deduce from Proposition 2 the existence of numbers h, k [not 0] in the center of F such that

$$(*) \qquad y\tau = \begin{cases} hy \text{ for } y \text{ in } J^{+}(\sigma) \\ ky \text{ for } y \text{ in } J^{-}(\sigma) \end{cases}.$$

Consider now some element x in A and let $X = \sum_{i=0}^{\infty} Fx\tau^{i}$. Since σ is an involution, we have $A = J^{+}(\sigma) \oplus J^{-}(\sigma)$; and consequently there exist uniquely determined elements x' and x'' in $J^{+}(\sigma)$ and $J^{-}(\sigma)$ respectively such that $x = x' + x''$. It follows from (*) that $x\tau^{i} = h^{i}x' + k^{i}x''$. Consequently $X \leqslant Fx' + Fx''$ so that $r(X) \leqslant 2$. It follows from the definition of X that $X\tau \leqslant X$; and X and $X\tau$ are subspaces of equal finite rank, since τ is a linear transformation. Hence $X = X\tau$ is a consequence of the Rank Formulas [II.2]. This completes the proof.

The following proposition is in some respects a kind of converse to Proposition 1 and will be needed in the future.

Proposition 3: *Suppose that σ is an involution and S a subspace of the linear manifold (F,A). Then $S = S\tau$ for every involution τ in $Z(\sigma)$ if, and only if, $S = 0$ or $J^{+}(\sigma)$ or $J^{-}(\sigma)$ or A.*

PROOF: The sufficiency of our conditions is an immediate consequence of Proposition 1. Thus we assume next that $S = S\tau$ for every involution τ in $Z(\sigma)$. Let $T = S \cap J^{+}(\sigma)$. It follows from Proposition 1 that

$$J^{+}(\sigma) = J^{+}(\sigma)\tau$$

for every involution τ in $Z(\sigma)$; and this implies

(*) $\quad T = T\tau$ for every involution τ in $Z(\sigma)$.

Suppose now, by way of contradiction, that $0 < T < J^{+}(\sigma)$. Then there exists an element $v \neq 0$ in T and an element w in $J^{+}(\sigma)$ which does not belong to T. There exist subspaces V,W such that $T = Fv \oplus V$ and $J^{+}(\sigma) = T \oplus Fw \oplus W$. Then

$$J^{+}(\sigma) = Fv \oplus V \oplus Fw \oplus W = F(v + w) \oplus V \oplus Fw \oplus W$$

and consequently there exists one and only one involution τ such that $(v + w)\tau = -v - w$ and $x\tau = x$ for every x in $V \oplus Fw \oplus W \oplus J^{-}(\sigma)$. It follows from Proposition 1 that $\sigma\tau = \tau\sigma$. But

$$v\tau = v\tau + w\tau - w = -v - 2w.$$

Since w is in $J^{+}(\sigma)$, but not in $T = S \cap J^{+}(\sigma)$, whereas v belongs to T, the element w is not in S and neither is $-v-2w$. Hence $S\tau$ is not part of S, contradicting our hypothesis. Thus we have shown that $S \cap J^{+}(\sigma)$ can only be 0 or $J^{+}(\sigma)$; and likewise we see that $S \cap J^{-}(\sigma) = 0$ or $J^{-}(\sigma)$. But these facts imply that $S = 0$ or $J^+(\sigma)$ or $J^-(\sigma)$ or A, as we intended to show.

VI.3. Transformations of Class 2

A linear transformation σ is said to be *of finite class,* if $0 = (\sigma - 1)^i$ for some positive i; and the smallest positive integer i with this property is *the class of* σ. Thus a linear transformation σ *is of class* 2 if, and only if, $\sigma \neq 1$ and $(\sigma - 1)^2 = 0$.

Noting that $x(\sigma - 1) = 0$ and $x\sigma = x$ are equivalent properties we see that a linear transformation σ is of class 2 if, and only if,

$$\sigma \neq 1 \textit{ and } A(\sigma - 1) \leqslant J^+(\sigma);$$

and the second condition is equivalent to saying that σ leaves invariant every element in the quotient space $A/J^+(\sigma)$. It is really this latter property which makes the transformations of class 2 significant, not the rather formal requirement that they satisfy some special equation.

Proposition 1: *Suppose that σ and τ are linear transformations and that σ is of class 2. Then $Z(\sigma) \leqslant Z(\tau)$ if, and only if, there exist numbers $h \neq 0$ and k in the center of F such that $\tau = k[h(\sigma - 1) + 1]$.*

Here we denote by t, for t in F, also the correspondence which maps the element x in A upon the element tx in A. We recall that this correspondence is a linear transformation if, and only if, t is a number, not 0, in the center of F.

PROOF: The sufficiency of our condition is practically obvious; and thus we assume immediately that $Z(\sigma) \leqslant Z(\tau)$. Since σ belongs to $Z(\sigma)$, σ likewise belongs to $Z(\tau)$ and consequently

$$\sigma\tau = \tau\sigma. \tag{1.1}$$

From this equation one deduces by the customary arguments that

$$[A(\sigma - 1)]\tau = A(\sigma - 1) \text{ and } J^+(\sigma)\tau = J^+(\sigma). \tag{1.2}$$

From $A(\sigma - 1) \leqslant J^+(\sigma)$ and the Complementation Principle [II.1] we deduce the existence of subspaces U and V such that

$$J^+(\sigma) = A(\sigma - 1) \oplus U, \qquad A = J^+(\sigma) \oplus V.$$

Next we note that $A(\sigma - 1) = V(\sigma - 1)$, since $J^+(\sigma)(\sigma - 1) = 0$; and that $v'(\sigma - 1) = v''(\sigma - 1)$ for v' and v'' in V implies $v' = v''$. It follows that

(1.3) $\sigma - 1$ induces a linear transformation ν of the linear manifold (F, V) upon the linear manifold $[F, A(\sigma - 1)]$.

We show next:

(1.4) There exists a number $k \neq 0$ in the center of F such that $x\tau = kx$ for every x in $A(\sigma - 1)$.

To prove this we observe first that τ induces [by (1.2)] a linear transformation τ' in $A(\sigma - 1)$. If ω is some linear transformation of $A(\sigma - 1)$, then there exists one and only one linear transformation ω^* of A with the following properties:

$$x\omega^* = \begin{cases} x\omega & \text{for } x \text{ in } A(\sigma - 1) \\ x & \text{for } x \text{ in } U \\ x\nu\omega\nu^{-1} & \text{for } x \text{ in } V \end{cases}$$

where we remember that the linear transformation ν of V upon $A(\sigma - 1)$ naturally possesses an inverse. Since every $x\omega$ for x in $A(\sigma - 1)$ belongs to $A(\sigma - 1)$, and since σ leaves invariant every element in $J^+(\sigma) = A(\sigma - 1) \oplus U$, we have $y\omega\sigma = y\omega = y\sigma\omega$ for every y in $J^+(\sigma)$. Since $x\omega^*$ belongs to V for x in V, and since ν coincides with $\sigma - 1$ on V, we find that

$$\begin{aligned} x\omega^*\sigma &= x\omega^*\nu + x\omega^* = x\nu\omega + x\omega^* = x(\sigma - 1)\omega + x\omega^* \\ &= x(\sigma - 1)\omega^* + x\omega^* = x\sigma\omega^* \end{aligned}$$

for x in V where we have to remember that ω and ω^* coincide on $A(\sigma - 1)$. Thus we have shown that $\omega^*\sigma = \sigma\omega^*$, proving that ω^* belongs to $Z(\sigma) \leqslant Z(\tau)$. But then τ and ω^* commute; and consequently τ' and ω commute. We have shown that τ' belongs to the center of $T[F, A(\sigma - 1)]$ and our contention is a consequence of IV.1, Proposition 1 and VI.1, Remark 1.

Consider next some linear transformation ω of the linear manifold (F, U). There exists one and only one linear transformation ω^* of A such that $u\omega^* = u$ for u in U and $x\omega^* = x$ for x in $A(\sigma - 1) \oplus V$. Since σ leaves invariant every element in $A(\sigma - 1) \oplus U$, and since ω^* leaves invariant every element in $A(\sigma - 1) \oplus V$, it follows that σ and ω^* commute; and this implies that ω^* belongs to $Z(\tau)$ or $\tau\omega^* = \omega^*\tau$.

We apply this result first on $\omega = -1$. We note first that $u\tau$ for u in U belongs to $J^+(\sigma) = A(\sigma - 1) \oplus U$. Hence there exist elements u', u'' in U and $A(\sigma - 1)$ respectively such that $u\tau = u' + u''$. Consequently

$$-u' - u'' = -u\tau = u(-1)^*\tau = u\tau(-1)^* = (u' + u')(-1)^* = -u' + u''$$

and $u'' = 0$ is a consequence of the fact that the characteristic of F is not 2. Thus we have shown that $U\tau \leqslant U$; and this implies $U = U\tau$, since τ maps the subspaces $A(\sigma - 1)$ and $A(\sigma - 1) \oplus U = J^+(\sigma)$ upon themselves. But the linear transformation which τ induces in U commutes with every linear transformation ω, since τ commutes with every ω^*; and now we infer again from VI.1, Proposition 1 and Remark 1 that there exists a number $k' \neq 0$ in the center of F such that $u\tau = k'u$ for every u in U.

Since $\sigma \neq 1$, $A(\sigma - 1) \neq 0$. Consequently there exists an element $a \neq 0$ in $A(\sigma - 1)$. Consider some element $u \neq 0$ in U. Then $U = Fu \oplus U'$

[Complementation Principle]; and there exists one and only one linear transformation α of A such that

$$u\alpha = u + a,\ x\alpha = x \text{ for } x \text{ in } A(\sigma - 1) \oplus U' \oplus V.$$

Since σ transforms this latter group into itself and leaves invariant the element a, we have $\sigma\alpha = \alpha\sigma$; and this implies as usual that $\tau\alpha = \alpha\tau$. Now we infer from (1.4) and from the result of the preceding paragraph that

$$ka + k'u = a\tau + u\tau = (a + u)\tau = u\alpha\tau = u\tau\alpha = k'u\alpha = k'a + k'u;$$

and $k = k'$ is a consequence of $a \neq 0$. Combining this with the result of the preceding paragraph and with (1.4) we see that we have obtained the following result.

(1.5) $\qquad x\tau = kx$ for every x in $A(\sigma - 1) \oplus U = J^{+}(\sigma)$.

Since k is a center element, not 0, we may form the linear transformation $\gamma = k^{-1}\tau$ which clearly satisfies $Z(\tau) = Z(\gamma)$. We note furthermore that $J^{+}(\sigma) \leqslant J^{+}(\gamma)$ by (1.5). Since γ and σ commute, and since γ leaves invariant every element in $A(\sigma - 1) \leqslant J^{+}(\sigma)$ we find that

$$[x - x\gamma](\sigma - 1) = x(\sigma - 1) - x\gamma(\sigma - 1) = x(\sigma - 1) - x(\sigma - 1)\gamma = 0$$

and this implies

$$A(\gamma - 1) \leqslant K(\sigma - 1) = J^{+}(\sigma) \leqslant J^{+}(\gamma) = K(\gamma - 1).$$

Consequently we have either $\gamma = 1$ or else γ is of class 2.
Next we prove

(1.6) $\qquad Fx(\gamma - 1) \leqslant Fx(\sigma - 1)$ for every x in A.

If x is any element in A, then $x = x' + x''$ with x' in V and x'' in $J^{+}(\sigma) \leqslant J^{+}(\gamma)$. It follows that $x(\sigma - 1) = x'(\sigma - 1)$ and $x(\gamma - 1) = x'\gamma(- 1)$; and thus it suffices to verify (1.6) for every element in V.

Assume now that x is an element in V such that $y = x(\gamma - 1)$ does not belong to $A(\sigma - 1)$. Then there exists a subspace Y such that

$$J^{+}(\sigma) = A(\sigma - 1) \oplus Fy \oplus Y,$$

since $A(\gamma - 1) \leqslant J^{+}(\sigma)$. There exists one and only one linear transformation η of A such that $y\eta = - y$ and $a\eta = a$ for every a in $A(\sigma - 1) \oplus Y \oplus V$. One verifies in the usual fashion that $\eta\sigma = \sigma\eta$ and that therefore $\eta\gamma = \gamma\eta$. Consequently

$$- y = y\eta = x(\gamma - 1)\eta = x\eta(\gamma - 1) = x(\gamma - 1) = y;$$

and this would imply $y = 0$ contradicting our assumption that y is not in $A(\sigma - 1)$. Thus we have shown that $V(\gamma - 1) \leqslant A(\sigma - 1)$.

Suppose next that x is an element in V such that $x(\sigma - 1)$ and $x(\gamma - 1)$ are independent elements. Both these elements belong to $A(\sigma - 1)$; and it is easy to construct a linear transformation of $A(\sigma - 1)$ which leaves invariant $x(\sigma - 1)$ and which maps $x(\gamma - 1)$ upon $- x(\gamma - 1)$. But we have shown before that this linear transformation of $A(\sigma - 1)$ is induced by a linear transformation $\varkappa$ of A such that $x\varkappa = x$ and $\varkappa\sigma = \sigma\varkappa$ [see the proof of (1.4)]. As usual we see that $\varkappa$ commutes with τ and γ; and now we find that

$$x(\gamma - 1) = x\varkappa(\gamma - 1) = x(\gamma - 1)\varkappa = - x(\gamma - 1).$$

Hence $x(\gamma - 1) = 0$ which certainly contradicts our assumption that $x(\gamma - 1)$ and $x(\sigma - 1)$ are independent elements.

If $x \neq 0$ is an element in V, then $x(\sigma - 1) \neq 0$; and it follows from the result of the preceding paragraph that $x(\gamma - 1)$ depends on $x(\sigma - 1)$. But this is equivalent to saying that $Fx(\gamma - 1) \leqslant Fx(\sigma - 1)$, as we intended to show.

Consider now any element x in $A(\sigma - 1)$. Then $x\nu^{-1}$ is a well determined element in V and $x\beta = x\nu^{-1}(\gamma - 1)$ is an element in $A(\gamma - 1)$. But it follows from (1.6) that

$$Fx\beta = Fx\nu^{-1}(\gamma - 1) \leqslant Fx\nu^{-1}(\sigma - 1) = Fx\nu^{-1}\nu = Fx,$$

if we remember that ν is exactly the mapping of V upon $A(\sigma - 1)$ which is induced by $\sigma - 1$. Thus β is an endomorphism of the linear manifold $A(\sigma - 1)$ on which we may apply III.1, Proposition 3. There exists therefore a number h in F such that $hx = x\beta$ for every x in $A(\sigma - 1)$. If $h \neq 0$, then we may infer from $A(\sigma - 1) \neq 0$ and $(fx)\beta = f(x\beta)$ for every f in F and x in $A(\sigma - 1)$ that h belongs to the center of F. Consider now any element d in V. Then $d(\sigma - 1)$ belongs to $A(\sigma - 1)$. Hence

$$hd(\sigma - 1) = d(\sigma - 1)\beta = d(\sigma - 1)\nu^{-1}(\gamma - 1) = d(\gamma - 1);$$

and consequently we find that

$$A[h(\sigma - 1) - (\gamma - 1)] = [J^+(\sigma) \oplus V]h\,[(\sigma - 1) - (\gamma - 1)] = V[h(\sigma - 1) - (\gamma - 1)] = 0$$

or $h(\sigma - 1) = \gamma - 1$. Hence

$$\tau = k\gamma = k[h(\sigma - 1) + 1]$$

where h and $k \neq 0$ are numbers in the center of F; and this completes the proof of Proposition 1.

Proposition 2: *Suppose that σ is a linear transformation of class 2, and that σ' is an involution such that $J^+(\sigma) = J^+(\sigma')$. Then the following properties of the linear transformation τ are equivalent.*

(I) *There exists a center element $h \neq 0$ in F such that $\tau = h(\sigma - 1) + 1$.*

(II) *τ is of class 2 and $Z(\sigma) = Z(\tau)$.*

(III) *There exists an element γ in $Z_2(\sigma')$ such that $\tau = \gamma^{-1}\sigma\gamma$.*

(IV) *The linear transformations σ and τ are conjugate in the group $T(F,A)$ of all linear transformations; and $Z(\sigma) \leqslant Z(\tau)$.*

PROOF: If (I) is true, then we have $\tau - 1 = h(\sigma - 1)$ and $\sigma - 1 = h^{-1}(\tau - 1)$. Since h is in the center of F, we find that $(\tau - 1)^2 = h^2(\sigma - 1)^2 = 0$ so that τ is of class 2. Application of Proposition 1 shows that $Z(\sigma) \leqslant Z(\tau) \leqslant Z(\sigma)$. Hence (II) is a consequence of (I). If conversely (II) is true, then we deduce from Proposition 1 the existence of numbers h,k in the center of F such that $\tau = k[h(\sigma - 1) + 1]$. Since τ and σ are of class 2, and since h,k are in the center of F, we find that

$$0 = (\tau - 1)^2 = [kh(\sigma - 1) + k - 1]^2 = 2k(k - 1)h(\sigma - 1) + (k - 1)^2.$$

This is equivalent to $(k-1)^2 = 2kh(1-k)(\sigma-1)$. But σ is of class 2; and so it follows that $(k - 1)^4 = 0$ which implies $k = 1$. Thus we have verified the equivalence of (I) and (II).

Assume next the validity of the equivalent properties (I) and (II). Denote by γ the uniquely determined linear transformation of A such that $x\gamma = hx$ for x in $J^+(\sigma') = J^+(\sigma)$ and $x\gamma = x$ for x in $J^-(\sigma')$. Since $h \neq 0$ is in the center of F, it follows from VI.2, Proposition 2 that γ belongs to $Z_2(\sigma')$. If x is an element in A, then $x = x' + x''$ with x' in $J^+(\sigma') = J^+(\sigma)$ and x'' in $J^-(\sigma')$. Hence $x'(\sigma - 1) = 0$ and consequently

$$\begin{aligned} x\gamma^{-1}(\sigma - 1)\gamma &= (x'\gamma^{-1} + x''\gamma^{-1})(\sigma - 1)\gamma = (h^{-1}x' + x'')(\sigma - 1)\gamma \\ &= x''(\sigma - 1)\gamma = hx''(\sigma - 1) = h[x' + x''](\sigma - 1) = hx(\sigma - 1) \end{aligned}$$

since $x''(\sigma - 1)$ belongs to $J^+(\sigma)$. Therefore

$$\gamma^{-1}\sigma\gamma - 1 = \gamma^{-1}(\sigma - 1)\gamma = h(\sigma - 1) = \tau - 1 \text{ or } \gamma^{-1}\sigma\gamma = \tau;$$

and this shows that (III) is a consequence of (I), (II).

Assume next the validity of (III). Then there exists an element γ in $Z_2(\sigma')$ such that $\tau = \gamma^{-1}\sigma\gamma$; and we infer from VI.2, Proposition 2 the existence of numbers h',h'' in the center of F neither of which is 0 such that

$$x\gamma = \begin{cases} h'x \text{ for } x \text{ in } J^+(\sigma') = J^+(\sigma) \\ h''x \text{ for } x \text{ in } J^-(\sigma'). \end{cases}$$

Since $x(\sigma - 1)$ belongs to $J^+(\sigma)$, we find for x in $J^-(\sigma')$ that

$$x\gamma^{-1}(\sigma - 1)\gamma = h''^{-1}x(\sigma - 1)\gamma = h''^{-1}h'x(\sigma - 1);$$

and for x in $J^+(\sigma') = J^+(\sigma)$ we find

$$x\gamma^{-1}(\sigma - 1)\gamma = 0 = h''^{-1}h'x(\sigma - 1).$$

Hence $\gamma^{-1}(\sigma - 1)\gamma = h''^{-1}h'(\sigma - 1)$ where $h''^{-1}h'$ is a center element in F.

Thus (I) is a consequence of (III) and we have shown the equivalence of (I) to (III).

In the presence of all three conditions (I) to (III) we are naturally assured of the validity of (IV). If conversely (IV) is valid, then τ is of class 2, since τ is conjugate to the linear transformation σ of class 2. From $Z(\sigma) \leqslant Z(\tau)$ and the argument used when deducing (I) from (II) we infer now the validity of (I); and this completes the proof of the equivalence of the conditions (I) to (IV).

Proposition 3: *The linear transformation σ is of class* 2 [*or* 1] *if, and only if, there exist involutions $\sigma'\sigma''$ such that $J^+(\sigma') = J^+(\sigma'')$ and $\sigma = \sigma'\sigma''$.*
PROOF: Assume first the existence of involutions σ', σ'' such that $J^+(\sigma') = J^+(\sigma'')$ and $\sigma = \sigma'\sigma''$. Let $S = J^+(\sigma') = J^+(\sigma'')$. If x is an element in A, then there exist uniquely determined elements x' and x'' in $J^-(\sigma')$ and $J^-(\sigma'')$ respectively such that $x - x'$ and $x - x''$ are both in S [since $A = S \oplus J^-(\sigma') = S \oplus J^-(\sigma'')$]. But then

$$x\sigma' = (x - x')\sigma' + x'\sigma' = (x - x') - x' = 2(x - x') - x \equiv -x \text{ modulo } S$$

and $x\sigma'' \equiv - x$ modulo S is seen likewise. Hence $x\sigma'\sigma'' \equiv x$ modulo S for every x in A; and for every x in S we have clearly $x\sigma'\sigma'' = x$. Consequently $A(\sigma'\sigma'' - 1) \leqslant S \leqslant J^+(\sigma'\sigma'')$; and this is equivalent to saying that $\sigma = \sigma'\sigma''$ is of class 2 [or 1 in case $\sigma = 1$].

Assume conversely that σ is of class 2. Then $A(\sigma - 1) \leqslant J^+(\sigma)$. There exists a subspace V such that $A = J^+(\sigma) \oplus V$; and there exists one and only one endomorphism σ' of A such that

$$x\sigma' = \begin{cases} -x + x(\sigma - 1) & \text{for } x \text{ in } V \\ x & \text{for } x \text{ in } J^+(\sigma). \end{cases}$$

If x is in V, then

$$x\sigma'^2 = [-x + x(\sigma - 1)]\sigma' = -[-x + x(\sigma - 1)] + x(\sigma - 1) = x,$$

since $x(\sigma - 1)$ belongs to $J^+(\sigma)$. Hence $\sigma'^2 = 1$ so that σ is an involution. Since $x(\sigma - 1) = 0$ for x in V implies $x = 0$, it follows that $J^+(\sigma') = J^+(\sigma)$. Let next $\sigma'' = \sigma'\sigma$. Then we have for x in V

$$x\sigma'' = x\sigma'\sigma = [-x + x(\sigma - 1)]\sigma = -x\sigma + x(\sigma - 1) = -x,$$

since $x(\sigma - 1)$ belongs to $J^+(\sigma)$; and now one sees that σ' is the uniquely determined involution such that $J^+(\sigma'') = J^+(\sigma)$ and $J^-(\sigma'') = V$. Thus we have represented $\sigma = \sigma'\sigma''$ in the desired form; and we have actually proved slightly more, namely the following

Corollary 1: *If the linear transformation σ is of class* 2, *if $A = J^+(\sigma) \oplus V$, and if σ'' is the uniquely determined involution such that $J^+(\sigma) = J^+(\sigma'')$ and $J^-(\sigma'') = V$, then $\sigma' = \sigma\sigma''$ is an involution such that $J^+(\sigma) = J^+(\sigma')$.*

We have now assembled all the facts which are actually needed for our desired multiplicative characterization of the linear transformations of class 2.

Theorem 1: *The linear transformation σ is of class 2 if, and only if, σ has the following properties.*

(*a*) *The linear transformation τ belongs to the center of $T(F,A)$ if, and only if, $Z(\sigma) < Z(\tau)$.*

(*b*) *The totality $Z^*(\sigma)$ composed of 1 and those transformations in the center of $Z(\sigma)$ which are conjugate to σ is a subgroup.*

(*c*) *There exists an involution σ' such that $\sigma'\tau\sigma' = \tau^{-1}$ for every τ in $Z^*(\sigma)$.*

(*d*) *There exists an involution α and a linear transformation σ'' in $Z_2(\alpha)$ such that $\sigma''^{-1}\tau\sigma'' = \tau^2$ for every τ in $Z^*(\sigma)$.*

(*e*) *If $Z^*(\sigma)$ has order 3, then the center of $T(F,A)$ has order 2.*

PROOF: We assume first that σ is of class 2. We are going to derive a number of properties of $Z(\sigma)$ which contain the above conditions (*a*) to (*e*), but which go a little further thus giving a better picture of the structure of $Z(\sigma)$.

We note first that σ is not in the center of T, since $\sigma \neq 1$ [VI.1, Proposition 1]; and that $Z(\tau) = T$ if, and only if, τ belongs to the center of T. Thus $Z(\sigma) < Z(\tau)$ for τ in the center of T. Assume conversely that $Z(\sigma) < Z(\tau)$. From Proposition 1 we deduce the existence of numbers $k \neq 0$ and h in the center of F such that $\tau = k[h(\sigma - 1) + 1]$. If $h \neq 0$, then $\sigma = h^{-1}[k^{-1}\tau - 1] + 1$ so that $Z(\tau) \leqslant Z(\sigma)$ which is impossible. Hence $h = 0$ and $\tau = k$ is in the center of T. This completes the proof of the validity of (*a*).

If τ is in the center of $Z(\sigma)$, then $Z(\sigma) \leqslant Z(\tau)$. If furthermore τ is conjugate to σ, then Proposition 2, (IV) is satisfied. Applying the equivalence of conditions (I) and (IV) of Proposition 2 we see the validity of the following statement.

(A.1) *$Z^*(\sigma)$ is the totality of linear transformations of the form $h(\sigma - 1) + 1$ with h in the center of F.*

Note that for $h = 0$ we obtain just the identity.

Suppose now that h', h'' are numbers in the center of F. Then we deduce from $(\sigma - 1)^2 = 0$ that

$$[h'(\sigma - 1) + 1][h''(\sigma - 1) + 1] = (h' + h'')(\sigma - 1) + 1.$$

[This equation contains as special cases

$$[h(\sigma - 1) + 1]^{-1} = (-h)(\sigma - 1) + 1 \text{ and } \sigma^2 = 2(\sigma - 1) + 1.]$$

From this equation one infers that

(A.2) *an isomorphism of the additive group of the center of F upon*

$Z^*(\sigma)$ is obtained by mapping the number h in the center of F upon the transformation $h(\sigma - 1) + 1$.

This implies in particular that $Z^*(\sigma)$ is a subgroup of the center of $Z(\sigma)$; and it follows from VI.1, Proposition 1 that the number of elements, not 1, in $Z^*(\sigma)$ is exactly the number of elements in the center of $T(F,A)$. Thus conditions (*b*) and (*e*) are verified.

We infer from Corollary 1 the existence of an involution σ' such that $J^+(\sigma) = J^+(\sigma') = J^+(\sigma\sigma')$ and such that $\sigma'' = \sigma\sigma'$ is an involution. Then

$$\sigma'\sigma\sigma' = \sigma'\sigma'' = (\sigma''\sigma')^{-1} = \sigma^{-1}.$$

Since σ is of class 2, we have $0 = (\sigma - 1)^2 = \sigma^2 - 2\sigma + 1$ so that

$$(2 - \sigma)\sigma = 1 \text{ or } \sigma^{-1} = 2 - \sigma \text{ and } \sigma^{-1} - 1 = 1 - \sigma.$$

If h is a number in the center of F, then

$$\begin{aligned}\sigma'[h(\sigma - 1) + 1]\sigma' &= h(\sigma'\sigma\sigma' - 1) + 1 = h(\sigma^{-1} - 1) + 1 \\ &= h(1 - \sigma) + 1 = (-h)(\sigma - 1) + 1 \\ &= [h(\sigma - 1) + 1]^{-1}\end{aligned}$$

by (A.2); and this proves the validity of (*c*).

Consider now some number $h \neq 0$ in the center of F. It follows from Proposition 2 and the choice of σ' that there exists a linear transformation $\gamma = \gamma(h)$ in $Z_2(\sigma')$ such that $\gamma^{-1}\sigma\gamma = h(\sigma - 1) + 1$. If z is any number in the center of F, then we find

$$\gamma^{-1}[z(\sigma - 1) + 1]\gamma = z(\gamma^{-1}\sigma\gamma - 1) + 1 = zh(\sigma - 1) + 1.$$

Noting that, by Proposition 2, every γ in $Z_2(\sigma')$ transforms σ into an element of the form $h(\sigma - 1) + 1$ with $h \neq 0$ in the center of F, we see that we have verified the following facts.

(A.3) *To every $h \neq 0$ in the center of F there exists a γ in $Z_2(\sigma')$ such that $\gamma^{-1}\tau\gamma = h(\tau - 1) + 1$ for every τ in $Z^*(\sigma)$; and to every γ in $Z_2(\sigma')$ there exists a number $h \neq 0$ in the center of F such that $\gamma^{-1}\tau\gamma = h(\tau - 1) + 1$ for every τ in $Z^*(\sigma)$.*

If we apply the first part of (A.3) in particular on $h = 2$ [which is different from 0, since the characteristic of F is not 2], and if we remember that every $\tau \neq 1$ in $Z^*(\sigma)$ is of class 2, then we find an element $\gamma = \gamma(2)$ in $Z(\sigma)$ such that

$$\gamma^{-1}\tau\gamma = 2(\tau - 1) + 1 = (\tau - 1)^2 + 2(\tau - 1) + 1 = \tau^2$$

for every τ in $Z^*(\sigma)$. This proves the validity of (*d*).

(A.4) *$Z(\sigma) \cap Z_2(\sigma')$ is the center of T and $Z(\sigma)Z_2(\sigma')$ is the normalizer of $Z(\sigma)$ in T.*

The normalizer of a subgroup consists of all the elements in T which transform the subgroup into itself. The proof of (A.4) may be deduced easily from the facts already derived, Proposition 2 and VI.2, Proposition 2. We omit the details, as no use will be made of (A.4).

We assume next that the linear transformation σ has the properties (*a*) to (*e*). Since σ belongs to $Z(\sigma)$, it follows from (*a*) that σ does not belong to the center of T.

We infer from condition (*d*) the existence of a linear transformation γ with the following properties.

(*d*.1) There exists an involution α such that γ belongs to $Z_2(\alpha)$.

(*d*.2) $\gamma^{-1}\tau\gamma = \tau^2$ for every τ in $Z^*(\sigma)$.

We deduce from (*d*.1) and VI.2, Corollary 2 the validity of the following facts.

(*d*.3) $$r\Big(\sum_{i=0}^{\infty} Fx\gamma^i\Big) \leqslant 2 \text{ and } \Big(\sum_{i=0}^{\infty} Fx\gamma^i\Big)\gamma = \sum_{i=0}^{\infty} Fx\gamma^i.$$

Next we prove that

(B.1) $r\Big(\sum_{i=0}^{\infty} Fx\tau^i\Big)$ is finite and $\Big(\sum_{i=0}^{\infty} Fx\tau^i\Big)\tau = \sum_{i=0}^{\infty} Fx\tau^i$ for every x in A and every τ in $Z^*(\sigma)$.

To prove this let $X = \sum_{i=0}^{\infty} Fx\gamma^i$. It follows from (*d*.3) that $r(X) \leqslant 2$ and $X = X\gamma$. This implies also $X = X\gamma^j$ for every positive and negative integer j. Naturally the rank of $X\tau$ cannot exceed 2 either. If

$$X\tau = Fx' + Fx'',$$

then we let $X' = \sum_{i=0}^{\infty} Fx'\gamma^i$ and $X'' = \sum_{i=0}^{\infty} Fx''\gamma^i$. It follows again from (*d*.3) that the ranks of X' and X'' do not exceed 2 and that

$$X'\gamma = X', \; X''\gamma = X''.$$

From $\sum_{i=0}^{\infty} X\tau\gamma^i \leqslant X' + X''$ and from $X = X\gamma^{-i}$ we deduce now that the totality of elements $x\gamma^{-i}\tau\gamma^i = x\tau^{2^i}$ for $0 \leqslant i$ belongs to the subspace $X' + X''$ of rank not exceeding 4. The elements $x\tau^{2^i}$ are therefore not independent. Consequently the elements $x, x\tau, \cdots, x\tau^j, \cdots$ are not independent either; and this proves the existence of a relation

$$x\tau^k = \sum_{j=0}^{k-1} c_j x\tau^j \text{ with } c_j \text{ in } F.$$

It is clear now that the rank of $\sum_{i=0}^{\infty} Fx\tau^i = V$ is finite [cannot exceed k];

and that $\left(\sum_{i=0}^{\infty} Fx\tau^i\right)\tau \leqslant \sum_{i=0}^{\infty} Fx\tau^i$. Since τ is a linear transformation, the subspaces V and $V\tau$ have equal finite rank; and $V = V\tau$ is a consequence of $V\tau \leqslant V$ and the Rank Formulas. This proves (B.1).

(B.2) $J^-(\tau) = 0$ for every τ in $Z^*(\sigma)$.

If x belongs to $J^-(\tau^{2^i})$ for some non-negative i, then

$$x\tau^{2^{i+1}} = - x\tau^{2^i} = x$$

so that

$$J^-(\tau^{2^i}) \leqslant J^+(\tau^{2^{i+1}}) \leqslant J^+(\tau^{2^j}) \text{ for } i < j.$$

Since the characteristic of F is not 2, we have always

$$J^+(\nu) \cap J^-(\nu) = 0;$$

and thus it follows that

$$J^-(\tau^{2^{i+1}}) \cap \sum_{j=0}^{i} J^-(\tau^{2^j}) \leqslant J^-(\tau^{2^{i+1}}) \cap J^+(\tau^{2^{i+1}}) = 0.$$

Assume now that $w \neq 0$ belongs to $J^-(\tau)$. Then

$$w\gamma^i\tau^{2^i} = w\tau\gamma^i = - w\gamma^i \text{ for } 0 < i$$

so that $w\gamma^i$ belongs to $J^-(\tau^{2^i})$. Since none of the $w\gamma^i$ is 0, it follows from $J^-(\tau^{2^{i+1}}) \cap \sum_{j=0}^{i} J^-(\tau^{2^j}) = 0$ that the infinitely many elements $w\gamma^i$ are independent; and this contradicts (d.3). Hence $J^-(\tau)$ cannot contain elements different from 0, proving (B.2).

(B.3) If τ is in $Z^*(\sigma)$, then $\tau + 1$ is a linear transformation of A.

To prove this we note first that $K(\tau + 1) = J^-(\tau) = 0$ by (B.2) so that $\tau + 1$ is a linear transformation of A upon its subspace $A(\tau + 1)$. Consider now some element x in A and form the subspace $X = \sum_{i=0}^{\infty} Fx\tau^i$. It follows from (B.1) that $r(X)$ is finite and that $X = X\tau$. Consequently $\tau + 1$ induces a linear transformation of X upon its subspace $X(\tau + 1)$. Since these two subspaces have equal finite rank, it follows from the Rank Formulas that $X = X(\tau + 1)$. Thus the random element x in A belongs to $X = X(\tau + 1) \leqslant A(\tau + 1)$; and this proves $A = A(\tau + 1)$. Hence $\tau + 1$ is a linear transformation of A.

(B.4) $\tau + \tau^{-1} = 2$ for every τ in $Z^*(\sigma)$.

This is obvious, if $\tau = 1$; and thus we may assume that $\tau \neq 1$. It follows from (*b*) that τ^2 belongs to $Z^*(\sigma)$; and it follows therefore from (B.3) that $\tau^2 + 1$ is a linear transformation of A. Consequently $\tau + \tau^{-1} = (\tau^2 + 1)\tau^{-1}$ is a linear transformation of A.

It is clear that $Z(\sigma) \leqslant Z(\tau)$, since τ is in the center of $Z(\sigma)$. But τ is not in the center of T, since σ and τ are conjugate, and since σ is not in the center of T. It follows from (*a*) that $Z(\sigma) = Z(\tau)$.

It is clear that $Z(\tau) \leqslant Z(\tau + \tau^{-1})$. There exists by (*c*) an involution σ' such that $\sigma'\tau\sigma' = \tau^{-1}$. It follows from (*d*) that $\tau^2 \neq 1$, since otherwise τ itself would be 1. Hence $\tau \neq \tau^{-1}$ and σ' does not belong to $Z(\tau)$. But $\sigma'(\tau + \tau^{-1})\sigma' = \tau^{-1} + \tau$, since σ' is an involution. Hence σ' belongs to $Z(\tau + \tau^{-1})$; and we have shown that

$$Z(\sigma) = Z(\tau) < Z(\tau + \tau^{-1}).$$

Since $\tau + \tau^{-1}$ is a linear transformation of A [belongs to T], it follows from condition (*a*) that $\tau + \tau^{-1}$ belongs to the center of T.

Now we distinguish two cases.

Case 1: $Z^*(\sigma)$ does not have order 3.

Then there exists an element τ' in $Z^*(\sigma)$ which is different from 1, τ and τ^{-1} [since $Z^*(\sigma)$ contains certainly these three distinct elements]. It follows from (*b*) that τ, τ', $\tau\tau'$ and $\tau\tau'^{-1}$ are elements in $Z^*(\sigma)$; and they are all conjugate in T, since they are all different from 1. Hence

$$\tau + \tau^{-1},\ \tau' + \tau'^{-1},\ \tau\tau' + (\tau\tau')^{-1},\ \tau\tau'^{-1} + (\tau\tau'^{-1})^{-1}$$

are conjugate center elements in T; and as such they have all the same value e, a center element, not 0, in F. Remembering the commutativity of $Z^*(\sigma)$ we find consequently that

$$e^2 = (\tau + \tau^{-1})(\tau' + \tau'^{-1}) = \tau\tau' + (\tau\tau')^{-1} + \tau\tau'^{-1} + (\tau\tau'^{-1})^{-1} = 2e$$

or $e = 2$, as we wanted to show.

Case 2: $Z^*(\sigma)$ has order 3.

Then $\tau^2 = \tau^{-1}$; and it follows from (*e*) that the center of T has order 2. But by VI.1, Proposition 2 the center of T is essentially the same as the multiplicative group of the center of F so that the center of F contains exactly 3 elements. Then F has characteristic 3. The element $\tau + \tau^{-1}$ in the center of T satisfies consequently

$$(\tau + \tau^{-1})^3 = \tau^3 + \tau^{-3} = 2;$$

and this implies that $\tau + \tau^{-1} = 2$, since the center of F consists of 0, 1 and 2 only. This completes the proof of (B.4).

Since σ belongs to $Z^*(\sigma)$, it follows from (B.4) that $\sigma + \sigma^{-1} = 2$ or $\sigma^2 + 1 = 2\sigma$; and this is clearly equivalent to $(\sigma - 1)^2 = 0$. Since $\sigma \neq 1$, this completes the proof of the fact that σ is of class 2.

Remark 1: If σ is an isomorphism of $T(F,A)$ upon $T(G,B)$, then σ maps elements in $T(F,A)$ which have properties (*a*) to (*e*) of Theorem 1 upon elements in $T(G,B)$ having these same properties. Thus it follows from Theorem 1 that

the isomorphism σ maps linear transformations of class 2 *upon linear transformations of class* 2.

It is an interesting problem to decide whether it is possible to omit any of the five conditions (*a*) to (*e*) of Theorem 1. Not much seems to be known at present. Thus the following criteria may be of some interest since they permit to simplify the five conditions at least in special situations.

Proposition 4: *The linear transformation σ is of class* 2 *if, and only if, σ is of finite class and satisfies condition* (*a*) *of Theorem* 1.

Proof: The necessity of the conditions is obvious [Theorem 1]. Thus we assume that σ satisfies these two conditions. It is clear that σ does not belong to the center of T so that in particular σ is not of class 1. Since σ is of finite class, there exists an integer m such that $1 < m$, $(\sigma - 1)^{m-1} \neq 0$, but $(\sigma - 1)^m = 0$.

Form $\tau = 1 + (\sigma - 1)^{m-1}$. Then $\tau \neq 1$, $(\tau - 1)^2 = (\sigma - 1)^{2m-2} = 0$, since $2 \leqslant m$. Consequently $1 = (2 - \tau)\tau$; and we have shown that τ is a linear transformation of class 2. This implies in particular that τ does not belong to the center of T.

Since we have obviously $Z(\sigma) \leqslant Z(\tau)$, and since τ is not in the center of T, we infer from condition (*a*) [of Theorem 1] that $Z(\sigma) = Z(\tau)$. Since τ is of class 2, we deduce from Proposition 1 the existence of numbers $k \neq 0$ and h in the center of F such that

$$\sigma = k[h(\tau - 1) + 1] = k[h(\sigma - 1)^{m-1} + 1].$$

Multiplying this equation by $\sigma - 1$ we deduce from $(\sigma - 1)^m = 0$ that

$$\sigma(\sigma - 1) = k(\sigma - 1) \text{ or } (\sigma - 1)^2 = (k - 1)(\sigma - 1);$$

and if we multiply this last equation by $(\sigma - 1)^{m-2}$, we find that

$$0 = (\sigma - 1)^m = (k - 1)(\sigma - 1)^{m-1}.$$

This would imply $0 = (\sigma - 1)^{m-1}$, if the number $k - 1$ in the center of F were not 0. Hence $k = 1$ implying $(\sigma - 1)^2 = (k - 1)(\sigma - 1) = 0$. Thus σ is of class 2, as we intended to show.

Remark 2: It should be noted that the condition: "σ is of finite class" is an additive condition and is therefore not an acceptable substitute for the multiplicative conditions (*b*) to (*f*) of Theorem 1.

Theorem 2: *Assume that the field F is of prime number characteristic p. Then the linear transformation σ of (F,A) is of class* 2 *if, and only if, $\sigma^p = 1$ and σ satisfies condition* (*a*) *of Theorem* 1.

PROOF: Since the prime number p is the characteristic of the field F, we have $(\sigma - 1)^p = \sigma^p - 1$. Hence $\sigma^p = 1$ if, and only if, σ is of class not exceeding p; and Theorem 2 is an immediate consequence of Proposition 4.

VI.4. Cosets of Involutions

If S is a subspace of the linear manifold A, then we denote by $\Delta(S)^+$ the totality of involutions σ of A such that $S = J^+(\sigma)$; and we denote by $\Delta(S)^-$ the totality of involutions σ such that $S = J^-(\sigma)$. We shall show in the present section that these systems Δ can be characterized in purely multiplicative terms within the group T, and that they reflect faithfully all the properties of the projective geometry of A [i.e. of the lattice of subspaces of A].

We note first $\Delta(S)^- = -\Delta(S)^+$. Every involution σ belongs to $\Delta[J^+(\sigma)]^+$ and $\Delta[J^-(\sigma)]^-$ and to no further systems of this type. Whenever we are in need of a short name, we shall refer to $\Delta(S)^+$ and $\Delta(S)^-$ as to Δ-systems; and we let $\Delta(S) = [\Delta(S)^+, \Delta(S)^-]$.

The following *notations* will be used throughout. If Φ is a system of involutions, then Φ^2 consists of all the linear transformations of the form $\sigma'\sigma''$ with σ' and σ'' in Φ; and $N\Phi$ consists of all the involutions σ which satisfy $\Phi = \sigma\Phi\sigma$. Note that $N\Phi$ consists of all the involutions in the soc. normalizer of Φ in T.

Proposition 1: *Every Δ-system Φ has the following properties.*

(*a*) *If α,β,γ belong to Φ, then $\alpha\beta\gamma = \gamma\beta\alpha$ belongs to Φ.*

(*b*) *If α,β are in Φ, then there exists one and only one γ in Φ such that $\gamma\beta\gamma = \alpha$.*

(*c*) *The involution σ belongs to $N\Phi$ if, and only if, there exists an involution α in Φ such that $\alpha\sigma = \sigma\alpha$.*

(*d*) *Every transformation in Φ^2 is of class* 1 *or* 2.

PROOF: If Φ is a Δ-system, then $\Phi = \Delta(S)^+$ or $\Phi = \Delta(S)^-$ for some subspace S of A. Since $\Delta(S)^- = -\Delta(S)^+$, there is no loss in generality, if we assume that $\Phi = \Delta(S)^+$.

We prove first:

(*a*.1) *If α,β,γ belong to Φ, then $\alpha\beta\gamma = \alpha - \beta + \gamma$.*

Clearly $S = J^+(\alpha) = J^+(\beta) = J^+(\gamma)$. Consequently the elements $x + x\alpha$, $x + x\beta$ and $x + x\gamma$ belong always to S and are therefore left invariant by α, β, and γ. Now we find that

$$\begin{aligned} x(\alpha - \beta + \gamma) &= x\alpha - x\beta + x\gamma = (x + x\alpha) - (x + x\beta) + x\gamma \\ &= [(x + x\alpha) - (x + x\beta) + x]\gamma = [x + x\alpha - x\beta]\gamma \\ &= x + x\alpha - x\beta\gamma = (x + x\alpha)\beta\gamma - x\beta\gamma = x\alpha\beta\gamma \end{aligned}$$

or $\alpha - \beta + \gamma = \alpha\beta\gamma$.

From $(a.1)$ we deduce that $\alpha\beta\gamma = \gamma\beta\alpha$ and that consequently

$$(\alpha\beta\gamma)^2 = \alpha\beta\gamma\gamma\beta\alpha = 1.$$

If x belongs to $J^+(\alpha\beta\gamma)$, then

$$x = x(\alpha + \gamma - \beta) \text{ or } x + x\beta = x\alpha + x\gamma;$$

and this shows that $x\alpha + x\gamma$ belongs to S. But

$$x\alpha + x\gamma = x + x\alpha + x\gamma - x;$$

and it follows that $x\gamma - x$ belongs to S, since $x + x\alpha$ certainly is in S. But $x\gamma - x$ belongs likewise to $J^-(\gamma)$; and so it belongs to the intersection 0 of $J^+(\gamma) = S$ and $J^-(\gamma)$. Hence $x = x\gamma$ belongs to S; and now it is fairly obvious that $S = J^+(\alpha\beta\gamma)$. This completes the proof of (a).

Suppose next that β,γ,η are involutions in Φ such that $\gamma\beta\gamma = \eta\beta\eta$. It follows from $(a.1)$ that $2\gamma - \beta = 2\eta - \beta$; and this implies $\gamma = \eta$, proving the uniqueness part of (b). If α,β are in Φ, then we let $\gamma = \frac{1}{2}(\alpha + \beta)$; and we deduce from $(a.1)$ that

$$\gamma^2 = 4^{-1}(2 + \alpha\beta + \beta\alpha) = 4^{-1}[2 + (\alpha\beta\alpha + \beta)\alpha] = 4^{-1}[2 + (2\alpha - \beta + \beta)\alpha] = 1$$

so that γ is an involution. It is clear that $S \leqslant J^+(\gamma)$; and if x is in $J^+(\gamma)$, then we find that

$$4x = 2x + 2x\gamma = 2x + x(\alpha + \beta) = (x + x\alpha) + (x + x\beta)$$

belongs to S, since the first summand is left invariant by α and the second by β so that they both belong to S. But with $4x$ also x belongs to the subspace S; and thus we have shown that $S = J^+(\gamma)$. Hence γ belongs to Φ. Now we deduce from $(a.1)$ that

$$\gamma\alpha\gamma = 2\gamma - \alpha = \alpha + \beta - \alpha = \beta;$$

and this completes the proof of (b).

Assume next that σ belongs to $N\Phi$ and that α is some involution in Φ. Then $\sigma\alpha\sigma$ belongs to Φ and we deduce from (b) the existence of one and only one involution β in Φ such that $\sigma\alpha\sigma = \beta\alpha\beta$. It follows from $(a.1)$ that

$$\sigma\alpha\sigma = \beta\alpha\beta = 2\beta - \alpha.$$

Pre- and post-multiplication of this equation by the involution σ results in

$$\alpha = 2\sigma\beta\sigma - \sigma\alpha\sigma = 2\sigma\beta\sigma - \beta\alpha\beta = 2\sigma\beta\sigma - (2\beta - \alpha),$$
$$2\beta = 2\sigma\beta\sigma \text{ or } \beta\sigma = \sigma\beta.$$

This shows the necessity of the condition given under (c).

Assume conversely that σ is an involution such that $\sigma\alpha = \alpha\sigma$ for some α in Φ. Then

$$S = J^+(\alpha) = J^+(\sigma\alpha\sigma) = J^+(\alpha)\sigma = S\sigma,$$
$$J^+(\sigma\beta\sigma) = S\sigma = S \text{ for every } \beta \text{ in } \Phi,$$
$$\sigma\Phi\sigma = \Phi,$$

since $\Phi = \Delta(S)^+$. This completes the proof of (c).

The validity of (*d*) is finally an immediate consequence of VI.3, Proposition 3.

Proposition 2: *Suppose that the set* Φ *of involutions has the properties* (*a*) *to* (*d*) *of Proposition* 1. *Then*

(1) Φ^2 *is an abelian subgroup of* T;

(2) $\sigma\sigma'\sigma = \sigma'^{-1}$ *for every* σ *in* Φ *and every* σ' *in* Φ^2.

(3) *If* Φ *contains at least two involutions, then* $\Phi \leqslant \Delta[J^+(\Phi^2)]^+$ *or* $\Phi \leqslant \Delta[J^+(\Phi^2)]^-$.

PROOF: If σ',σ'' belong both to Φ^2, then there exist elements $\alpha,\beta,\gamma,\delta$ in Φ such that $\sigma' = \alpha\beta$, $\sigma'' = \gamma\delta$. It follows from (*a*) that $\alpha\beta\gamma$ is in Φ. Hence $\sigma'\sigma'' = (\alpha\beta\gamma)\delta$ belongs to Φ^2. Since $\sigma'^{-1} = \beta\alpha$, we see that Φ^2 is a subgroup. Next we deduce from (*a*) that

$$\sigma'\sigma'' = \alpha\beta\gamma\delta = \gamma\beta\alpha\delta = \gamma\delta\alpha\beta = \sigma''\sigma'.$$

Hence Φ^2 is abelian.

If σ,α,β are in Φ, then it follows from (*a*) that

$$\sigma\alpha\beta\sigma = \beta\alpha\sigma^2 = \beta\alpha = (\alpha\beta)^{-1};$$

and this proves (2).

Next we show that

$$(3') \qquad J^+(\Phi^2) \neq 0.$$

It is a consequence of condition (*d*) that $(\sigma - 1)^2 = 0$ for every σ in Φ^2. If σ',σ'' belong both to Φ^2, then it follows from (1) that $\sigma'\sigma'' = \sigma''\sigma'$ belongs likewise to Φ^2. Hence

$$\begin{aligned} 0 &= (\sigma'\sigma'' - 1)^2 = (\sigma'\sigma'')^2 - 2\sigma'\sigma'' + 1 = \sigma'^2\sigma''^2 - 2\sigma'\sigma'' + 1 \\ &= (2\sigma' - 1)(2\sigma'' - 1) - 2\sigma'\sigma'' + 1 = 2(\sigma'\sigma'' - \sigma' - \sigma'' + 1) \\ &= 2(\sigma' - 1)(\sigma'' - 1) \end{aligned}$$

or $(\sigma' - 1)(\sigma'' - 1) = 0$ for every σ',σ'' in Φ^2. But this implies

$$A(\sigma' - 1) \leqslant K(\sigma'' - 1) = J^+(\sigma'') \text{ for every } \sigma',\sigma'' \text{ in } \Phi^2.$$

If every $A(\sigma - 1)$ for σ in Φ^2 is 0, then Φ^2 consists of 1 only so that $J^+(\Phi^2) = A$; and otherwise we derive from the preceding inequality that

$$(3'') \qquad \sum_{\sigma\in\Phi^2} A(\sigma - 1) \leqslant \bigcap_{\sigma\in\Phi^2} J^+(\sigma) = J^+(\Phi^2),$$

an inequality which is slightly sharper than the desired (3′).

Suppose now that σ belongs to $N\Phi$. Then clearly $\sigma\Phi^2\sigma = \Phi^2$ so that the involution σ belongs likewise to $N\Phi^2$. It follows that

$$(4') \qquad J^+(\Phi^2) = J^+(\sigma\Phi^2\sigma) = J^+(\Phi^2)\sigma$$

for every involution σ in $N\Phi$.

Consider now an involution α in Φ. If τ is an involution such that $\tau\alpha = \alpha\tau$, then it follows from condition (*c*) of Proposition 1 that τ belongs to $N\Phi$; and consequently it follows from (4′) that $J^+(\Phi^2) = J^+(\Phi^2)\tau$ for every involution τ in $Z(\alpha)$. We apply VI.2, Proposition 3 and find that

$$(4'') \qquad J^+(\Phi^2) = 0 \text{ or } J^+(\alpha) \text{ or } J^-(\alpha) \text{ or } A \text{ for every } \alpha \text{ in } \Phi.$$

It follows from (3′) that $J^+(\Phi^2) \neq 0$; and $J^+(\Phi^2) = A$ would imply that $\Phi^2 = 1$ in which case Φ would consist of one involution only, a possibility which we are going to exclude from now on [see the additional hypothesis of (3)].

Suppose now that $J^+(\Phi^2) = J^+(\alpha)$ for some α in Φ. If β is some further involution in Φ, then we infer from condition (*b*) of Proposition 1 the existence of an involution γ in Φ such that $\beta = \gamma\alpha\gamma$. Hence

$$J^+(\beta) = J^+(\gamma\alpha\gamma) = J^+(\alpha)\gamma = J^+(\Phi^2)\gamma = J^+(\Phi^2)$$

may be inferred from (4′), since γ as an involution in Φ belongs to $N\Phi$ too. This implies $\Phi \leqslant \Delta[J^+(\Phi^2)]^+$. If however $J^+(\Phi^2) \neq J^+(\alpha)$ for every α in Φ, then it follows from the preceding considerations [see (4″)] that $J^+(\Phi^2) = J^-(\alpha)$ for every α in Φ; and this implies $\Phi \leqslant \Delta[J^+(\Phi^2)]^-$. This completes the proof of (3).

We are now ready to give the desired multiplicative characterization of the Δ-systems.

Theorem 1: *The set Φ of involutions is a Δ-system if, and only if, Φ is a maximal system with properties (a) to (d) of Proposition* 1.

Proof: Assume first that S is a subspace and that $\Phi = \Delta(S)^+$. Then we deduce from Proposition 1 that Φ meets the requirements (*a*) to (*d*). Suppose now that $\Phi < \Phi'$ and that Φ' too meets the requirements (*a*) to (*d*). Let $S' = J^+(\Phi'^2)$. From $\Phi^2 \leqslant \Phi'^2$ we deduce $S' = J^+(\Phi'^2) \leqslant J^+(\Phi^2) = S$. Since $\Phi < \Phi'$, it follows that Φ' contains at least two involutions; and we deduce from Proposition 2, (3) that $\Phi' \leqslant \Delta(S')^+$ or $\Phi' \leqslant \Delta(S')^-$. In the second case we would deduce from $\Phi < \Phi'$ and $S' \leqslant S$ that we have for every σ in Φ the following equations:

$$J^-(\sigma) = S' \leqslant S = J^+(\sigma)$$

which clearly implies $J^-(\sigma) = 0$ and $\sigma = 1$. But the elements in Φ', not 1, which exist because of $\Phi < \Phi'$ cannot be conjugate to $\sigma = 1$, contradicting condition (*b*). Thus it follows that $\Phi' \leqslant \Delta(S')^+$. If σ belongs to Φ, then σ is in Φ' too; and it follows that $S = J^+(\sigma) = S'$. But this implies

$$\Delta(S)^+ = \Phi < \Phi' \leqslant \Delta(S')^+ = \Delta(S)^+,$$

a contradiction which proves the maximality of Φ.

Assume now conversely that Φ is a maximal set of involutions satisfying the conditions (*a*) to (*d*) of Proposition 1. If Φ consists of 1 only, then $\Phi = \Delta(A)^+$; and if Φ consists of - 1 only, then $\Phi = \Delta(A)^-$. Assume next that Φ consists of one involution σ only; and assume that $\sigma \neq \pm 1$. Then $0 < J^+(\sigma) < A$; and it is easy to construct an involution $\sigma' \neq \sigma$ such that $J^+(\sigma) = J^+(\sigma')$. Consequently $\Phi < \Delta[J^+(\sigma)]^+ = \Phi'$; and it follows from Proposition 1 that Φ' too has properties (*a*) to (*d*) of Proposition 1. This contradicts the assumed maximality of Φ; and thus we have shown that Φ contains exactly one involution if, and only if, Φ consists of $+1$ or -1 only. Assume finally that Φ contains at least two distinct involutions. We apply Proposition 3 and find that $\Phi \leqslant \Delta(S)^+$ or $\Phi \leqslant \Delta(S)^-$ where $S = J^+(\Phi^2)$. But Δ-systems have properties (*a*) to (*d*) of Proposition 1; and now it follows from the maximality of Φ that $\Phi = \Delta(S)^+$ or $\Phi = \Delta(S)^-$. Hence Φ is a Δ-system. This completes the proof of our theorem and of the following fact.

Corollary 1: *Φ is a maximal system of involutions with properties (a) to (d) of Proposition 1 which contains just one involution if, and only if, the involution in Φ is ± 1.*

We say that the subspace V is *between* the subspaces U and W, if $U \leqslant V \leqslant W$ *or* $W \leqslant V \leqslant U$. Our next objective is to characterize betweenness of subspaces in terms of the corresponding Δ-systems. The solution of this problem will be contained in Theorem 2 below whose statement and proof we precede by the proofs of several propositions.

Proposition 3: *The following properties of the subspaces S and T of A are equivalent.*

(I) $S \leqslant T$ *or* $T \leqslant S$.

(II) $\Delta(S)^+ \leqslant N\Delta(T)^+$.

(III) $\Delta(T)^+ \leqslant N\Delta(S)^+$.

REMARK: It is clear that we may substitute minus-signs for the plus-signs in conditions (II) and (III), since $\Delta(X)^- = -\Delta(X)^+$.

PROOF: Assume first that $S \leqslant T$. If σ belongs to $\Delta(S)^+$, then $S = J^+(\sigma)$; and it follows from $S \leqslant T$ and $A = J^+(\sigma) \oplus J^-(\sigma)$, that

$$T = S \oplus [T \cap J^-(\sigma)].$$

Denote by U some subspace such that

$$J^-(\sigma) = [T \cap J^-(\sigma)] \oplus U;$$

and denote by σ' the uniquely determined involution such that $T = J^+(\sigma')$, $U = J^-(\sigma')$. It is clear that σ' belongs to $\Delta(T)^+$; and it follows from VI.2, Proposition 1 that $\sigma'\sigma = \sigma\sigma'$. We deduce from Proposition 1, (*c*) that σ belongs to $N\Delta(T)^+$. Thus (II) is a consequence of $S \leqslant T$. Assume next that τ belongs to $\Delta(T)^+$. Then $T = J^+(\tau)$. There exists a subspace V

such that $T = S \oplus V$. Denote by τ' the uniquely determined involution such that

$$S = J^{+}(\tau'),\ V \oplus J^{-}(\tau) = J^{-}(\tau').$$

It is clear that τ' belongs to $\Delta(S)^{+}$; it follows from VI.2, Proposition 1 that $\tau\tau' = \tau'\tau$; and we deduce from Proposition 1, (c) that τ belongs to $N\Delta(S)^{+}$. Hence (III) is a consequence of $S \leqslant T$. An interchange of S and T effects an interchange of conditions (II) and (III); and now we see that (I) implies both (II) and (III).

Assume next that neither $S \leqslant T$ nor $T \leqslant S$. There exist subspaces S' and T' such that $S = (S \cap T) \oplus S'$ and $T = (S \cap T) \oplus T'$; and it follows from our hypothesis that neither S' nor T' is 0. Denote by s and t elements, different from 0, in S' and T' respectively; and denote by T'' some subspace satisfying $T' = Ft \oplus T''$. Finally denote by W some subspace such that $A = (S + T) \oplus W$. Then

$$A = (S \cap T) \oplus S' \oplus Ft \oplus T'' \oplus W = (S \cap T) \oplus S'' \oplus F(t + s) \oplus T'' \oplus W,$$

since $t \neq 0$ and s is in S'; and consequently there exists one and only one involution ω such that $J^{+}(\omega) = S$, $J^{-}(\omega) = F(t + s) \oplus T'' \oplus W$. It is clear that ω belongs to $\Delta(S)^{+}$. But

$$t\omega = (t + s)\omega - s\omega = -(t + s) - s = -t - 2s$$

does not belong to T, since $s \neq 0$ is in S' and hence in S, though not in $S \cap T$. Thus $T\omega \neq T$. If τ belongs to $\Delta(T)^{+}$, then $T = J^{+}(\tau)$; and it follows from VI.2, Proposition 1 that $\omega\tau \neq \tau\omega$. Now we deduce from Proposition 1, (c) that ω does not belong to $N\Delta(T)^{+}$ or $\Delta(S)^{+} \nleqslant N\Delta(T)^{+}$. Thus (II) is not true, if (I) fails to hold, so that (I) is a consequence of (II); and one sees similarly that (I) is a consequence of (III). This completes the proof.

If S is some subspace of A, and if σ', σ'' belong to $\Delta(S)^{+}$, then $-\sigma'$ and $-\sigma''$ belong to $\Delta(S)^{-}$; and we have $\sigma'\sigma'' = (-\sigma')(-\sigma'')$. Hence

$$[\Delta(S)^{+}]^{2} = [\Delta(S)^{-}]^{2};$$

and it is justified to denote this subgroup more shortly by $\Delta(S)^{2}$.

Lemma 1: *The linear transformation σ belongs to $\Delta(S)^{2}$ if, and only if, $A(\sigma - 1) \leqslant S \leqslant J^{+}(\sigma)$.*

Proof: If σ belongs to $\Delta(S)^{2}$, then there exist involutions σ', σ'' such that $\sigma = \sigma'\sigma''$ and $S = J^{+}(\sigma') = J^{+}(\sigma'')$. It is clear that $S \leqslant J^{+}(\sigma)$ and from $x(\sigma - 1) = (x\sigma' + x)\sigma'' - (x\sigma'' + x)$ we deduce that $A(\sigma - 1) \leqslant S$, since $x\sigma' + x$ and $x\sigma'' + x$ belongs to $J^{+}(\sigma') = S$ and $J^{+}(\sigma'') = S$ respectively.

Assume conversely that $A(\sigma - 1) \leqslant S \leqslant J^{+}(\sigma)$. There exist subspaces V, W such that $J^{+}(\sigma) = S \oplus V$ and $A = J^{+}(\sigma) \oplus W$. There exists one and only one involution σ'' such that $J^+(\sigma'') = S$, $J^-(\sigma'') = V \oplus W$. It is clear that σ'' belongs to $\Delta(S)^+$. Let $\sigma' = \sigma\sigma''$. Then

$$x\sigma' = x\sigma\sigma'' = x(\sigma - 1)\sigma'' + x\sigma'' = x(\sigma - 1) + x\sigma'',$$

since $A(\sigma - 1) \leqslant S = J^+(\sigma'')$; and hence $\sigma' = \sigma + \sigma'' - 1$. From $(\sigma - 1)^2 = 0$ and $\sigma''^2 = 1$ we deduce next that

$$\sigma'^2 = [(\sigma - 1) + \sigma'']^2 = (\sigma - 1)\sigma'' + \sigma''(\sigma - 1) + 1.$$

From

$$A(\sigma - 1) \leqslant S = J^+(\sigma'') = A(1 + \sigma'') \leqslant J^+(\sigma)$$

we deduce again that $(\sigma - 1)\sigma'' = \sigma - 1$ and $(\sigma'' + 1)(\sigma - 1) = 0$; and this implies $\sigma'^2 = 1$. It is clear that $S \leqslant J^+(\sigma')$. If conversely x belongs to $J^+(\sigma')$, then $x = x\sigma' = x(\sigma + \sigma'' - 1)$ so that $-x(\sigma - 1) = x(\sigma'' - 1)$ belongs to the intersection $A(\sigma - 1) \cap J^-(\sigma'') \leqslant S \cap (V + W) = 0$. Hence $x = x\sigma''$ so that x belongs to $J^+(\sigma'') = S$; and we have shown that σ' too belongs to $\Delta(S)^+$. Consequently $\sigma = \sigma'\sigma''$ belongs to $\Delta(S)^2$, as we intended to prove.

REMARK 1: Compare result and proof with VI.3, Proposition 3 and VI.3, Corollary 1.

Proposition 4: *Assume that S, T and X are subspaces of A.*

(*a*) $0 = S \cap T$ *or* $A = S + T$ *if, and only if,* $\Delta(S)^2 \cap \Delta(T)^2 = 1$.

(*b*) $0 < S \cap T \leqslant X \leqslant S + T < A$ *if, and only if,*

$$1 < \Delta(S)^2 \cap \Delta(T)^2 \leqslant \Delta(X)^2.$$

PROOF: It is an immediate consequence of Lemma 1 that the linear transformation σ belongs to $\Delta(S)^2 \cap \Delta(T)^2$ if, and only if,

$$A(\sigma - 1) \leqslant S \cap T \text{ and } S + T \leqslant J^+(\sigma).$$

Noting the equivalence of the assertions $\sigma = 1$, $A(\sigma - 1) = 0$ and $J^+(\sigma) = A$ one verifies immediately the validity of (*a*).

To prove (*b*) we make the [by (*a*) equivalent] hypotheses that

$$0 < S \cap T,\ S + T < A \text{ and } 1 \neq \Delta(S)^2 \cap \Delta(T)^2.$$

Assume first that $S \cap T \leqslant X \leqslant S + T$. Every element σ in $\Delta(S)^2 \cap \Delta(T)^2$ satisfies

$$A(\sigma - 1) \leqslant S \cap T \leqslant X \leqslant S + T \leqslant J^+(\sigma);$$

and it follows from Lemma 1 that σ belongs to $\Delta(X)^2$, proving the necessity of our condition. Assume next that $S \cap T \not\leqslant X$. Then there exists an element $w \neq 0$ in $S \cap T$, but not in X. Since $S + T < A$, there exists a hyperplane H and an element $a \neq 0$ such that

$$S + T \leqslant H \quad\text{and}\quad A = Fa \oplus H.$$

There exists furthermore one and only one linear transformation ω such that $x\omega = x$ for x in H and $a\omega = a + w$. Then

$$A(\omega - 1) = Fw \leqslant S \cap T \leqslant S + T \leqslant H = J^+(\omega)$$

so that ω belongs to $\Delta(S)^2 \cap \Delta(T)^2$. But $A(\omega - 1) \not\leqslant X$ so that ω does not belong to $\Delta(X)^2$ [by Lemma 1]. Assume finally that $X \not\leqslant S + T$. Then there exists an element $y \neq 0$ in X which does not belong to $S + T$. Denote by Y some subspace such that $A = (S + T) \oplus Fy \oplus Y$. Since $S \cap T \neq 0$ there exists an element $z \neq 0$ in $S \cap T$; and there exists one and only one linear transformation ν such that $x\nu = x$ for x in $S + T + Y$ and $y\nu = y + z$. As before one verifies that ν belongs to the cross cut of $\Delta(S)^2$ and $\Delta(T)^2$; but ν does not belong to $\Delta(X)^2$, since

$$X \not\leqslant J^+(\nu) \text{ [as } y \neq y\nu].$$

This completes the proof of (*b*).

Before stating our next results we introduce some simplification of our *notation*. Since $N\Delta(X)^+ = N\Delta(X)^-$, we term this system of involutions just $N\Delta(X)$; and the pair $[\Delta(X)^+, \Delta(X)^- = -\Delta(X)^+]$ we have denoted by $\Delta(X)$. Using this notation Proposition 3, (II) takes the simpler and more comprehensive form: $\Delta(S) \leqslant N\Delta(T)$.

Theorem 2: *Suppose that S, T and X are subspaces of A. Then S and T are both different from 0 and A and X is between S and T if, and only if,*

(*a*) $\Delta(S) \leqslant N\Delta(T)$ *and/or* $\Delta(T) \leqslant N\Delta(S)$,

(*b*) $1 < \Delta(S)^2 \cap \Delta(T)^2 \leqslant \Delta(X)^2$.

Proof: Assume first that X is between S and T and that neither S nor T is 0 or A. Then we may assume that $0 < S \leqslant X \leqslant T < A$. It follows from Proposition 3 that (*a*) is true and that the two inequalities stated under (*a*) are equivalent. Since $S = S \cap T$ and $T = S + T$, we may deduce the validity of (*b*) from Proposition 4, (*b*).

Assume next the validity of (*a*) and (*b*). It follows from (*a*) and Proposition 3 that $S \leqslant T$ or $T \leqslant S$; and we may assume that $S \leqslant T$ and hence $S = S \cap T$, $T = S + T$. It follows from (*b*) and Proposition 4, (*b*) that

$$0 < S \cap T = S \leqslant X \leqslant S + T = T < A.$$

Hence X is between S and T and neither S nor T is 0 or A.

We have mapped every subspace S of A upon a certain substructure of the group $\mathrm{T}(F,A)$ which consists of the basic systems

$$\Delta(S) = [\Delta(S)^+, \Delta(S)^-]$$

and the derived systems $\Delta(S)^2$ and $N\Delta(S)$. In Theorem 1 we have given a complete characterization of the class of substructures of T which are images of subspaces S under this mapping Δ; and in Theorem 2 we have shown how the relation of betweenness is reflected by properties of this mapping Δ. We have still to decide the question to which extent the mapping Δ is one to one.

Theorem 3: (*a*) *The following properties* (I) *to* (IV) *of the subspace S of A are equivalent.*

(I) $S = 0$ *or* A.

(II) $\Delta(S) = [+1, -1]$.

(III) $\Delta(S)^2 = 1$.

(IV) $N\Delta(S)$ *consists of all the involutions in* T.

(*b*) *If* $0 < S \leqslant A$, *then* $S = J^+[\Delta(S)^2]$.

(*c*) *If* Φ *is a maximal set of involutions with properties* (*a*) *to* (*d*) *of Proposition* 1, *then* $\Delta[J^+(\Phi^2)] = [\Phi, -\Phi]$.

Proof: It is obvious that the properties (I) and (II) are equivalent and that the properties (III) and (IV) are consequences of (II). If (III) is true, then $\Delta(S)^+$ consists of one involution σ only; and $\sigma = \pm 1$, since $\Delta(S)^+$ is a maximal system of involutions with properties (*a*) to (*d*) of Proposition 1 [by Theorem 1], and since therefore we may apply Corollary 1. Thus (II) is a consequence of (III). If (IV) is true, then we have $\Delta(X) \leqslant N\Delta(S)$ for every subspace X of A; and it follows from Proposition 3 that $S \leqslant X$ or $X \leqslant S$ for every subspace X of A. But the only subspaces S of A with this property are clearly 0 and A; and thus (I) is a consequence of (IV). This completes the proof of (*a*).

Assume next that $0 \neq S$. If x is an element in A, but not in S, then $A = S \oplus Fx \oplus X$ for some subspace X [Complementation Principle]. Since $S \neq 0$, there exists an element $s \neq 0$ in S; and since $x \neq 0$, we have $A = S \oplus F(x+s) \oplus X$. Consequently there exists one and only one linear transformation σ such that $y\sigma = y$ for y in $S \oplus X$ and $x\sigma = x + s$. Clearly $A(\sigma - 1) = Fs \leqslant S \leqslant J^+(\sigma)$ so that σ belongs to $\Delta(S)^2$ [Lemma 1]. But $x\sigma \neq x$, since $s \neq 0$. Hence x does not belong to $J^+[\Delta(S)^2]$. On the other hand it is an immediate consequence of Lemma 1 that $S \leqslant J^+[\Delta(S)^2]$; and thus we have shown that $S = J^+[\Delta(S)^2]$ for $S \neq 0$. This proves (*b*).

The validity of (*c*) finally is easily deduced from Proposition 2, (3), Theorem 1 and Corollary 1.

If Φ is a maximal system of involutions with properties (*a*) to (*d*) of Proposition 1, then we may term $[\Phi, -\Phi]$ a Δ-*system*. Theorems 1 and 3 imply then the validity of the following basic result.

Theorem 4: *Mapping S onto $\Delta(S)$ and mapping $[\Phi, -\Phi]$ upon $J^+(\Phi^2)$ constitute reciprocal* [*and therefore one-to-one*] *correspondences between the set of all subspaces, not* 0, *of A and the set of all Δ-systems in* $\mathrm{T}(F,A)$.

We conclude this section by showing how a Δ-system may be reconstructed from its derived subgroup Δ^2.

Proposition 5: *Suppose that $0 < S < A$. Then the involution α belongs to $\Delta(S)$ if, and only if, $\alpha\sigma\alpha = \sigma^{-1}$ for every σ in $\Delta(S^2)$.*

PROOF: The necessity of the condition is an immediate consequence of Proposition 2. Thus assume that $\alpha\sigma\alpha = \sigma^{-1}$ for every σ in $\Delta(S)^2$. Suppose first that $J^+(\alpha) \not\leqslant S$. Then there exists an element $w \neq 0$ in $J^+(\alpha)$ which does not belong to S. There exists a subspace W of A such that $A = S \oplus Fw \oplus W$. If s is any element in S, then $A = S \oplus F(w + s) \oplus W$, since $w \neq 0$. Consequently there exists one and only one linear transformation σ of A such that $w\sigma = w + s$ and $x\sigma = x$ for every x in $S + W$. It is clear that

$$A(\sigma - 1) = Fs \leqslant S \leqslant S + W = J^+(\sigma).$$

We deduce from Lemma 1 that σ belongs to $\Delta(S)^2$; and it follows from our hypothesis that $\alpha\sigma\alpha = \sigma^{-1}$. Hence

$$s = w(\sigma - 1) = w\alpha(\sigma^{-1} - 1)\alpha = w(1 - \sigma)\sigma^{-1}\alpha = - s\alpha,$$

since w belongs to $J^+(\alpha)$, and since $s\sigma = s$ as s is in S. Thus we have shown that

$$J^+(\alpha) \not\leqslant S \text{ implies } S \leqslant J^-(\alpha);$$

and one verifies likewise that

$$J^-(\alpha) \not\leqslant S \text{ implies } S \leqslant J^+(\alpha).$$

Combining these two implications one verifies that S equals $J^+(\alpha)$ or $J^-(\alpha)$, since $0 < S < A$; and α consequently belongs to $\Delta(S)$.

VI.5. The Isomorphisms of the Full Linear Group

The equivalence of the following properties of the linear manifold (F,A) is fairly evident:

(I) F has prime number characteristic p.

(II) $pA = 0$.

Thus we may term the characteristic of F likewise the characteristic of the linear manifold (F,A). Noting that ± 1 are the only elements in the field F whose square is 1 one deduces from VI.1, Proposition 2 that (F,A) has characteristic 2 if, and only if, the identity is the only involution in the center of $\mathrm{T}(F,A)$. As in the preceding three sections we shall assume in this section that the linear manifolds under consideration have characteristic different from 2; and this is equivalent to requiring that the center of T contains an involution, not 1.

To the hypothesis just stated we shall add the further hypothesis that the rank of the linear manifolds under consideration is at least three. It may be noted that a linear manifold (F,A) of rank 2 contains apart from

0 and A only points so that $0 < S \leqslant T < A$ implies $S = T$. Using VI.4, Proposition 3 and VI.4, Theorem 4 one may now find without too much difficulty an internal property of $\mathrm{T}(F,A)$ which is equivalent to the requirement $3 \leqslant r(A)$. We leave the actual formulation of this property to the reader.

Consider now the linear manifolds (F,A) and (G,B) neither of which has characteristic 2 and both of which have rank at least three. Our objective is to find a criterion for the isomorphy of their groups $\mathrm{T}(F,A)$ and $\mathrm{T}(G,B)$ and, in case these groups are isomorphic, to give a survey of all their isomorphisms. A solution of the first problem is contained in the following proposition.

Structure Theorem: *The groups* $\mathrm{T}(F,A)$ *and* $\mathrm{T}(G,B)$ *are isomorphic if, and only if, the linear manifolds* (F,A) *and* (G,B) *are projectively equivalent or are projective duals of each other.*

See III.1, Projective Structure Theorem for a characterization of projective equivalence and IV.1, Duality Theorem for a characterization of duality.

The proof of this Structure Theorem will be obtained as a corollary to our results giving a survey of the totality of all possible isomorphisms between $\mathrm{T}(F,A)$ and $\mathrm{T}(G,B)$. We begin by constructing some special classes of isomorphisms.

We recall first that $\mathrm{P}(F,A)$ is the ring of all endomorphisms of the linear manifold (F,A) and that $\mathrm{T}(F,A)$ consists of all those elements in $\mathrm{P}(F,A)$ which possess an inverse in the ring $\mathrm{P}(F,A)$. In other words: T *is the unit group of* P. [See V.1.]

INDUCED ISOMORPHISMS OF THE FIRST KIND:

Suppose that σ is an isomorphism of the ring $\mathrm{P}(F,A)$ upon the ring $\mathrm{P}(G,B)$. This isomorphism maps the unit group $\mathrm{T}(F,A)$ of $\mathrm{P}(F,A)$ upon the unit group $\mathrm{T}(G,B)$ of $\mathrm{P}(G,B)$; and consequently σ induces an isomorphism σ' of $\mathrm{T}(F,A)$ upon $\mathrm{T}(G,B)$. This isomorphism σ' we call an induced isomorphism of the first kind. [See V.4 for a discussion of the isomorphisms of the ring $\mathrm{P}(F,A)$.]

INDUCED ISOMORPHISMS OF THE SECOND KIND:

Suppose that σ is an anti-isomorphism of $\mathrm{P}(F,A)$ upon $\mathrm{P}(G,B)$. This anti-isomorphism induces an anti-isomorphism of the unit-group $\mathrm{T}(F,A)$ of $\mathrm{P}(F, A)$ upon the unit-group $\mathrm{T}(G,B)$ of $\mathrm{P}(G,B)$. If we define σ' by the rule:

$$\tau^{\sigma'} = (\tau^{\sigma})^{-1} \text{ for } \tau \text{ in } \mathrm{T}(F,A),$$

then one sees easily that σ' is an isomorphism of $\mathrm{T}(F,A)$ upon $\mathrm{T}(G,B)$; and we shall call σ' an induced isomorphism of the second kind. [See V.5 for a discussion of the anti-isomorphisms of the ring $\mathrm{P}(F,A)$.] •

These induced isomorphisms do not exhaust the possibilities; but we shall show that the most general isomorphism differs from the induced ones only by some rather special automorphisms which we are going to construct next.

SINGULAR AUTOMORPHISMS OF T(F,A):

An automorphism α of T(F,A) will be termed singular, if $\sigma^{\alpha}\sigma^{-1}$ belongs to the center Z of T for every linear transformation σ in T [such automorphisms are often called center automorphisms of T].

(1) *If α is a singular automorphism of* T, *then mapping σ in* T *upon $\sigma^{\alpha}\sigma^{-1}$ constitutes a homomorphism α^* of* T *into* Z,

since

$$(\sigma'\sigma'')^{\alpha^*} = (\sigma'\sigma'')^{\alpha}(\sigma'\sigma'')^{-1} = \sigma'^{\alpha}\sigma''^{\alpha^*}\sigma'^{-1} = \sigma'^{\alpha^*}\sigma''^{\alpha^*}$$

as σ^{α^*} belongs to the center Z of T.

(2) *Suppose that η is a homomorphism of* T *into* Z. *Then there exists a [singular] automorphism α of* T *such that $\alpha^* = \eta$ if, and only if, mapping ζ in* Z *upon $\zeta\zeta^{\eta}$ constitutes an automorphism of* Z.

The necessity of this condition is a consequence of the fact that every automorphism of T induces an automorphism in the center of T. If conversely our condition is satisfied, then denote by ν the inverse automorphism to the automorphism of Z which maps ζ onto $\zeta\zeta^{\eta}$. Then we have

$$\zeta = \zeta^{\nu}\zeta^{\eta\nu} = \zeta^{\nu}\zeta^{\nu\eta}$$

for every ζ in Z. Define the mapping α of T into itself by the rule $\sigma^{\alpha} = \sigma\sigma^{\eta}$ for every σ; and define the mapping β of T into itself by the rule $\sigma^{\beta} = \sigma\sigma^{-\eta\nu}$ which latter mapping is well defined, since η maps T into Z. It is easy to see that α and β are endomorphisms of T and that $\alpha\beta = \beta\alpha = 1$, proving that α is an automorphism of T such that $\alpha^* = \eta$.

EXAMPLE: Suppose that F is the field of complex numbers and that the linear manifold (F,A) has finite rank n. Every linear transformation σ of (F,A) may be represented by an n by n matrix with complex coefficients whose determinant does not depend on the special representation of the linear transformations by matrices [see V.1]. Thus the determinant of σ is a uniquely determined number in F; and the absolute value $|\sigma|$ of the determinant is a well determined non-negative real number. Let $\sigma^{\eta} = |\sigma|^2$. Then it is fairly obvious that η is a homomorphism of T into the multiplicative group of positive real numbers which is essentially the same as a subgroup of the center Z of T. Every center element in T is a multiplication by a complex number c and the determinant of this linear transformation is c^n. Now the reader will find it not too difficult to verify that $c^{\eta} = |c|^{2n}$, and that therefore the criterion of (2) is satisfied by η. Mapping the linear transformation σ onto $\sigma^{\alpha} = \sigma|\sigma|^2$ constitutes therefore a non-trivial singular automorphism of T.

We want to show that this singular automorphism α does not commute with all the induced automorphisms of the first kind. There exist automorphisms ω of the field F of complex numbers which do not map real numbers onto real numbers [see anyone of the standard works on algebra treating the theory of formally real fields]. If we consider a fixed representation of T by matrices with coefficients in F, then we may map the linear transformation represented by the matrix (a_{ik}) upon the linear transformation represented by the matrix (a_{ik}^{ω}); and this mapping constitutes an induced automorphism β of the first kind of T. Then

$$\sigma^{\beta\alpha} = \sigma^{\beta} \mid \sigma^{\beta} \mid^2, \ \sigma^{\alpha\beta} = \sigma^{\beta} \mid \sigma \mid^{2\beta} = \sigma^{\beta} \mid \sigma \mid^{2\omega}.$$

It follows that $\beta\alpha \neq \alpha\beta$, since $\mid \sigma^{\beta} \mid^2$ is always a positive real number whereas $\sigma^{2\omega}$ is sometimes not even real.

The reader will not find it difficult to produce further examples of singular automorphisms. It might be an interesting problem to obtain a survey of all the singular automorphisms or at least to decide the question whether there exist always non-trivial singular automorphisms of the full linear group.

A complete description of all possible isomorphisms of T(F,A) is contained in the following

Isomorphism Theorem: *Every isomorphism of* T(F,A) *upon* T(G,B) *may be represented in one and only one way in the form* $\sigma'\sigma''$ *where* σ' *is a singular automorphism of* T(F,A) *and where* σ'' *is an induced isomorphism [of the first or second kind] of* T(F,A) *upon* T(G,B).

The proofs of these fundamental theorems will be effected in a number of steps, some of interest in themselves. We begin by giving a characterization of the singular automorphisms.

Proposition 1: *The following properties of the automorphism* σ *of* T(F,A) *are equivalent.*

(I) σ *is a singular automorphism.*

(II) σ *leaves invariant every linear transformation of class* 2.

(III) $\Delta(S)^{\sigma} = \Delta(S)$ *for every subspace* S *of* A.

Proof: Suppose first that σ is a singular automorphism of T. Then $\tau^* = \tau^{-1}\tau^{\sigma}$ belongs for every τ in T to the center Z of T. If α,β are linear transformations, the $(\alpha^{-1}\beta\alpha)^* = (\alpha^*)^{-1}\beta^*\alpha^* = \beta^*$, since mapping τ on τ^* is [by (1)] a homomorphism of T into the center Z of T.

If α is an involution, so is α^{σ}. Hence

$$1 = (\alpha^{\sigma})^2 = (\alpha\alpha^*)^2 = \alpha^2\alpha^{*2} = \alpha^{*2} \text{ or } \alpha^* = \pm 1 \text{ for } \alpha^2 = 1.$$

Consider now a transformation ω of class 2. It follows from VI.3, Proposition 3 that there exist involutions ω',ω'' such that $J^+(\omega') = J^+(\omega'')$

and $\omega = \omega'\omega''$. The involutions ω' and ω'' are conjugate in T [see, for instance, VI.4, Proposition 1, (*b*)]; and thus it follows from what we have shown already that $\omega'^* = \omega''^* = \pm 1$. Hence

$$\omega^\sigma = \omega'^\sigma\omega''^\sigma = \omega'\omega'^*\omega''\omega''^* = \omega'\omega'' = \omega.$$

Therefore (II) is a consequence of (I). That (III) is a consequence of (II), may be inferred from VI.4, Proposition 5.

Assume finally the validity of (III). If S is any subspace of A, then we deduce $[\Delta(S)^2]^\sigma = \Delta(S)^2$ from (III); and if τ is a linear transformation, then we infer $\Delta(S\tau)^2 = \tau^{-1}\Delta(S)^2\tau$ from VI.4, Lemma 1. Consequently we have for every subspace U and for every linear transformation α

$$\Delta(S\alpha^\sigma)^2 = \alpha^{-\sigma}\Delta(S)^2\alpha^\sigma = \alpha^{-\sigma}[\Delta(S)^2]^\sigma\alpha^\sigma = [\alpha^{-1}\Delta(S)^2\alpha]^\sigma = [\Delta(S\alpha)^2]^\sigma = \Delta(S\alpha)^2$$

and this implies $S\alpha^\sigma = S\alpha$ for every subspace S and every linear transformation α, as follows from VI.4, Theorem 3, (*b*). The linear transformation $\alpha' = \alpha^\sigma\alpha^{-1}$ satisfies consequently $S\alpha' = S$ for every subspace S of A; and it follows from VI.1, Proposition 1 that every α' belongs to the center of T. Thus σ is a singular automorphism of T; and (I) has been shown to be a consequence of (III). This completes the proof.

Next we need a lemma relating the ring and the group of a linear manifold. Concerning the notations used the reader is referred to V.2. We note that $J - 1$ denotes the totality of endomorphisms $j - 1$ for j in the subset J of the endomorphism ring P.

Lemma 1: *If S is a subspace of the linear manifold (F,A), and if $0 < S < A$, then*

(*a*) $\Delta(S)^2 - 1 = N(S) \cap Q(S)$,

(*b*) $L[N(S) \cap Q(S)] = Q(S)$;

(*c*) $R[N(S) \cap Q(S)] = N(S)$.

PROOF: An endomorphism σ of A belongs to the intersection $N(S) \cap Q(S)$ if, and only if, $A\sigma \leqslant S \leqslant K(\sigma)$. This implies $\sigma^2 = 0$ so that

$$(1 + \sigma)(1 - \sigma) = 1 - \sigma^2 = 1.$$

Hence $1 + \sigma$ is a linear transformation of class 2; and it follows from VI.4, Lemma 1 that $1 + \sigma$ belongs to $\Delta(S)^2$. That all the $\tau - 1$ for τ in $\Delta(S)^2$ belong to $N(S) \cap Q(S)$, is likewise deduced from VI.4, Lemma 1. This proves (*a*).

It follows from V.2, Proposition 2 and from VI.4, Theorem 3, (*b*) that

$$\begin{aligned} L[N(S) \cap Q(S)] &= Q(K[N(S) \cap Q(S)]) = Q(K[\Delta(S)^2 - 1]) \\ &= Q(J^+[\Delta(S)^2]) = Q(S), \end{aligned}$$

proving (*b*).

From $S < A$ and VI.4, Lemma 1 one deduces readily that

$$S = A[\Delta(S)^2 - 1];$$

and thus it follows from (*a*) that $S = A[N(S) \cap Q(S)]$. Now we deduce from V.2, Proposition 2 that

$$N(S) = N(A[N(S) \cap Q(S)]) = R[N(S) \cap Q(S)];$$

and this proves (*c*).

Proposition 2: *The identity is the only automorphism of* $\mathrm{T}(F,A)$ *which is at the same time singular and induced [of first or second kind].*

PROOF: If σ is a singular automorphism of T, then it follows from Proposition 1 that

$$\Delta(S)^2 = [\Delta(S)^2]^\sigma \text{ for every subspace } S \text{ of } A. \tag{2.1}$$

Assume now that there exists an automorphism α of $\mathrm{P}(F,A)$ which induces σ in the subgroup T of P. It follows from Lemma 1 and (2.1) that, for $0 < S < A$,

$$[N(S) \cap Q(S)]^\alpha = [\Delta(S)^2]^\alpha - 1 = [\Delta(S)^2]^\sigma - 1 = \Delta(S)^2 - 1 = N(S) \cap Q(S),$$
$$Q(S)^\alpha = L[N(S) \cap Q(S)]^\alpha = L[(N(S) \cap Q(S))^\alpha] = L[N(S) \cap Q(S)] = Q(S),$$

since the automorphic image of a left-annihilator is the left-annihilator of the automorphic image. But now it follows from V.1, Corollary 1 that $\alpha = 1$. Hence $\sigma = 1$, since σ is induced by α.

Assume next that there exists an anti-automorphism β of P such that

$$\tau^\sigma \tau^\beta = 1 \text{ for every } \tau \text{ in T}.$$

It follows from Lemma 1 and (2.1) that, for $0 < S < A$,

$$[N(S) \cap Q(S)]^\beta = [\Delta(S)^2]^\beta - 1 = [\Delta(S)^2]^{-\sigma} - 1 = \Delta(S)^2 - 1 = N(S) \cap Q(S),$$

since $\Delta(S)^2$ is a subgroup and consequently is equal to the totality of inverses of its elements,

$$N(S)^\beta = [R[N(S) \cap Q(S)]]^\beta = L([N(S) \cap Q(S)]^\beta)$$
$$= L[N(S) \cap Q(S)] = Q(S).$$

It follows from V.5, Lemma 2 that an auto-duality of A is defined by mapping the subspace S of A upon the subspace $AN(S)^\beta = AQ(S)$ of A. But it follows from V.2, Proposition 1 that $S = AQ(S)$; and thus we have shown that there exists an auto-duality of A which leaves invariant every subspace of A except 0 and A. This is patently impossible, since $2 < r(A)$. This completes the proof of Proposition 2. Actually we have shown slightly more, namely the

Corollary 1: *The identity is the only automorphism of the ring* $\mathrm{P}(F,A)$ *which induces a singular automorphism of* $\mathrm{T}(F,A)$ *and induced automorphisms of the second kind are never singular.*

Finally we have to relate with projectivities and dualities the special kind of betweenness preserving transformations that appear in our discussion.

Lemma 2: *Suppose that the one-to-one correspondence σ maps the totality of subspaces, not* 0 *or A, of (F,A) upon the totality of subspaces, not* 0 or B, *of (G,B), and that σ preserves betweenness. Then σ is induced by a projectivity or by a duality.*

[Note that a slight complication arises from the fact that 0σ and $A\sigma$ are undefined; and the main question is whether 0σ and $A\sigma$ can be defined in such a way that betweenness is still preserved.]

PROOF: If $0 < S \leqslant T < A$, then S is between S and T. Hence $S\sigma$ is between $S\sigma$ and $T\sigma$ and consequently

$$0 < S\sigma \leqslant T\sigma < B \text{ or } 0 < T\sigma \leqslant S\sigma < B.$$

Next we note that a subspace [not 0 or the total space] is a point or a hyperplane if, and only if, it is not properly between subspaces different from 0 and the total space. It follows that σ maps points on points or hyperplanes and that σ maps likewise hyperplanes upon points or hyperplanes. Now we distinguish two cases.

Case 1: There exists a point P in A such that $P\sigma$ is a point.

If H is a hyperplane through P, then $H\sigma \neq P\sigma$ and $H\sigma$ is either part of $P\sigma$ or else $H\sigma$ contains $P\sigma$. But the point $P\sigma$ does not contain subspaces different from 0 and $P\sigma$; and thus it follows that $P\sigma < H\sigma$. Hence $H\sigma$ is not a point and consequently $H\sigma$ is a hyperplane through $P\sigma$.

Consider now some point Q in A. There exists a hyperplane H through P and Q, since $2 < r(A)$. We have seen in the preceding paragraph of the proof that $H\sigma$ is a hyperplane in B. Applying the dual argument it follows that $Q\sigma$ is a point on $H\sigma$. Thus we have seen that σ maps points on points and hyperplanes on hyperplanes.

Suppose now that $0 < S \leqslant T < A$. There exists a point V such that $V \leqslant S$. It follows from our hypothesis that $S\sigma$ is between $V\sigma$ and $T\sigma$. But $V\sigma$ has been shown to be a point, implying that $0 < V\sigma \leqslant S\sigma \leqslant T\sigma$. The mapping σ preserves order; and now one sees that σ is induced by a projectivity of A upon B.

Case 2: There exists a point Q in A such that $Q\sigma$ is not a point.

From the result of Case 1 we deduce that $X\sigma$ is not a point whenever X is a point in A; and thus we see that σ maps points onto hyperplanes and hyperplanes onto points. Suppose now that $0 < S \leqslant T < A$. There exists a point V such that $V \leqslant S$. Since $V\sigma$ is a hyperplane and $S\sigma$ between $V\sigma$ and $T\sigma$, it follows that $T\sigma \leqslant S\sigma \leqslant V\sigma$. Consequently σ is induced by a duality as we intended to show.

PROOF OF THE ISOMORPHISM THEOREM:

Suppose that σ is an isomorphism of the group $\mathrm{T}(F,A)$ upon the group

$T(G,B)$. It follows from VI.3, Theorem 1 [VI.3, Remark 1] that σ maps transformations of class 2 upon transformations of class 2. If a system Φ of involutions in $T(F,A)$ has the properties (a) to (d) of VI.4, Proposition 1, then Φ^σ has the same properties, since σ maps transformations of class 2 upon transformations of class 2. It follows from VI.4, Theorem 1 that σ maps Δ-systems in $T(F,A)$ upon Δ-systems in $T(G,B)$.

Consider now a subspace S of A such that $0 < S < A$. Then we form the Δ-system $\Delta(S)$; and this we map onto the Δ-system $\Delta(S)^\sigma$ in $T(G,B)$. The latter we map upon the subspace

$$S^\nu = J^+([\Delta(S)^\sigma]^2)$$

of B; and it follows from VI.4, Theorem 4 [and the fact that σ is a one-to-one and exhaustive mapping] that ν constitutes a one-to-one mapping of the subspaces, not 0 nor A, of A onto the totality of subspaces, not 0 nor B, of B. It follows from VI.4, Theorem 2 that ν preserves betweenness; and now we deduce from Lemma 2 that ν is induced by a projectivity or by a duality which we shall denote by ν too. We note that

$$\Delta(S^\nu) = \Delta(S)^\sigma \text{ for } 0 < S < A$$

as a consequence of VI.4, Theorem 3.

Case 1: ν is a projectivity.

It follows from the First Fundamental Theorem of Projective Geometry [III.1] that the projectivity ν is induced by a semi-linear transformation ω of (F,A) upon (G,B) so that $S^\nu = S^\omega$. This semi-linear transformation ω induces an isomorphism σ' of $P(F,A)$ upon $P(G,B)$ which satisfies

$$N(S)^{\sigma'} = \omega^{-1}N(S)\omega = N(S^\omega) = N(S^\nu),$$
$$Q(S)^{\sigma'} = \omega^{-1}Q(S)\omega = Q(S^\omega) = Q(S^\nu)$$

for every subspace S of A, as follows from V.4, Lemma 3. We deduce from Lemma 1 that

$$[\Delta(S)^2]^{\sigma'} - 1 = [\Delta(S)^2 - 1]^{\sigma'} = [N(S) \cap Q(S)]^{\sigma'}$$
$$= N(S)^{\sigma'} \cap Q(S)^{\sigma'} = N(S^\nu) \cap Q(S^\nu) = \Delta(S^\nu)^2 - 1;$$

and now it follows from VI.4, Theorem 4 [or VI.4, Proposition 5] that

$$\Delta(S)^{\sigma'} = \Delta(S)^\sigma \text{ for every } 0 < S < A.$$

An immediate application of Proposition 1 shows that $\sigma\sigma'^{-1} = \sigma''$ is a singular automorphism of $T(F,A)$. Hence $\sigma = \sigma''\sigma'$ is the product of a singular automorphism and of the isomorphism of the first kind which has been induced by σ'; and the uniqueness of this representation is a consequence of Proposition 2.

Case 2: ν is a duality.

It follows from V.5, Theorem 2 that there exists an anti-isomorphism σ' of P(F,A) upon P(G,B) which induces ν according to V.5, Lemma 2. Hence

$$S^{\nu} = K[Q(S)^{\sigma'}] = BN(S)^{\sigma'} \text{ for every subspace } S \text{ of } A.$$

It follows from V.2, Proposition 4 that

$$N(S^{\nu}) = Q(S)^{\sigma'},\ Q(S^{\nu}) = N(S)^{\sigma'},$$

since the anti-isomorphism σ' interchanges right and left annulets. Now we apply Lemma 1 and find that

$$\begin{aligned}\Delta(S^{\nu})^2 - 1 &= N(S^{\nu}) \cap Q(S^{\nu}) = Q(S)^{\sigma'} \cap N(S)^{\sigma'} = [Q(S) \cap N(S)]^{\sigma'} \\ &= [\Delta(S)^2 - 1]^{\sigma'} = [\Delta(S)^2]^{\sigma'} - 1.\end{aligned}$$

Applying the definition of ν and VI.4, Theorem 4 [or VI.4, Proposition 5] we find that

$$\Delta(S)^{\sigma} = \Delta(S)^{\sigma'} \text{ for every subspace } S \text{ with } 0 < S < A.$$

We denote by σ'' the isomorphism of the second kind which the anti-isomorphism $\sigma\sigma'$ induces in T(F,A). Then $\sigma\sigma''^{-1}$ is an automorphism of T(F,A) which leaves invariant all the Δ-systems; and it follows from Proposition 1 that $\sigma\sigma''^{-1}$ is a singular automorphism of T(F,A). Thus we have represented σ as the product of a singular automorphism and of an induced isomorphism of the second kind; and this representation is unique by Proposition 2. This completes the proof of the Isomorphism Theorem.

PROOF OF THE STRUCTURE THEOREM:

If the groups T(F,A) and T(G,B) are isomorphic, then it follows from the Isomorphism Theorem that there exists an isomorphism or an anti-isomorphism of the ring P(F,A) upon P(G,B). If there exists an isomorphism of P(F,A) upon P(G,B), then (F,A) and (G,B) have, by V.4, Structure Theorem, the same structure and are therefore in particular projectively equivalent. If there exists an anti-isomorphism between the rings, then it follows from V.5, Lemma 2 that there exists a duality of (F,A) upon (G,B) and the linear manifolds are duals of each other.

Conversely if there exists a projectivity of (F,A) upon (G,B), then we deduce from the First Fundamental Theorem of Projective Geometry [III.1] the existence of a semi-linear transformation of (F,A) upon (G,B). This semi-linear transformation induces [by V.4] an isomorphism between the rings which in turn induces an isomorphism between the groups. If there exists a duality of (F,A) upon (G,B), then it follows from V.5, Theorem 2 that there exists an anti-isomorphism between their rings which in turn induces an isomorphism [of the second kind] between the groups $\mathrm{T}(F,A)$ and $\mathrm{T}(G,B)$. This completes the proof of the Structure Theorem.

The Automorphism Group of the Group $\mathrm{T}(F,A)$

This group A may be analyzed as follows. It contains the subgroup A_s of all the singular automorphisms which is a normal subgroup, since it consists of exactly those automorphisms of T which induce the identity automorphism in T/Z [which latter group is essentially the same as the group of collineations; see III.2]. There is furthermore the subgroup A_i of the induced automorphisms of the first or second kind. It follows from Corollary 1 that this subgroup A_i is essentially the same as the group of automorphisms and anti-automorphisms of the ring $\mathrm{P}(F,A)$ which latter group we termed in V.5 the "extended group of the ring" and which group was shown [in V.4 and V.5] to be essentially the same as the group of auto-projectivities and auto-dualities of the basic linear manifold (F,A). It is a consequence of the Isomorphism Theorem and Corollary 1 that A splits into these two subgroups:

$$\mathrm{A} = \mathrm{A}_s\mathrm{A}_i, \; 1 = \mathrm{A}_s \cap \mathrm{A}_i.$$

We note that, in general, A_i is not going to be a normal subgroup of A. For, if this were the case, then every induced automorphism of T would commute with every singular automorphism of T; and we have shown by an example that this is not always the case.

The group A_i contains all the inner automorphisms of T; they are induced automorphisms of the first kind which are induced by linear transformations. The group A_i equals the group of inner automorphisms if, and only if, there do not exist any automorphisms $\neq 1$ of the field F [which then is commutative] and $r(A)$ is infinite [so that there do not exist any dualities of A]. Naturally the inner automorphisms form a normal subgroup of the group of all the automorphisms of T so that in this particular case A is the direct product of its group of inner automorphisms and its group of singular automorphisms; we leave the details of this discussion to the reader.

APPENDIX I

Groups of Involutions

If a subgroup of the group $\mathrm{T}(F,A)$ consists of involutions only, then we speak of *a group of involutions*. This concept has been made the basis of the treatment of the isomorphisms of T by various authors, notably Dieudonné [1,2] and " Sur les systèmes maximaux d'involutions conjuguées permutables dans les groupes projectifs " [*Summa Brasil. Math.*, vol. 2 (1950), pp. 59-94], and Mackey [*Annals of Mathematics*, vol. 43 (1942), pp. 250-260]. We have not made any use of this concept; but the groups of involutions have an interesting relation with the direct decompositions of the linear manifold (F,A) into points and this relation shall be explored in the present appendix. We assume as before that the characteristic of F is not 2.

Lemma 1: *The set Φ of involutions in* T *is contained in a group of involutions if, and only if, $\sigma'\sigma'' = \sigma''\sigma'$ for every σ',σ'' in Φ.*

PROOF: If Φ is part of a group of involutions, then $\sigma'\sigma''$ is an involution whenever σ' and σ'' are involutions in Φ. But this implies

$$1 = (\sigma'\sigma'')^2 = \sigma'\sigma''\sigma'\sigma'' \text{ or } \sigma'\sigma'' = \sigma''^{-1}\sigma'^{-1} = \sigma''\sigma',$$

proving the necessity of our condition. Assume conversely that any two involutions in Φ commute. If $\sigma_1, \cdots, \sigma_n$ belong to Φ, then

$$(\sigma_1 \cdots \sigma_n)^2 = \sigma_1^2 \cdots \sigma_n^2 = 1;$$

and this shows that the subgroup of T which is generated by Φ consists of involutions only, is a group of involutions.

Suppose now that A is the direct sum of the points in the set D—this is equivalent to saying that A is the sum of the points in D and that none of the points in D lies on the hyperplane spanned by the other points in D. Then we denote by $\Theta(D)$ *the totality of involutions σ such that $P\sigma = P$ for every P in D.* We shall also speak of D as a decomposition of A into points and of $\Theta(D)$ as the system of involutions belonging to the decomposition D.

Proposition 1: *If D is a decomposition of A into points, then $\Theta(D)$ is a maximal group of involutions.*

PROOF: If P is a point in D, and if $P = Fp$, then we have $p\sigma = ep$ for every σ in $\Theta(D)$. But σ is an involution; and consequently we have

$$p = p\sigma^2 = e^2p.$$

This implies $e = \pm 1$. If σ',σ'' are any two involutions in $\Theta(D)$, then it follows from this remark that $p\sigma' = \pm p$ and $p\sigma'' = \pm p$; and this implies $p\sigma'\sigma'' = p\sigma''\sigma'$ for every p in every P in D. Since A is the sum of the points P in D, it follows that any two involutions in $\Theta(D)$ commute; and it follows from Lemma 1 that products of involutions in $\Theta(D)$ are themselves involutions. Now it is clear that $\Theta(D)$ is a group of involutions.

Suppose now that Θ' is a group of involutions such that $\Theta(D) \leqslant \Theta'$. If σ' belongs to Θ', then it follows from Lemma 1 that $\sigma\sigma' = \sigma'\sigma$ for every σ in $\Theta(D)$. It follows from VI.2, Proposition 1 that

$$J^+(\sigma)\sigma' = J^+(\sigma) \text{ and } J^-(\sigma)\sigma' = J^-(\sigma) \text{ for every } \sigma \text{ in } \Theta(D).$$

If P is a point in D, P' the hyperplane spanned by the remaining points in D, then $A = P \oplus P'$ and there exists one and only one involution σ such that $P = J^+(\sigma)$ and $P' = J^-(\sigma)$. Naturally σ belongs to $\Theta(D)$; and it follows from the preceding remark that

$$P\sigma' = J^+(\sigma)\sigma' = J^+(\sigma) = P.$$

Thus $P = P\sigma'$ for every P in D and σ' belongs to $\Theta(D)$. Hence $\Theta' = \Theta(D)$ showing the maximality of the group $\Theta(D)$ of involutions.

Proposition 2: *The following properties of the maximal group Θ of involutions are equivalent.*

(I) *Θ belongs to a direct decomposition D of A into points: $\Theta = \Theta(D)$.*

(II) *If x is an element in A, then the set $x\Theta$ [of images of x under involutions in Θ] spans a subspace of finite rank.*

(III) *A is the sum of the points P such that $P = P\Theta$.*

(IV) *If S is a subspace of A such that $\Theta \leqslant N\Delta(S)$, then Θ and $\Delta(S)^+$ have one [and only one] involution in common.*

Proof: Assume first that $\Theta = \Theta(D)$ for a direct decomposition D of A into points. If x is an element in A, then there exist finitely many points P_i in D such that x belongs to $\sum_{i=1}^{n} P_i$. There exist elements p_i in P_i such that $x = \sum_{i=1}^{n} p_i$. If σ is an involution in Θ, then $P_i\sigma = P_i$ so that $p_i\sigma$ belongs to P_i. Consequently $x\sigma = \sum_{i=1}^{n} (p_i\sigma)$ belongs to $\sum_{i=1}^{n} P_i$; and we have shown that $x\Theta$ is part of the subspace $\sum_{i=1}^{n} P_i$ of finite rank n. Hence (II) is a consequence of (I).

Assume next the validity of (II) and denote by S the sum of all the points P such that $P = P\Theta$ [the Θ-invariant points]. We prove first:

(III*) *If M is a Θ-invariant subspace of finite rank, then $M \leqslant S$.*

This is certainly true for points; and thus we may make the inductive hypothesis that $1 < r(M)$ and that all Θ-invariant subspaces of rank smaller than $r(M)$ are contained in S. If σ is in Θ, then $M = M\sigma$ and M contains therefore with x also the elements $\frac{1}{2}(x + x\sigma)$ and $\frac{1}{2}(x - x\sigma)$. Since x is their sum, and since the first of these is in $J^+(\sigma) \cap M$, the second one in $J^-(\sigma) \cap M$, it follows that

$$M = [J^+(\sigma) \cap M] \oplus [J^-(\sigma) \cap M] \text{ for every } \sigma \text{ in } M.$$

Now we distinguish two cases.

Case 1: There exists an involution σ in Θ such that $0 < J^+(\sigma) \cap M < M$.

Since any two involutions in Θ commute by Lemma 1, we deduce from VI.2, Proposition 1 that $J^+(\sigma)$ and $J^-(\sigma)$ are Θ-invariant subspaces of S. Since intersections of Θ-invariant subspaces are likewise Θ-invariant, we have represented M as the direct sum of its Θ-invariant subspaces $M \cap J^+(\sigma)$ and $M \cap J^-(\sigma)$. Since the first of these has positive lower rank than M, so has the second one and it follows from the inductive hypothesis that they both are part of S; and consequently M itself is part of S.

Case 2: There does not exist any involution σ in Θ such that

$$0 < J^+(\sigma) \cap M < M.$$

If σ is an involution in Θ, then either $M = M \cap J^+(\sigma) \leqslant J^+(\sigma)$ or else $0 = M \cap J^+(\sigma)$. But M is the direct sum of $M \cap J^+(\sigma)$ and $M \cap J^-(\sigma)$; and thus it follows in the latter case that $M = M \cap J^-(\sigma) \leqslant J^-(\sigma)$. Thus $M \leqslant J^+(\sigma)$ or $M \leqslant J^-(\sigma)$ for every σ in Θ; and this implies that every subspace of M is Θ-invariant. Hence all the points in M are Θ-invariant so that every point in M and therefore M itself is part of S. This completes the inductive proof of (III*).

Consider now any element x in A. Denote by X the subspace of A spanned by x and all the elements $x\sigma$ for σ in Θ. Clearly X is a Θ-invariant subspace of A; and it follows from condition (II) that X has finite rank. It follows from (III*) that X is part of S; and so x itself is contained in S. Hence $S = A$; and we have shown that (III) is a consequence of (II).

Assume next the validity of (III). Then A is the sum of all its Θ-invariant points, and there exists therefore a set D' of Θ-invariant points whose direct sum is A. Form the group $\Theta(D')$ of involutions belonging to this direct decomposition D' of A into points. Every involution in Θ leaves invariant every point in D' [by construction]. Hence $\Theta \leqslant \Theta(D')$. Since Θ is a maximal group of involutions, and since $\Theta(D')$ is a group of involutions [by Proposition 1], it follows that $\Theta = \Theta(D')$; and thus we have shown that (I) is a consequence of (III). This proves the equivalence of the first three conditions. [The reader may find it a useful exercise to show that $D = D'$.]

Assume again that $\Theta = \Theta(D)$ for some direct decomposition D of A into points; and suppose that S is a subspace such that $\Theta \leqslant N\Delta(S)$. If σ is in Θ, then we infer from VI.4, Proposition 1, (*a*) the existence of an involution τ in $\Delta(S)^+$ such that $\tau\sigma = \sigma\tau$. From $S = J^+(\tau)$ and VI.2, Proposition 1 we infer $S = S\sigma$. Thus S is a Θ-invariant subspace; and this implies [as before] that

$$S = [J^+(\sigma) \cap S] \oplus [J^-(\sigma) \cap S] \text{ for every involution } \sigma \text{ in } \Theta.$$

If P is a point in D, then we denote by P' the hyperplane spanned by the remaining points in D. There exists one and only one involution σ in $\Theta(D)$ such that $P = J^+(\sigma)$, $P' = J^-(\sigma)$; and this implies that either $P \leqslant S$ or else $S \leqslant P'$ [for every P in D]. Denote now by D_S the totality of points P in D which are contained in S; and denote by D'_S the remaining points in D. If U is the sum of the points P in D_S, V the intersection of all the hyperplanes P' for P in D'_S, then $U \leqslant S \leqslant V$. But A is the direct sum of the points in D; and so it follows that $U = V$. Thus we have shown that

$$S \text{ is the direct sum of the points in } D_S.$$

Denote now by S' the direct sum of the points in D'_S. Then $A = S \oplus S'$ and there exists one and only one involution ν such that $S = J^+(\nu)$, $S' = J^-(\nu)$. It is clear that ν leaves invariant every point in D, since every point in D is either on S or on S'. Thus ν belongs to the intersection of Θ and $\Delta(S)^+$. [That ν is uniquely determined by this property, may be verified in various ways; we leave this to the reader.] Hence (IV) is a consequence of (I).

Assume next the validity of (IV). We prove first:

(IV.a) If S is a Θ-invariant subspace of A, then there exists one and only one involution τ in Θ such that $S = J^+(\tau)$; and there exists one and only one Θ-invariant subspace S' of A such that $A = S \oplus S'$.

If σ is in Θ, then $\sigma^{-1}\Delta(S)^+\sigma = \Delta(S\sigma)^+ = \Delta(S)^+$ so that σ belongs to $N\Delta(S)$. Hence $\Theta \leqslant N\Delta(S)$; and we deduce from (IV) the existence of an involution τ in the intersection $\Theta \cap \Delta(S)^+$. Thus we have found an involution τ in Θ such that $S = J^+(\tau)$. We note that $J^-(\tau) = S'$ is Θ-invariant, as follows from VI.2, Proposition 1, since τ commutes with every σ in Θ. Thus we have found a Θ-invariant subspace S' such that $A = S \oplus S'$.

Assume now that S'' is a further Θ-invariant subspace such that $A = S \oplus S''$. If x is an element in S'', then $x\tau$ and $x + x\tau$ belong to S''. But $x + x\tau$ belongs also to $J^+(\tau) = S$. Hence $x + x\tau = 0$ is an element in the intersection 0 of S and S''. Consequently x belongs to $J^-(\tau) = S'$; and we have shown that $S'' \leqslant S'$. Since $S \oplus S' = S \oplus S''$ this inequality implies $S' = S''$. If τ'' is another involution in Θ such that $S = J^+(\tau')$, then $A = S \oplus J^-(\tau')$; and $J^-(\tau')$ is Θ-invariant. It follows from what we have already shown that $J^-(\tau) = J^-(\tau')$; and this implies $\tau = \tau'$. This completes the proof of (IV.a).

(IV.b) $r(S) \leqslant 1$, if S is Θ-invariant and if S has the property:

(*) $$S \leqslant J^+(\sigma) \text{ or } S \leqslant J^-(\sigma) \text{ for every } \sigma \text{ in } \Theta.$$

To prove this assume that $1 < r(S)$. Then $S = U \oplus V$ where neither U nor V is 0 [or S]. We infer from (IV.a) the existence of a uniquely determined Θ-invariant subspace S' such that $A = S \oplus S'$. There exists one and only one involution ω such that $J^+(\omega) = S' \oplus U$ and $J^-(\omega) = V$. It is clear that ω commutes with every σ in Θ on S'; and it follows from (*) that ω commutes with every σ in Θ on S. Hence ω commutes with every σ in Θ; and it follows from Lemma 1 and the maximality of Θ that ω belong to Θ. But this contradicts (*), since U and V are both different from 0. Hence $r(S) \leqslant 1$, as we intended to show.

(IV.c) If $S \neq 0$ is a Θ-invariant subspace of A, then S contains a Θ-invariant point.

S contains an element $v \neq 0$. Denote by V the subspace spanned by $v\Theta$. Since Θ is a group, V is Θ-invariant and is actually the smallest Θ-invariant subspace containing v. If there is given an ordered [by inclusion] set of Θ-invariant subspaces, then their join is also a Θ-invariant subspace. It follows from the Maximum Principle of Set Theory [see Appendix S] that there exists a maximal Θ-invariant subspace M of V which does not contain the element $v \neq 0$. We deduce from (IV.a) the existence of one and only one Θ-invariant subspace M' of A such that $A = M \oplus M'$. From $M \leqslant V$ we deduce that $V = M \oplus (M' \cap V)$; and $M' \cap V \neq 0$, since $M < V$.

Let $V'' = M' \cap V$. Then V'' is Θ-invariant as the intersection of the Θ-invariant subspaces M' and V. If σ is an involution in Θ, then it follows as usual that $V'' = [V'' \cap J^+(\sigma)] \oplus [V'' \cap J^-(\sigma)]$ where $V'' \cap J^+(\sigma)$ and $V'' \cap J^-(\sigma)$ are both Θ-invariant. Since $V'' \neq 0$, at least one of these components is not 0. Suppose, for instance, $0 \neq V'' \cap J^+(\sigma)$. Then $M < M \oplus [V'' \cap J^+(\sigma)] \leqslant V$. Since M is a maximal Θ-invariant subspace of V which does not contain v, and since $V'' \cap J^+(\sigma)$ is Θ-invariant, it follows that the Θ-invariant subspace $M \oplus [V'' \cap J^+(\sigma)]$ of V contains v. But V is the smallest Θ-invariant subspace which contains V. Hence $M \oplus V'' = V = M \oplus [V'' \cap J^+(\sigma)]$ so that $V'' = V'' \cap J^+(\sigma)$ or $V'' \leqslant J^+(\sigma)$. If $V'' \cap J^-(\sigma)$ were different from 0, then we would see likewise that $V'' \leqslant J^-(\sigma)$. Thus we see that the Θ-invariant subspace $V'' \neq 0$ of V and S meets requirement (*) of (IV.b); and it follows from (IV.b) that $r(V'') = 1$. Thus (IV.c) has been verified.

Now denote by B the sum of all the Θ-invariant points in A. Then B itself is Θ-invariant; and it follows from (IV.a) that there exists a Θ-invariant subspace B' such that $A = B \oplus B'$. If B' were different from 0, then we would infer from (IV.c) the existence of a Θ-invariant point $P \leqslant B'$ which is impossible, since B is the sum of all Θ-invariant points. Hence $B' = 0$ and $A = B$ is the sum of all Θ-invariants points. Thus we have shown that (III) is a consequence of (IV); and this completes the proof of the equivalence of the four properties (I) to (IV).

That the four equivalent properties (I) to (IV) are not always satisfied, is the content of our next result.

Proposition 3: *The following properties of the linear manifold (F,A) are equivalent.*

(I) *A has finite rank.*

(II) *Every maximal group of involutions belongs to a direct decomposition of A into points.*

(III) *Any two maximal groups of involutions are conjugate in* $\mathrm{T}(F,A)$.

PROOF: If A has finite rank, then every maximal group of involutions meets requirement (II) of Proposition 2 and belongs therefore to a direct decomposition into points. Hence (I) implies (II).

Assume next the validity of (II). If D and D' are direct decompositions of A into points, then D and D' contain both $r(A)$ points [II.2, Uniqueness Theorem]. It is now easy to construct a linear transformation σ mapping D upon D'; and we have $\sigma^{-1}\Theta(D)\sigma = \Theta(D')$. This shows that (III) is a consequence of (II).

Assume now the validity of (III). Suppose that Θ is some maximal group of involutions. There exists a direct decomposition D of A into points; and Θ and $\Theta(D)$ are conjugate [by (III) and Proposition 1]. Consequently there exists a linear transformation σ such that

$$\Theta = \sigma^{-1}\Theta(D)\sigma = \Theta(D\sigma);$$

and this shows that (II) is a consequence of (III).

Assume finally that A has infinite rank. Then $A = U \oplus V$ where U possesses a countably infinite basis $b_1, b_2, \cdots, b_i, \cdots$. Since U is the direct sum of the points $F(b_1 - b_2), \cdots, F(b_i - b_{i+1}), Fb_{i+1}, \cdots, Fb_{i+j}, \cdots$, there exists one and only one involution σ_i such that

$$J^+(\sigma_i) = \sum_{j=1}^{i} F(b_j - b_{j+1}),\ J^-(\sigma_i) = V + \sum_{i<j} Fb_j.$$

One deduces from VI.2, Proposition 1 that $\sigma_i\sigma_k = \sigma_k\sigma_i$ for every i and k. It follows from Lemma 1 that the involutions σ_i are contained in some group of involutions; and applying the Maximum Principle of Set Theory we may find a maximal group Θ of involutions which contains all the involutions σ_i. But

$$b_1\sigma_i = \left[\sum_{j=1}^{i}(b_j - b_{j+1}) + b_{i+1}\right]\sigma_i = b_1 - 2b_{i+1}$$

so that $b_1\Theta$ contains all the elements $b_1, b_2, \cdots, b_i, \cdots$. Thus the group Θ of involutions does not meet requirement (II) of Proposition 2 and does consequently not belong to a direct decomposition of A into points. Hence (II) is not true, if $r(A)$ is infinite. Consequently (I) is a consequence of (II); and this completes the proof.

Remark 1: If D is a direct decomposition of A into points, then one may verify that the point P belongs to D if, and only if, P is $\Theta(D)$-invariant. This shows that the mapping Θ constitutes a one-to-one correspondence between the direct decompositions into points and the maximal groups of involutions which meet the requirements (I) to (IV) of Proposition 2.

Remark 2: If D is a direct decomposition into points, then $\Theta(D)$ contains exactly $2^{r(A)}$ involutions, since to every subset of D there belongs one and only one involution in $\Theta(D)$ which leaves invariant the elements in the points of the given subset of D. If $r(A)$ is finite, then this permits to determine the rank of A from the order of the maximal groups of involutions; but in case of infinite $r(A)$, this leads to questions connected with the soc. continuum hypothesis.

VI.6. Characterization of the Full Linear Group within the Group of Semi-linear Transformations

In III.2 we have pointed out that the full linear group $\Gamma(F,A)$ of the linear manifold (F,A) is a normal subgroup of the group $\Lambda(F,A)$ of all semi-linear transformations of A. It is our objective in this section to improve upon this result and to find a characterization of Γ as a subgroup of Λ which is invariant under isomorphisms of Λ. This will make it possible to derive [in VI.7] from the results of the first part of this chapter results on the isomorphisms of Λ which are comparable with those obtained in VI.5 for Γ.

We assume in this and the next section that all linear manifolds under consideration have *characteristic different from* 2 *and rank not less than* 3, though some of our considerations remain valid without these restrictions.

We recall that every semi-linear transformation σ of (F,A) upon itself is a pair consisting of an automorphism σ of the field F and an automorphism σ of the additive group A which are related by the condition: $(fa)^\sigma = f^\sigma a^\sigma$ for f in F and a in A. It will prove convenient to refer to the first of these automorphisms as to the component field automorphism of the semi-linear transformation σ. [Note our return to exponential notation.]

Lemma 1: *The semi-linear transformation σ is involutorial, but not linear if, and only if,*

(a) *there exists a basis of A whose elements are left invariant by σ and*

(b) *the component field automorphism of σ is involutorial, but not* 1.

Proof: The sufficiency of these conditions is fairly obvious. Thus assume that $\sigma^2 = 1$, but that σ is not linear. Then the necessity of (*b*) follows almost immediately from the fact that a semi-linear transformation is linear if, and only if, its component field automorphism is 1. Since the characteristic of F is not 2, every element a in A and every element f in F has the form:

$$a = \tfrac{1}{2}[a + a^\sigma] + \tfrac{1}{2}[a - a^\sigma],\ f = \tfrac{1}{2}[f + f^\sigma] + \tfrac{1}{2}[f - f^\sigma]$$

where the first component is σ-invariant whereas the second one is σ-skew. Since the component field automorphism is not 1, there exists at least one element $i \neq 0$ in F such that $i^\sigma = -i$. If a is a σ-invariant element in A, then $(ia)^\sigma = i^\sigma a^\sigma = -ia$ so that ia is σ-skew. Consequently every element x in A has the form: $x = x' + ix''$ where x' and x'' are both σ-invariant [let $x' = \tfrac{1}{2}(x + x^\sigma)$ and $x'' = \tfrac{1}{2}i^{-1}(x - x)^\sigma$]. Now it is clear that a basis of A over F may be found which consists of σ-invariant elements only. This shows the necessity of (*a*).

Lemma 2: *Involutorial semi-linear transformations which are not linear are conjugate* [*in the group* Λ] *if, and only if, their component field automorphisms are conjugate* [*in the group of all automorphisms of the field* F].

PROOF: The necessity of our condition is an immediate consequence of the fact that the product of two semi-linear transformations is obtained by multiplication of their components [so that in particular the component field automorphism of the product is the product of the component field automorphisms of the factors]. Assume now that σ' and σ'' are involutorial semi-linear transformations, that neither σ' nor σ'' is linear and that there exists an automorphism α of F such that their component field automorphisms are related by the formula: $\sigma'' = \alpha^{-1}\sigma'\alpha$. It follows from Lemma 1 that there exists a basis B' of (F,A) whose elements are σ'-invariant and a basis B'' which is σ''-invariant. Since B' and B'' contain the same number of elements [namely $r(A)$], there exists a one-to-one correspondence mapping B' upon B''; and consequently there exists a semi-linear transformation β whose component field automorphism is α and which maps B' upon B''. One verifies then that $\beta^{-1}\sigma'\beta$ has the same component field automorphism as σ'' and leaves invariant every element in the basis B'' as does σ''. Hence $\sigma'' = \beta^{-1}\sigma'\beta$, as we intended to show.

Proposition 1: *Suppose that* σ *is an involutorial semi-linear transformation. Then* σ *and* $-\sigma$ [$= (\sigma, -\sigma)$] *are conjugate semi-linear transformations if, and only if,*

(*) σ *is not linear or else* $r[J^+(\sigma)] = r[J^-(\sigma)]$.

REMARK 1: If σ is not linear, then it follows from Lemma 1 that neither the totality of σ-invariant elements nor the totality of σ-skew elements in A is a subspace of A and that they both span A itself.

PROOF: Suppose first that σ is linear and that $-\sigma = \alpha^{-1}\sigma\alpha$ where α is some semi-linear transformation. Then

$$J^-(\sigma) = J^+(-\sigma) = J^+(\alpha^{-1}\sigma\alpha) = J^+(\sigma)^\alpha;$$

and this implies the validity of (*), since semi-linear transformations preserve rank [see III.1].

Assume conversely the validity of (*). If σ is not linear, then σ and $-\sigma$ are involutorial semi-linear transformations with the same component field automorphisms and neither σ nor $-\sigma$ is linear. It follows from Lemma 2 that σ and $-\sigma$ are conjugate in Λ. If σ is linear, then it follows from (*) that $r[J^+(\sigma)] = r[J^-(\sigma)]$. Since $A = J^+(\sigma) \oplus J^-(\sigma)$, there exists an involution τ of A which interchanges $J^+(\sigma)$ and $J^-(\sigma)$; and it follows from

$$J^+(\sigma) = J^-(\sigma)^\tau = J^-(\tau\sigma\tau),\ J^-(\sigma) = J^+(\sigma)^\tau = J^+(\tau\sigma\tau)$$

that $\tau\sigma\tau = -\sigma$. This completes the proof.

REMARK 2: Note that involutorial linear transformations [= involutions] with property (*) exist if, and only if, either $r(A)$ is infinite or else $r(A)$ is an even integer.

REMARK 3: If we denote as usual by -1 the linear transformation of A which maps a in A upon $-a$, then the notation $-\sigma$ used in Proposition 1 is quite justified, since $-\sigma = (-1)\sigma$. We note that -1 is uniquely determined in Λ as its *only center element of order* 2, as follows from VI.1, Proposition 2, since the center of Λ is necessarily part of the center of T.

We denote by N^* *the totality of involutorial semi-linear transformations* σ *which are not conjugate to* $-\sigma$. It follows from Remark 3 that this class N^* is a group theoretical invariant of Λ; and it follows from Proposition 1 that $\mathrm{N}^* \leqslant \mathrm{T}$.

We recall that a semi-linear transformation σ has been termed a *multiplication*, if there exists a number $f \neq 0$ in F such that $x^\sigma = fx$ for every x in A—the component field automorphism is then the inner automorphism induced by f. The totality N of multiplications is a normal subgroup of Λ[see III.1 and III.2].

Proposition 2: N *is the centralizer of* N^* *in* Λ.

PROOF: It has been pointed out before that [as a consequence of Proposition 1] $\mathrm{N}^* \leqslant \mathrm{T}$. The centralizer of T in Λ is consequently part of the centralizer of N^* in Λ. But we have shown before [III.2, Survey of the Groups of a Linear Manifold] that T is the centralizer of N in Λ. Thus every element in N commutes with every element in N^*; and this shows that N is part of the centralizer of N^*.

Suppose now that the semi-linear transformation σ commutes with every transformation in N^*. Consider a point P in A. Then there exists an involution ν in T such that $P = J^+(\nu)$. Since $3 \leqslant r(A)$, we have

$1 = r(P) < r[J^-(\nu)]$; and it follows from Proposition 1 that ν belongs to N^*. Hence $\nu\sigma = \sigma\nu$ and therefore

$$P = J^+(\nu) = J^+(\sigma^{-1}\nu\sigma) = J^+(\nu)^\sigma = P^\sigma.$$

The semi-linear transformation σ induces therefore the identity projectivity; and it follows from III.1, Proposition 3 that σ is a multiplication. Hence σ is in N; and N is therefore exactly the centralizer of N^* in Λ.

Theorem 1: *The full linear group* T *is the second centralizer of* N^* *in* Λ.

PROOF: N is the centralizer of N^* in Λ [Proposition 2]. T is the centralizer of N in Λ [III.2, Survey of Groups of a Linear Manifold]. Hence T is the second centralizer of N^* in Λ.

Theorem 2: *Every isomorphism of* $\Lambda(F,A)$ *upon* $\Lambda(G,B)$ *maps* $N^*(F,A)$, $N(F,A)$ *and* $T(F,A)$ *upon* $N^*(G,B)$, $N(G,B)$ *and* $T(G,B)$ *respectively.*

PROOF: Since -1 is the only involution in the center of $\Lambda(F,A)$, an isomorphism σ of $\Lambda(F,A)$ upon $\Lambda(G,B)$ is going to map -1 [of $\Lambda(F,A)$] upon -1 [of $\Lambda(G,B)$]. It follows from the definition of N^* that σ maps $N^*(F,A)$ upon $N^*(G,B)$; and now it follows from Proposition 2 and Theorem 1 that σ maps $N(F,A)$ upon $N(G,B)$ and $T(F,A)$ upon $T(G,B)$.

VI.7. The Isomorphisms of the Group of Semi-linear Transformations

Consider linear manifolds (F,A) and (G,B) neither of which has characteristic 2 nor rank less than 3. Then we are going to prove the

Structure Theorem: $\Lambda(F,A)$ *and* $\Lambda(G,B)$ *are isomorphic groups if, and only if,* (F,A) *and* (G,B) *are either projectively equivalent or duals of each other.*

Combining this result with VI.5, Structure Theorem it follows that $T(F,A)$ and $T(G,B)$ are isomorphic if, and only if, $\Lambda(F,A)$ and $\Lambda(G,B)$ are isomorphic.

The proof of this Structure Theorem will be obtained as a consequence of a theorem which gives an explicit construction of the isomorphisms [if any] between $\Lambda(F,A)$ and $\Lambda(G,B)$. We begin by enumerating some of these possibilities. They are similar to, though more involved than, the corresponding constructions in VI.5.

SINGULAR AUTOMORPHISMS OF $\Lambda(F,A)$:

An automorphism α of $\Lambda(F,A)$ will be termed *singular*, if $\sigma^\alpha\sigma^{-1}$ belongs to N for every semi-linear transformation σ in Λ.

Every semi-linear transformation in N, i.e. every multiplication of A may be identified with the corresponding number, not 0, in F; and this we are going to do. If σ is a semi-linear transformation, a an element in A and $f \neq 0$ in F, then we have

$$(1) \qquad a^{\sigma^{-1}f\sigma} = (fa^{\sigma^{-1}})^{\sigma} = f^{\sigma}a \text{ or } \sigma^{-1}f\sigma = f^{\sigma}.$$

If α is a singular automorphism of Λ, then we let $\sigma^* = \sigma^{\alpha}\sigma^{-1}$ for every semi-linear transformation σ in Λ; and we find that

$$(2) \qquad (\sigma\tau)^* = \sigma^{\alpha}\tau^{\alpha}\tau^{-1}\sigma^{-1} = \sigma^{\alpha}\sigma^{-1}\sigma\tau^{\alpha}\tau^{-1}\sigma^{-1} = \sigma^*\tau^{*\sigma^{-1}}.$$

Mapping σ onto σ^* is consequently a soc. *crossed homomorphism of* Λ *into* N *or* F; and it is easily seen that such a crossed homomorphism of Λ into F comes from a singular automorphism of Λ if, and only if, mapping f in F onto f^*f effects an automorphism of F [note our identification of F and N].

Proposition 1: *The following properties of the automorphism* α *of* Λ *are equivalent.*

(I) α *is a singular automorphism of* Λ.

(II) α *induces a singular automorphism in* T.

(III) $\Delta(S)^{\alpha} = \Delta(S)$ *for every subspace* S *of* A.

PROOF: Assume first that α is a singular automorphism of Λ. It follows from VI.6, Theorem 2 that α induces an automorphism in T. If σ is a linear transformation, then σ and σ^{α} belong to T. Hence $\sigma^{\alpha}\sigma^{-1}$ belongs to the intersection of T and N. But T is the centralizer of N in Λ [III.2, Survey of Groups of a Linear Manifold] so that the intersection of N and T is exactly the center of N. It follows from VI.1, Proposition 2 that the center of N is just the center Z of T; and thus we have shown that α induces a singular automorphism in T; (I) implies (II).

That (II) implies (III), is an immediate consequence of VI.5, Proposition 1.

Assume finally the validity of (III). If σ is a semi-linear transformation in Λ, then we let $\sigma^* = \sigma^{\alpha}\sigma^{-1}$. It follows from (III) that

$$\Delta(S^{\sigma^*}) = \sigma^{*-1}\Delta(S)\sigma^* = \sigma\sigma^{-\alpha}\Delta(S)\sigma^{\alpha}\sigma^{-1} = \sigma[\sigma^{-1}\Delta(S)\sigma]^{\alpha}\sigma^{-1}$$
$$= \sigma[\Delta(S)^{\sigma}]^{\alpha}\sigma^{-1} = \sigma\Delta(S^{\sigma})\sigma^{-1} = \Delta(S)$$

for every subspace S of A; and now we deduce from VI.4, Theorem 3, (*b*) that $S = S^{\sigma^*}$ for every subspace S of A. But the only semi-linear transformations inducing the identity projectivity are multiplications [III.1, Proposition 3]; and thus we have shown that σ^* belongs to N for every σ. Hence α is a singular automorphism of Λ, proving that (I) is a consequence of (III).

This proposition shows in particular that our present definition of singular automorphism is in complete agreement with that used in VI. 5.

INDUCED ISOMORPHISMS OF THE FIRST KIND:

Suppose that σ is a semi-linear transformation of the linear manifold (F,A) upon the linear manifold (G,B). If τ is a semi-linear transformation in $\Lambda(F,A)$, then $\sigma^{-1}\tau\sigma = \tau^{\sigma}$ is a semi-linear transformation in $\Lambda(G,B)$; and mapping τ onto τ^{σ} constitutes an isomorphism of $\Lambda(F,A)$ upon $\Lambda(G,B)$. We shall refer to this isomorphism of $\Lambda(F,A)$ upon $\Lambda(G,B)$ as to *the isomorphism induced by the semi-linear transformation* σ [of (F,A) upon (G,B)], an induced isomorphism of the first kind.

Induced automorphisms of the first kind are just the inner automorphisms of $\Lambda(F,A)$. An element in $\Lambda(F,A)$ induces the identity automorphism if, and only if, it belongs to the center of $\Lambda(F,A)$. We prove the following simple fact.

Lemma 1: *The semi-linear transformation* σ *belongs to the center of* $\Lambda(F,A)$ *if, and only if, it is a multiplication by a center element in F which is left invariant by every automorphism of the field F.*

PROOF: It is clear that center elements in Λ belong to the centralizer of T in Λ which is just the group of multiplications. If f is a number, not 0, in F and σ a semi-linear transformation, then the multiplication f has the property $f^{\sigma} = \sigma^{-1}f\sigma$ [by (1)]. Thus this multiplication belongs to the center of Λ if, and only if, $f = f^{\sigma}$ for every automorphism σ of F [as every automorphism of F occurs in at least one semi-linear transformation]. Lemma 1 is an immediate consequence.

Every automorphism of Λ induces an automorphism in T. If we consider an element z in the center Z of T, then it follows from VI.1, Proposition 1 that z is just the same as a center element in F. Thus center elements in F lead to inner automorphisms of Λ which induce the identity automorphism in T, though they will, in general, not induce the identity automorphism in Λ, as follows from Lemma 1.

The construction of the induced isomorphisms of the second kind is somewhat more involved and is closely related to similar constructions in V.5. We begin with the construction of the

NATURAL ANTI-ISOMORPHISM AND NATURAL ISOMORPHISM OF $\Lambda(F,A)$ INTO $\Lambda(A^*,F)$

[where (A^*,F) is the adjoint space to (F,A); see II.3].

Consider any semi-linear transformation σ in $\Lambda(F,A)$. If f is a linear form over A, then we define a single-valued mapping $f^{\sigma*}$ of A into F by the rule

$$af^{\sigma*} = (a^{\sigma}f)^{\tau^{-1}} \text{ for every } a \text{ in } A. \tag{3}$$

It is clear that $f^{\sigma*}$ is additive; and we have

$$(xa)f^{\sigma*} = [(xa)^{\sigma}f]^{\sigma^{-1}} = [x^{\sigma}(a^{\sigma}f)]^{\sigma^{-1}} = x(af^{\sigma*})$$

so that $f^{\sigma*}$ is a linear form over A too. It is fairly clear now that σ^* constitutes a one-to-one and additive mapping of A^* into itself. We verify

(3a) $$(fy)^{\sigma^*} = f^{\sigma^*}y^{\sigma^{-1}} \text{ for } f \text{ in } A^* \text{ and } y \text{ in } F.$$

This is true, since we have for every a in A

$$a(fy)^{\sigma^*} = [a^{\sigma}(fy)]^{\sigma^{-1}} = [(a^{\sigma}f)y]^{\sigma^{-1}} = (a^{\sigma}f)^{\sigma^{-1}}y^{\sigma^{-1}} = (af^{\sigma^*})y^{\sigma^{-1}} = a(f^{\sigma^*}y^{\sigma^{-1}}).$$

Next we prove

(4) $$(\sigma\tau)^* = \tau^*\sigma^* \text{ for } \sigma, \tau \text{ in } \Lambda(F,A).$$

This is true, since we have for every a in A, f in A^*

$$af^{(\sigma\tau)^*} = (a^{\sigma\tau}f)^{\tau^{-1}\sigma^{-1}} = ([(a^{\sigma})^{\tau}f]^{\tau^{-1}})^{\sigma^{-1}} = [a^{\sigma}f^{\tau^*}]^{\sigma^{-1}} = a(f^{\tau^*})^{\sigma^*} = af^{\tau^*\sigma^*}.$$

Since $1^* = 1$, it is clear now that every σ^* has an inverse and that therefore σ^* is a semi-linear transformation in $\Lambda(A^*,F)$ whose component field automorphism is the inverse of the component field automorphism of σ; and that the mapping of σ onto σ^* constitutes the *natural anti-isomorphism of* $\Lambda(F,A)$ *into* $\Lambda(A^*,F)$. But mapping group elements onto their inverses constitutes an anti-automorphism; and thus we obtain an isomorphism of $\Lambda(F,A)$ into $\Lambda(A^*,F)$, if we map σ in $\Lambda(F,A)$ upon σ^{*-1} in $\Lambda(A^*,F)$. This isomorphism will be called *the natural isomorphism of* $\Lambda(F,A)$ *into* $\Lambda(A^*,F)$. [Note that σ and σ^{*-1} have the same component field automorphism.]

Proposition 2: *The following properties of the linear manifold* (F,A) *are equivalent.*

(I) *The natural isomorphism maps* $\Lambda(F,A)$ *upon the whole group* $\Lambda(A^*,F)$.

(II) $\Lambda(F,A)$ *and* $\Lambda(A^*,F)$ *are isomorphic groups.*

(III) $r(A)$ *is finite.*

PROOF: It is clear that (I) implies (II). Assume next the validity of (II). Then it follows from VI.6, Theorem 2 that $T(F,A)$ and $T(A^*,F)$ are likewise isomorphic groups. But this implies by VI.5, Structure Theorem that (F,A) and (A^*F) are either projectively equivalent or duals of each other [the reader ought to verify the irrelevance of the fact that the numbers in F act as left-multipliers on A and as right-multipliers on A^*]. In either case we have $r(A) = r(A^*)$; and this implies the finiteness of $r(A)$ by II.3, Corollary 1. Thus (III) is a consequence of (II).

Assume finally that $r(A)$ is finite. To prove (I) it suffices to show that the natural anti-isomorphism maps $\Lambda(F,A)$ upon the whole of $\Lambda(A^*,F)$; and this we are going to do. Thus assume that τ is some semi-linear transformation in $\Lambda(A^*,F)$. If a is an element in A, then we define a single-valued mapping a' of A^* into F by the rule:

(3*) $$a'f = (af^{\tau})^{\tau^{-1}} \text{ for every } f \text{ in } A^*.$$

One verifies by a direct computation that a' is a linear form over (A^*,F). Since $r(A)$ is finite, it follows from II.3, Theorem 2 that there exists one and only one element a^σ in A such that $a^\sigma f = a'f$ for every f in A^*. Thus we have mapped τ in $\Lambda(A^*,F)$ upon a definite single-valued mapping $\sigma = \sigma(\tau)$ of A into itself. This mapping satisfies by its very definition the following rule which is just a restatement of (3*):

(3 +) $\quad a^{\sigma(\tau)}f = (af^\tau)^{\tau^{-1}}$ for every a in A and every f in A^*.

One verifies as before that

$$(xa)^{\sigma(\tau)} = x^{\tau^{-1}}a^{\sigma(\tau)} \text{ for } x \text{ in } F \text{ and } a \text{ in } A;$$
$$\sigma(\tau'\tau'') = \sigma(\tau'')\sigma(\tau'); \ \sigma(1) = 1;$$

and from these two facts one deduces in particular that $\sigma(\tau)$ is a semi-linear transformation in $\Lambda(F,A)$, whose component field automorphism is τ^{-1}. Remembering the definition (3) of the natural anti-isomorphism of $\Lambda(F,A)$ into $\Lambda(A^*,F)$ we find now that

$$af^{\sigma(\tau)^*} = (a^{\sigma(\tau)}f)^{\sigma(\tau)^{-1}} = (a^{\sigma(\tau)}f)^\tau = [(af^\tau)^{\tau^{-1}}]^\tau = af^\tau$$

for every a in A and every f in A^*. But linear forms over A are equal if they have the same value on every a in A. Hence $f^\tau = f^{\sigma(\tau)^*}$ for every f in A^*; and consequently $\tau = \sigma(\tau)^*$. Thus τ is the image of $\sigma(\tau)$ under the natural anti-isomorphism of $\Lambda(F,A)$ into $\Lambda(A^*,F)$. Hence

$$[\Lambda(F,A)]^* = \Lambda(A^*,F);$$

and this proves (i). Hence (i) is a consequence of (iii); and this completes the proof.

Remark 1: The mapping $\sigma(\tau)$ of $\Lambda(A^*,F)$ is essentially the same as the natural anti-isomorphism of $\Lambda(A^*,F)$, if we only remember that A is the adjoint space of A^* whenever $r(A)$ is finite [II.3, Corollary 1].

We are now ready to construct the

Induced Isomorphisms of the Second Kind:

Consider any anti-semi-linear transformation σ of (A^*,F) upon the linear manifold (G,B). If α is a semi-linear transformation in $\Lambda(A^*,F)$, then $\sigma^{-1}\alpha\sigma = \alpha^\sigma$ is easily seen to be a semi-linear transformation in $\Lambda(G,B)$ —note in the proof mainly that $\sigma^{-1}\alpha\sigma$ is an automorphism of the field G, if σ is an anti-automorphism of F upon G and α is an automorphism of F; and it is clear that mapping α upon α^σ constitutes an isomorphism of $\Lambda(A^*,F)$ upon $\Lambda(G,B)$. Denote by ν the natural isomorphism of $\Lambda(F,A)$ into $\Lambda(A^*,F)$ so that $\tau^\nu = \tau^{*-1}$ for every τ in $\Lambda(F,A)$. Then $\nu\sigma$ is an isomorphism of $\Lambda(F,A)$ into $\Lambda(G,B)$; and we shall refer to this isomorphism $\nu\sigma$ as to *the isomorphism induced by* σ, *an induced isomorphism of the second kind.*

Since mapping α in $\Lambda(A^*,F)$ upon α^σ in $\Lambda(G,B)$ constitutes an isomorphism of $\Lambda(A^*,F)$ upon $\Lambda(G,B)$, the isomorphism $\nu\sigma$ is an isomorphism of $\Lambda(F,A)$ upon $\Lambda(G,B)$ if, and only if, ν is an isomorphism of $\Lambda(F,A)$ upon $\Lambda(A^*,F)$. It follows from Proposition 2 that this is equivalent to the finiteness of $r(A)$. We restate this result as follows.

Corollary 1: *The induced isomorphisms of the second kind are "isomorphisms upon" if, and only if, the rank is finite.*

REMARK 2: There exist always induced isomorphisms of the second kind, since there exist always anti-semi-linear transformations of (A^*,F) [see, for instance, IV.1, Construction of Canonical Dual Space]. This shows the indispensability of the condition that the rank be finite.

REMARK 3: The reader ought to verify the following fact. If the isomorphism σ of $\Lambda(F,A)$ upon $\Lambda(G,B)$ is an induced isomorphism of the first [second] kind, then σ induces an isomorphism of $\mathrm{T}(F,A)$ upon $\mathrm{T}(G,B)$ which is an induced isomorphism of the first [second] kind in the sense of VI. 5.

We are now ready to state the following comprehensive result.

Isomorphism Theorem: *Every isomorphism of $\Lambda(F,A)$ upon $\Lambda(G,B)$ may be represented in one and essentially only one way in the form $\sigma'\sigma''$ where σ' is a singular automorphism of $\Lambda(F,A)$ and where σ'' is an induced isomorphism [of the first or second kind] of $\Lambda(F,A)$ upon $\Lambda(G,B)$.*

Here we say that the representations $\sigma = \sigma'\sigma'' = \tau'\tau''$ of the isomorphism σ as a product of singular automorphisms σ', τ' and induced isomorphisms σ'', τ'' are *essentially the same,* if σ'' and τ'' are of the same kind and $\tau'^{-1}\sigma' = \tau''\sigma''^{-1}$ is an inner automorphism induced by a multiplication [an element in N].

The proof of this fundamental theorem will be preceded by the proofs of various facts.

Proposition 3: *If σ is an isomorphism of $\Lambda(F,A)$ upon $\Lambda(G,B)$, then there exists a uniquely determined projectivity or duality σ' of A upon B such that*

$$\Delta(S^{\sigma'}) = \Delta(S)^{\sigma} \text{ and } S^{\sigma'} = J^{+}([\Delta(S)^{2}]^{\sigma})$$

for every subspace S, not 0 nor A, of A.

PROOF: That σ' is uniquely determined by σ, is an almost immediate consequence of the second formula. To prove the existence of σ' we procede as follows. The isomorphism σ of $\Lambda(F,A)$ upon $\Lambda(G,B)$ induces an isomorphism of $\mathrm{T}(F,A)$ upon $\mathrm{T}(G,B)$ [VI.6, Theorem 2]. Now we continue as in the proof of VI.5, Isomorphism Theorem. If S is a subspace of A such that $0 < S < A$, then we let $S^{\nu} = J^{+}([\Delta(S)^{2}]^{\sigma}) = J^{+}([\Delta(S)^{\sigma}]^{2})$. Since σ induces an isomorphism of $\mathrm{T}(F,A)$ upon $\mathrm{T}(G,B)$, it follows from VI.4, Theorem 4 that ν is a one-to-one mapping of the subspaces, not 0 nor A, of A upon the totality of subspaces, not 0 nor B, of B; and it follows from VI.4, Theorem 2 that ν preserves betweenness. We apply VI.5, Lemma 2; and find that ν is induced by a projectivity or by a duality which we denote by σ'. That σ' satisfies also the first requirement

$$\Delta(S^{\sigma'}) = \Delta(S)^{\sigma},$$

is easily deduced from VI.4, Theorem 3, (*c*).

PROOF OF THE STRUCTURE THEOREM:

If there exists an isomorphism of $\Lambda(F,A)$ upon $\Lambda(G,B)$, then it follows from Proposition 3 that there exists a projectivity or a duality of A upon B. Assume conversely the existence of a projectivity or a duality of A upon B. If there exists a projectivity of A upon B, then there exists a semi-linear transformation of A upon B [III.1, First Fundamental Theorem of Projective Geometry]; and consequently there exists an induced isomorphism of the first kind of $\Lambda(F,A)$ upon $\Lambda(G,B)$. If there exists a duality of A upon B, then there exists an anti-semi-linear transformation of (A^*,F) upon (G,B) [IV.1, Lemma 2]; and $r(A)$ is finite [IV.1, Existence Theorem]. It follows from Corollary 1 that there exists an induced isomorphism of the second kind of $\Lambda(F,A)$ upon $\Lambda(G,B)$. Thus $\Lambda(F,A)$ and $\Lambda(G,B)$ are isomorphic groups in either case.

Lemma 2: *If $r(A)$ is finite, and if ν is the natural isomorphism of $\Lambda(F,A)$ upon $\Lambda(A^*,F)$, then $\Delta(S)^{\nu} = \Delta[E(S)]$ for every subspace S of A.*

PROOF: We recall firstly that $E(S)$ is the totality of linear forms f in A^* such that $Sf = 0$. We note next that ν maps every semi-linear transformation of A upon a semi-linear transformation of A^* with the same component field automorphism. Thus ν maps in particular linear transformations upon linear transformations and involutions upon involutions. If σ is an involution in $\mathrm{T}(F,A)$, then $\sigma^{\nu} = \sigma^*$ where the * indicates the natural anti-isomorphism of $\Lambda(F,A)$ upon $\Lambda(A^*,F)$ which has been defined by rule (3).

The involution σ belongs to $\Delta(S)^+$ if, and only if, $S = J^+(\sigma)$; and this is equivalent to the following two properties:

$$s^{\sigma} = s \text{ for every } s \text{ in } S;\ x^{\sigma} \equiv -x \text{ modulo } S \text{ for every } x \text{ in } A.$$

If f belongs to $E(S)$, then we have consequently

$$0 = (x + x^{\sigma})f = xf + x^{\sigma}f = xf + xf^{\sigma^*} = x(f + f^{\sigma^*}) \text{ for every } x;$$

and this implies $f^{\sigma^*} = -f$ for f in $E(S)$. If next f is any element in A^*, and if s is an element in S, then we have

$$s(f - f^{\sigma^*}) = sf - sf^{\sigma^*} = sf - s^{\sigma}f = sf - sf = 0,$$

since $S = J^+(\sigma)$; and consequently $f - f^{\sigma^*}$ belongs to $E(S)$. This proves that $E(S) = J^-(\sigma^*)$ if $S = J^+(\sigma)$.

Assume conversely that $E(S) = J^-(\sigma^*)$. This is equivalent to the properties:

$$f^{\sigma^*} = -f \text{ for } f \text{ in } E(S) \text{ and } g^{\sigma^*} \equiv g \text{ modulo } E(S) \text{ for every } g \text{ in } A^*.$$

If s is some element in S and g some element in A^*, then we have consequently

$$0 = s(g - g^{\sigma^*}) = sg - sg^{\sigma^*} = sg - s^\sigma g = (s - s^\sigma)g.$$

Thus $s - s^\sigma$ is annihilated by every linear form over A; and it follows from II.3, Proposition 2 that $s - s^\sigma = 0$ or $s = s^\sigma$ for every s in S. If a is in A and f is in $E(S)$, then we have

$$(a + a^\sigma)f = af + a^\sigma f = af + af^{\sigma^*} = a(f + f^{\sigma^*}) = a0 = 0$$

so that $a + a^\sigma$ belongs to $S[E(S)] = S$ by II.3, Proposition 2. Hence $J^+(\sigma) = S$.

We have shown therefore that σ is an involution satisfying $S = J^+(\sigma)$ if, and only if, σ^* is an involution satisfying $E(S) = J^-(\sigma^*)$. This is equivalent to saying that

$$[\Delta(S)^+]^* = \Delta[E(S)]^-;$$

and this statement is equivalent to the assertion of the lemma, since the natural isomorphism and the natural anti-isomorphism have the same effect on involutions.

Proof of Isomorphism Theorem:

We begin by proving the uniqueness part of this theorem. Thus assume that α,β are singular automorphisms of $\Lambda(F,A)$, that γ,δ are induced isomorphisms of $\Lambda(F,A)$ upon $\Lambda(G,B)$, and that $\alpha\gamma = \beta\delta = \sigma$. It is a consequence of Propositions 1 and 3 that there exists a projectivity or duality σ' such that

$$\Delta(S^{\sigma'}) = \Delta(S)^\sigma = \Delta(S)^\gamma = \Delta(S)^\delta \text{ for every subspace } S \text{ of } A.$$

If γ is an induced isomorphism of the first kind, then γ is induced by a semi-linear transformation τ of (F,A) upon (G,B); and it follows that

$$\Delta(S^{\sigma'}) = \Delta(S)^\gamma = \tau^{-1}\Delta(S)\tau = \Delta(S^\tau)$$

so that σ' is the projectivity induced by the semi-linear transformation τ. If γ is an induced isomorphism of the second kind, then $r(A)$ is finite and there exists an anti-semi-linear transformation ω of (A^*,F) upon (G,B) such that γ maps the semi-linear transformation η in $\Lambda(F,A)$ upon $\eta^{\nu\omega} = \omega^{-1}\eta^\nu\omega$ where ν is the natural isomorphism of $\Lambda(F,A)$ upon $\Lambda(A^*,F)$. Now we deduce from Lemma 2 that

$$\Delta(S^{\sigma'}) = \Delta(S)^\gamma = \omega^{-1}\Delta(S)^\nu\omega = \omega^{-1}\Delta[E(S)]\omega = \Delta[E(S)^\omega]$$

so that σ' is the duality mapping S upon $E(S)^{\omega}$ [II.3, Theorem 3]. We restate these results for future application as follows:

(I) *If α is a singular automorphism of $\Lambda(F,A)$, and if the isomorphism γ of $\Lambda(F,A)$ upon $\Lambda(G,B)$ is induced by the semi-linear transformation τ of A upon B, then*

$$\Delta(S^{\tau}) = \Delta(S)^{\alpha\gamma} \textit{ for every subspace } S \textit{ of } A.$$

(II) *If α is a singular automorphism of $\Lambda(F,A)$, and if the isomorphism γ of $\Lambda(F,A)$ upon $\Lambda(G,B)$ is an induced isomorphism of the second kind which is derived from the anti-semi-linear transformation ω of A^* upon B, then*

$$\Delta[E(S)^{\omega}] = \Delta(S)^{\alpha\gamma} \textit{ for every subspace } S \textit{ of } A.$$

We apply these results on $\sigma = \alpha\gamma = \beta\delta$; and the corresponding mapping σ' which is either a projectivity or a duality. It follows that σ' is a projectivity if, and only if, γ and δ are induced isomorphisms of the first kind which derive from the semi-linear transformations inducing σ'; and that σ' is a duality if, and only if, γ and δ are induced isomorphisms of the second kind which derive from anti-semi-linear transformations inducing σ'.

If γ and δ are induced by semi-linear transformations γ' and δ' respectively, then we have

$$\Delta(S^{\gamma'}) = \Delta(S)^{\alpha\gamma} = \Delta(S)^{\beta\delta} = (\Delta S^{\delta'}) \text{ for every subspace } S \text{ of } A.$$

The same projectivity of A upon B is consequently induced by the semi-linear transformations γ' and δ'; and they differ therefore by a multiplication of A only [III.1, Proposition 3]. Hence $\beta^{-1}\alpha = \delta\gamma^{-1}$ is an inner automorphism of $\Lambda(F,A)$ induced by an element in N; and this fact constitutes the essential uniqueness of the representation of σ. If γ and δ are induced isomorphisms of the second kind, then we show the essential uniqueness of the representation of σ by a very similar argument based on (II) whose details we leave to the reader.

Assume finally that there is given some isomorphism σ of $\Lambda(F,A)$ upon $\Lambda(G,B)$. We infer from Proposition 3 the existence of a projectivity or duality σ' such that

$$\Delta(S^{\sigma'}) = \Delta(S)^{\sigma} \text{ for every subspace } S \text{ of } A.$$

If σ' is a projectivity, then we deduce from III.1, First Fundamental Theorem of Projective Geometry the existence of a semi-linear transformation τ of A upon B such that $S^{\tau} = S^{\sigma'}$ for every subspace S. This transformation τ induces an isomorphism γ of $\Lambda(F,A)$ upon $\Lambda(G,B)$; and it follows from (I) that

$$\Delta(S)^{\sigma} = \Delta(S^{\sigma'}) = \Delta(S^{\tau}) = \Delta(S)^{\gamma} \text{ for every subspace } S \text{ of } A.$$

But then it follows from Proposition 1 that $\sigma\gamma^{-1}$ is a singular automorphism of $\Lambda(F,A)$.

If next σ' is a duality, then $r(A)$ is finite [IV.1, Existence Theorem]; and we infer from IV.1, Lemma 2 the existence of an anti-semi-linear transformation ω of A^* upon B such that $E(S)^\omega = S^{\sigma'}$ for every subspace S of A. This transformation ω induces an isomorphism δ of the second kind of $\Lambda(F,A)$ upon $\Lambda(G,B)$; and it follows from (II) that

$$\Delta(S)^\sigma = \Delta(S^{\sigma'}) = \Delta[E(S)^\omega] = \Delta(S)^\delta \text{ for every subspace } S \text{ of } A.$$

But then it follows from Proposition 1 that $\sigma\delta^{-1}$ is a singular automorphism of $\Lambda(F,A)$; and this completes the proof.

The reader ought to explore the possibilities of constructing a proof of the Isomorphism Theorem which makes more use of the results of VI.5 than our proof; cp. the proof of the Structure Theorem.

The Automorphism Group of the Group $\Lambda(F,A)$ will be denoted by $A(\Lambda)$. It contains the subgroup $A_s(\Lambda)$ of singular automorphisms which is a normal subgroup, since the singular automorphisms are exactly those automorphisms of Λ which induce the identity automorphism in Λ/N [note that N is, by VI.6, Theorem 2, a characteristic subgroup of Λ]. There is furthermore the subgroup $A_i(\Lambda)$ of all the induced automorphisms of Λ and the normal subgroup $A_0(\Lambda)$ of the inner automorphisms of Λ [which equals the group of induced automorphisms of the first kind]. Clearly $A_s(\Lambda) \cap A_0(\Lambda)$ is a normal subgroup too; it consists of all those inner automorphisms which are induced by elements in N [uniqueness part of Isomorphism Theorem].

It follows from VI.6, Theorem 2 that every automorphism α of Λ induces an automorphism α^τ of T. Clearly τ is a [natural] homomorphism of $A(\Lambda)$ into the automorphism group $A(T)$ of T. It is a consequence of Proposition 1 that the kernel of τ consists of singular automorphisms only; and $A_s(\Lambda) \cap A_0(\Lambda)$ is part of the kernel of τ, since multiplications in N commute with linear transformations. One verifies easily that τ maps singular automorphisms onto singular automorphisms, induced automorphisms of the first or second kind upon induced automorphisms of the same kind; and that every induced automorphism of T is the τ-image of an induced automorphism of Λ.

Consider now a singular automorphism α and an inner automorphism β of Λ. Their commutator is naturally in $A_s(\Lambda) \cap A(_0\Lambda)$ so that $\alpha^\tau\beta^\tau = \beta^\tau\alpha^\tau$. Thus we have shown that τ maps singular automorphisms of Λ upon singular automorphisms of T which commute with every induced automorphism of the first kind of T. We have given an example in VI.5, showing that not all the singular automorphisms of T have this property. Thus, in general, τ maps $A(\Lambda)$ upon a proper part of $A(T)$ and in particular $A_s(\Lambda)$ upon a proper part of $A_s(T)$.

CHAPTER VII

Internal Characterization of the System of Subspaces

The totality $S = S(A) = S(F,A)$ of subspaces of the linear manifold (F,A) is the projective geometry defined by this linear manifold and it derives its structure from such relations and operations connecting subspaces as $U < V$, $U \cap V$, $U + V$. In the present chapter we shall enumerate a number of properties of these relations which suffice to characterize this system as the system of subspaces of a linear manifold.

The problem we are treating here is known under various names like "principles of geometry", "introducing coordinates in a projective space" etc. It is the one chapter in projective geometry which may be treated by synthetic methods only, since its prime objective is the construction of the algebraical machinery [= the underlying linear manifold]. Once this construction has been effected, we may use the algebraical tools wherever convenient.

In the preceding chapters we had to assume fairly often that the linear manifold had rank greater than 2. In the present chapter we shall have to assume that the rank is at least 4 [dimension at least 3]. Our arguments will apply to a certain class of planes too [Desarguesian planes]; but the highly interesting and somewhat sophisticated problem of the projective planes in general we shall not touch. Concerning these the reader is advised to consult the literature.

A Short Bibliography of the Principles of Geometry

E. Artin: Coordinates in Affine Geometry. *Reports of a Math. Coll.*, University of Notre Dame (2), vol. 2 (1940), pp. 15-20.

R. Baer: Homogeneity of Projective Planes. *Amer. Journal of Math.*, vol. 64 (1942), pp. 137-152.

A Unified Theory of Projective Spaces and Finite Abelian Groups. *Trans. Amer. Math. Soc.*, vol. 52 (1942), pp. 283-343.

G. Birkhoff: Combinatorial Relations in Projective Geometry. *Ann. of Math.*, vol. 36 (1935), pp. 743-748.

Lattice Theory, 2nd ed. *Amer. Math. Soc. Coll. Pub.*, 25, New York, 1948.

O. Bottema: De elementaire meetkunde van het platte vlak. Groningen-Batavia, 1938.
H. S. M. Coxeter: The Real Projective Plane. New York, 1949.
M. Hall: Projective Planes. *Trans. Amer. Math. Soc.*, vol. 54 (1943), pp. 229-277.
L. Heffter: Grundlagen und analytischer Aufbau der Geometrie, 2. Aufl. Leipzig, 1950.
H. Hermes and G. Köthe: Die Theorie der Verbände. *Enzyklopädie der math. Wiss.*, I, 1, 13.
G. W. Hessenberg: Grundlagen der Geometrie. Herausgegeben von W. Schwan, Leipzig, 1930.
A. Heyting: Intuitionistische axiomatiek der projectieve meetkunde. Groningen, 1925.
D. Hilbert: Grundlagen der Geometrie, 7. Aufl. Leipzig, 1930.
J. Hjelmslev: Grundlag for den projektive Geometrie. Kjobenhavn, 1933.
G. Köthe: Die Theorie der Verbände, ein neuer Versuch zur Grundlegung der Algebra und der projektiven Geometrie. *Jahresberichte der Deutschen Math. Ver.*, vol. 47 (1937), pp. 125-144.
F. Levi: Geometrische Konfigurationen. Leipzig, 1929.
H. Liebmann: Synthetische Geometrie. Leipzig and Berlin, 1934.
K. Menger: New Foundations of Projective and Affine Geometry. *Ann. of Math.*, vol. 37 (1936), pp. 456-482.
R. Moufang: Alternativkörper und der Satz vom vollständigen Vierseit (D_9). *Abhandlungen aus dem mathematischen Seminar der Hamburgischen Universität*, vol. 9 (1933), pp. 207-222.
J. von Neumann: Continuous Geometry. Princeton, N. J., 1936-1937.
M. Pasch: Vorlesungen über neuere Geometrie mit einem Anhang: Die Grundlegung der Geometrie in historischer Entwicklung von Max Dehn. Berlin, 1926.
K. Reidemeister: Vorlesungen über Grundlagen der Geometrie. Berlin, 1930.
G. de B. Robinson: The Foundations of Geometry. Toronto, 1940.
F. Schur: Grundlagen der Geometrie. Leipzig and Berlin, 1909.
G. Thomsen: Grundlagen der Elementargeometrie in gruppentheoretischer Behandlung. Berlin and Leipzig, 1933.
O. Veblen and J. W. Young: Projective Geometry, vol. 1, 2, Boston, 1918-1938.

VII.1. Basic Concepts, Postulates and Elementary Properties

The framework of our discussion is provided by the concept: *partially ordered set*. This is a set S of elements together with a relation "$\leqslant$". For obvious reasons we shall denote the elements in S by capital letters; and instead of $U \leqslant V$ we shall use such phrases as U is part of V, U is contained in V, U is on V, V contains U, V passes through U. If $U \leqslant V$ and $U \neq V$ hold at the same time, then we shall also write $U < V$.

This containedness relation is subject to the following natural postulates.

I. *$U \leqslant V$ and $V \leqslant U$ if, and only if, $U = V$.*

II. *$U \leqslant V$ and $V \leqslant W$ imply $U \leqslant W$.*

These are the postulates of partial order. Note that it is entirely possible that two elements U and V in S satisfy neither $U \leqslant V$ nor $V \leqslant U$; and that consequently $U \nleqslant V$ just excludes this one possibility [that $U \leqslant V$] without implying $V < U$.

DEFINITION 1: *Suppose that Θ is a set of elements in S. If the element*

U in S has the properties	*V in S has the properties*
(a') $U \leqslant X$ for every X in Θ;	*(a'') $X \leqslant V$ for every X in Θ;*
(b') if M is an element in S such that $M \leqslant X$ for every X in Θ then $M \leqslant U$;	*(b'') if N is an element in S such that $X \leqslant N$ for every X in Θ then $V \leqslant N$;*
then we let $$U = \bigcap_{X \in \Theta} X.$$	*then we let* $$V = \sum_{X \in \Theta} X.$$

V is often called the least upper bound of Θ and U the greatest lower bound. For our purposes it is more convenient to term U the intersection and V the sum of Θ. In case Θ contains only a few elements, then we shall use notations like

$H \cap K$, $H \cap K \cap L$ etc for U $\qquad$ $H + K$, $H + K + L$ etc for V.

It is immediately obvious that there exists at most one intersection and at most one sum of Θ. But the existence of at least one such element cannot be proved: and so we impose the following further postulate.

III. *If Θ is a [non vacuous] subset of S, then there exists the intersection $\bigcap_{X \in \Theta} X$ and the sum $\sum_{X \in \Theta} X$ of Θ.*

This postulate implies in particular the existence of the intersection 0 of all the elements in S and of the sum A of all the elements in S. [Note that the symbol customary in the theory of partially ordered sets for the sum of all elements is 1; but again it is obvious why we select the symbol A in accordance with usage in preceding chapters.]

One proves easily that the sum of Θ is the intersection of the totality of elements which contain every element in Θ; and that the intersection of Θ is the sum of all the elements which are contained in every element in Θ. On the basis of this remark one may show that it suffices to postulate the existence of A and all the intersections or the existence of 0 and all the sums [in III].

We make the obvious remark that the set of all the subspaces of a linear manifold satisfies Postulates I to III with respect to the containedness relation and that our definitions of intersection and sum are in accordance with those used in the theory of linear manifolds.

IV. *If U, V, W are elements in S, and if $U \leqslant V$, then*

$$V \cap (U + W) = U + (V \cap W).$$

This is exactly Dedekind's Law [II.1]. This law is often called the modular law. We shall usually refer to Postulate IV as to Dedekind's Law. One verifies now the equivalence of the following relations:

$$X \cap Y = X, \qquad X \leqslant Y, \qquad X + Y = Y.$$

V. *If U is an element in S, then there exists an element V in S such that $0 = U \cap V$, $A = U + V$.*

This is the Complementation Theorem [II.1]; and we shall refer often to Postulate V as to the Complementation Principle. A system meeting requirements I to V is called usually a "complete, complemented, modular lattice."

From V we deduce now the following *General Complementation Principle*:

If $U \leqslant V \leqslant W$, then there exists an element T in S such that $U = V \cap T$, $W = V + T$.

PROOF: From V we deduce the existence of an element H in S such that $0 = V \cap H$, $A = V + H$. Let $T = U + (H \cap W)$. Then we deduce from Dedekind's Law and $U \leqslant V$ that

$$V \cap T = V \cap [U + (H \cap W)] = U + (V \cap H \cap W) = U + 0 = U;$$

and from Dedekind's Law and $U \leqslant V \leqslant W$ we deduce that

$$V + T = V + U + (H \cap W) = V + (H \cap W) = W \cap (V + H) = W \cap A = W.$$

DEFINITION 2: *If $U < V$, and if $U \leqslant X < V$ implies $X = U$, then V is a point modulo U.*

Instead of saying that V is a point modulo U we shall usually say that V/U is a point; and if V is a point modulo 0, then we shall say shorter that V is a point. Notation and definition are in accordance with those used before [see in particular II.1 and II.2].

VI. *If U is an element, not 0, in S, then U contains a point.*

It is a fairly trivial remark that the set $S(A)$ of all the subspaces of the linear manifold A meets the requirement VI.

Lemma 1: *If $U < V$, then there exists a point P such that $P \leqslant V$, $P \not\leqslant U$, and such that consequently $U + P$ is part of V and a point modulo U.*

PROOF: Since $0 \leqslant U < V$ we may deduce from the General Complementation Principle the existence of an element W such that $0 = U \cap W$, $V = U + W$. From $U < V$ we infer $W \neq 0$. Consequently there exists by VI a point $P \leqslant W$. If P were part of U, then $P \leqslant U \cap W = 0$ which is impossible. Thus we have shown that $P \leqslant V$, $P \not\leqslant U$; and this implies clearly that $U < U + P \leqslant V$. Suppose finally that X is an element in S such that $U \leqslant X < U + P$. Then $P \not\leqslant X$ so that $P \cap X < P$. But P is a point; and so we find that $P \cap X = 0$. Now we deduce from Dedekind's Law that

$$X = X \cap (U + P) = U + (X \cap P) = U + 0 = U;$$

and this proves that $(U + P)/U$ is a point.

VII. *If Θ is a set of elements in S, if P is a point in S, and if $P \leqslant \sum_{X \in \Theta} X$, then there exists a finite number of elements $X_1, \cdots, X_k$ in Θ such that $P \leqslant \sum_{i=1}^{k} X_i$.*

VERIFICATION OF VII IN $S(F,A)$:

A point P in $S(F,A)$ has the form Fp; and P is part of the sum of the subspaces X in Θ if, and only if, p belongs to this sum. But the sum of the X in Θ consists of all the finite sums $x_1 + \cdots + x_k$ with x_i in X_i and X_i in Θ. Now it is evident how to complete the verification of VII.

Lemma 2: *If Θ is a set of elements in S, if $U < V \leqslant \sum_{X \in \Theta} X$, then there exists a finite number of elements $X_1, \cdots, X_k$ in Θ such that*

$$U \cap \sum_{i=1}^{k} X_i < V \cap \sum_{i=1}^{k} X_i.$$

PROOF: We infer from Lemma 1 the existence of a point P which is part of V, but not of U. We deduce from Postulate VII the existence of a finite number of elements $X_1, \cdots, X_k$ in Θ such that $P \leqslant \sum_{i=1}^{k} X_i$. Then $U \cap \sum_{i=1}^{k} X_i < V \cap \sum_{i=1}^{k} X_i$ is a consequence of the fact that P is part of the second intersection, but not of the first one.

Corollary 1: *If Θ is a set of elements in S, if $U \leqslant X$ for every X in Θ and $V \leqslant \sum_{X \in \Theta} X$, and if V/U is a point, then there exists a finite number of elements $X_1, \cdots, X_k$ in Θ such that $V \leqslant \sum_{i=1}^{k} X_i$.*

PROOF: We deduce from Lemma 2 the existence of a finite number of elements $X_1, \cdots, X_k$ in Θ such that

$$U = U \cap \sum_{i=1}^{k} X_i < V \cap \sum_{i=1}^{k} X_i \leqslant V.$$

But V/U is a point and so $V = V \cap \sum_{i=1}^{k} X_i$ or $V \leqslant \sum_{i=1}^{k} X_i$ is a consequence of the preceding inequality.

VIII. *If P and Q are distinct points in S, then there exists a point R, different from P and Q, such that $R \leqslant P + Q$.*

It is easy to see that the requirements imposed upon the "third" point R on the line $P + Q$ are equivalent to the following conditions:

$$P + Q = Q + R = R + P.$$

VERIFICATION OF VIII IN $S(A)$:

If P and Q are distinct points in the F-space A, then there exist elements p and q neither of which is 0 such that $P = Fp$, $Q = Fq$; and $R = F(p+q)$ is the desired third point on the line $P + Q$.

The imposition of Postulate VIII does not constitute an essential loss in generality, since a system with properties I to VII may be "decomposed" in a unique [and natural] fashion into components with properties I to VIII whose structures determine that of the original system. For details of this reduction cf.

Orrin Frink: Complemented Modular Lattices and Projective Spaces of Infinite Dimension. *Trans. Amer. Math. Soc.*, vol. 60 (1946), pp. 452-467, where further references may be found.

Lemma 3: *If $U < V < W$, then there exist two different elements T' and T'' in S such that $U = V \cap T' = V \cap T''$, $W = V + T' = V + T''$.*

PROOF: We deduce from the General Complementation Principle the existence of an element T in S such that $U = V \cap T$, $W = V + T$; and it follows from $U < V < W$ that $U < T < W$. We deduce now from Lemma 1 the existence of points P, Q such that $P \leqslant T$, $P \not\leqslant U$ and $Q \leqslant V$, $Q \not\leqslant U$. Since $U < U + P \leqslant T$, we may deduce from the General Complementation Principle the existence of an element H in S such that $U = H \cap (U + P)$, $T = H + U + P = H + P$.

Since P and Q are distinct points, we deduce from Postulate VIII the existence of a third point R such that $P + Q = Q + R = R + P$. Then we let $T^* = H + R$.

We note first that Q is not on T, since otherwise Q were on $T \cap V = U$. If P or Q were on T^*, then both would be on T^* [since $P + R = R + Q$]. Consequently we would have $T = H + P \leqslant H + R = T^*$; and since Q is certainly not on T, this would imply $H \leqslant T = H + P < H + R$. This implies in particular that $H + R$ is a point modulo H; see Lemma 1 and its proof. Consequently $H + P = H$ or $P \leqslant H$. But this implies

$$P \leqslant H \cap (U + P) = U$$

which is impossible. Thus we have been led to a contradiction which shows that neither P nor Q is on T^*.

It follows in particular that $T \neq T^*$. We have furthermore

$$V + T^* = V + H + R = V + Q + R + H = V + Q + P + H = V + T = W;$$

and we deduce from Dedekind's Law that

$$\begin{aligned} V \cap T^* &= V \cap (H + R) = V \cap (H + R) \cap (T + R) \\ &= V \cap (H + P + R) \cap (H + R) = V \cap (H + Q + P) \cap (H + R) \\ &= [Q + (V \cap T)] \cap (H + R) = (Q + U) \cap (H + R) \\ &= U + [Q \cap (H + R)] = U, \end{aligned}$$

since the point Q is not on $T^* = H + R$. This completes the proof.

DEFINITION 3: *If U, V are elements in S such that $U \leqslant V$, then V/U is the totality of elements X in S such that $U \leqslant X \leqslant V$.*

It is clear that V/U is a partially ordered set with respect to the same relation "$\leqslant$" as used in S. The null element of V/U is U and the "A" element of V/U is V. On the basis of Lemmas 1 to 3 and Corollary 1 one verifies now easily:

Proposition 1: *V/U satisfies all the Postulates* I *to* VIII.

REMARK ON THE PRINCIPLE OF DUALITY:

We have shown [Chapter IV] that every linear manifold of finite rank possesses a dual, though it need not be self-dual, and that the linear manifolds of infinite rank do not possess any duals. The theory of linear manifolds consequently lacks duality; and in order to obtain this duality we would have to exclude the linear manifolds of infinite rank. Consequently it is impossible to give a self-dual system of postulates for a projective geometry that comprises the linear manifolds of infinite rank. For a self-dual system of postulates for projective geometry see

Karl Menger: The Projective Space. *Duke Math. Journal*, vol. 17 (1950), pp. 1-14;

M. Esser: Self-dual Postulates for n-Dimensional Projective Geometry. *Duke Math. Journal*, vol. 18 (1951), pp. 475-480.

VII.2. Dependent and Independent Points

Throughout we shall assume the validity of the Postulates I to VIII which we enumerated in § 1; and we shall refer to them only by their roman numerals. Not all of them will be needed on every occasion; and the reader should check which of them have actually been used.

Proposition 1: *Every element X in S is the sum of the points contained in X.*

Here we may make use of the convention that 0 is the sum of the empty set.

PROOF: Denote by Y the sum of all the points contained in X. We deduce from the General Complementation Principle the existence of an element Z in S such that $X = Y + Z$ and $0 = Y \cap Z$. If Z were not 0, then we would infer from Postulate VI the existence of a point $P \leqslant Z$. Clearly P is contained in X too. Consequently it follows from the definition of Y that $P \leqslant Y$. Hence $P \leqslant Y \cap Z = 0$, a contradiction. Thus $Z = 0$; and $X = Y$ is the sum of the points contained in X.

DEFINITION 1: *The point P is dependent on the set Φ of points, if $P \leqslant \sum_{X \in \Phi} X$.*

Proposition 2: *Dependency of points has the following properties:*

(*a*) *No point depends on the empty set.*

(*b*) *If the point P depends on the set Φ of points, and if Φ is a subset of the set Θ of points, then P depends on Θ.*

(*c*) *If the point P depends on the set Φ of points, then P depends on a finite subset of Φ.*

(*d*) *If the point P depends on the set Φ of points, and if every point in Φ depends on the set Θ of points, then P depends on Θ.*

(*e*) *If the point P depends on the finite set of points $P_1, \cdots, P_k$, but on no proper subset, then P_1 depends on $P, P_2, \cdots, P_k$.*

(*g*) *Every point depends on itself.*

PROOF: If P depends on Φ, and if every point in Φ depends on Θ, then we have the inequalities

$$P \leqslant \sum_{X \in \Phi} X \leqslant \sum_{Y \in \Theta} Y,$$

since every X in Φ satisfies $X \leqslant \sum_{Y \in \Theta} Y$; and this proves that P depends on Θ. Hence (*d*) is true. (*c*) is an immediate consequence of Postulate VII. (*a*), (*b*) and (*g*) are obvious.

Suppose finally that P depends on the finitely many points $P_1, \cdots, P_k$, but not on any proper subset of them. Then we have

$$P \leqslant \sum_{i=1}^{k} P_i, \quad P \not\leqslant \sum_{i=2}^{k} P_i.$$

We deduce from Dedekind's Law that

$$P + \sum_{i=2}^{k} P_i = [P + \sum_{i=2}^{k} P_i] \cap \sum_{i=1}^{k} P_i = \sum_{i=2}^{k} P_i + [P_1 \cap (P + \sum_{i=2}^{k} P_i)].$$

We have

$$\sum_{i=2}^{k} P_i < P + \sum_{i=2}^{k} P_i,$$

since P is not part of the summation from 2 to k; and this implies in conjunction with the preceding equation that

$$0 \neq P_1 \cap [P + \sum_{i=2}^{k} P_i].$$

But P_1 is a point so that the above intersection equals P_1. Hence

$$P_1 \leqslant P + \sum_{i=2}^{k} P_i;$$

and this proves $\sum_{i=1}^{k} P_i = P + \sum_{i=2}^{k} P_i$. This shows the validity of (*e*).

REMARK 1: The rules (*a*) to (*e*) are the soc. Postulates of Algebraical Dependency which we have used in an implicite fashion in II.2. For a detailed theory of algebraical dependency the reader ought to consult

S. S. MacLane: A Lattice Formulation for Transcendence Degrees and *p*-Bases. *Duke Math. Journal*, vol. 4 (1938), pp. 455-468.

DEFINITION 2: *The set Φ of points is independent, if none of the points in Φ depends on the other points.*

DEFINITION 3: *The set Φ of points is a basis of the element E in S, if Φ is independent and E is the sum of the points in Φ.*

Theorem 1: *Every element in S possesses a basis.*

Theorem 2: *Any two bases of the element E in S contain the same number of elements; this [finite or infinite] number is the rank $r(E)$ of E.*

The proofs of these two theorems are quite similar to the corresponding proofs in II.2 and may consequently be left to the reader.

We note that an element L in S is called a line, if its rank is 2. A line L is the sum of any two different points, contained in L. Elements of rank 3 are called planes; they are sums of three independent points.

Theorem 3: *If the element E has finite rank, then every element contained in E has finite rank. If E' and E'' are elements of finite rank, then*

$$r(E') + r(E'') = r(E' + E'') + r(E' \cap E'').$$

The proofs of these facts are quite similar to the proofs of the corresponding facts in II.2. We leave them to the reader.

Again we note that distinct lines in a plane meet in a point, and that distinct planes in an element of rank 4 meet in a line.

We remark finally that the name *hyperplane* will be used for elements H such that A/H is a point. If A has finite rank, then $r(A) = r(H) + 1$; and if A has infinite rank, then $r(A) = r(H)$. Note that the rank condition defines hyperplanes in the finite case only.

Whenever possible, we shall make use of the appropriate geometrical terminology. Thus we shall speak of lines passing through a point, carrying a point, meeting in a point etc.

VII.3. The Theorem of Desargues

The considerations of the present section are almost identical with those customary in ordinary three dimensional projective geometry. Thus we shall omit some of the details.

Proposition 1: *Assume that the three lines L, L', L'' pass through the point P, but are not contained in a plane. Assume furthermore that Q, R are points*

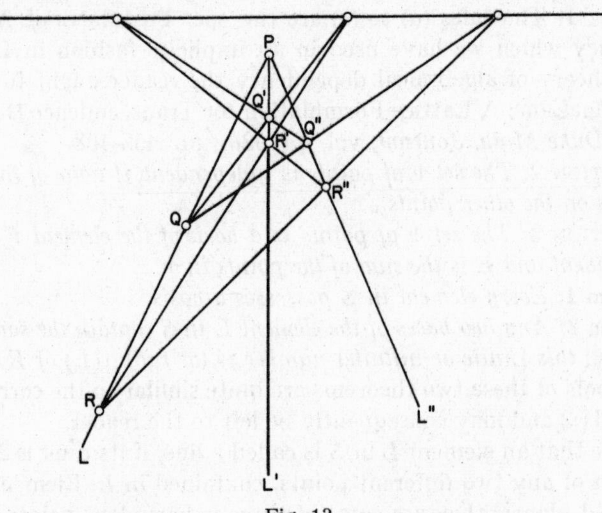

Fig. 13

on L, that Q', R' are points on L' and that Q'', R'' are points on L'' and that none of these points equals P. Assume finally that $Q + Q' \neq R + R'$, $Q' + Q'' \neq R' + R''$ and $Q'' + Q \neq R'' + R$. Then $(Q + Q') \cap (R + R')$, $(Q' + Q'') \cap (R' + R'')$, $(Q'' + Q) \cap (R'' + R)$ are collinear points.

Proof: We note first that $L + L'$, $L' + L''$ and $L'' + L$ are three distinct planes. Next we note that $Q + Q' + Q''$ and $R + R' + R''$ are two distinct planes. Finally we remark that $Q + Q'$ and $R + R'$ are two distinct lines meeting in the point $(Q + Q') \cap (R + R')$; and likewise we see that $(Q' + Q'') \cap (R' + R'')$ and $(Q'' + Q) \cap (R'' + R)$ are points. But all these points are on the intersection of the distinct planes $Q + Q' + Q''$ and $R + R' + R''$. But their intersection is clearly a line, since they are contained in the element $L + L' + L''$ of rank 4. This completes the proof.

Proposition 2: *Assume that $3 < r(A)$. Suppose that the three distinct lines L,L',L'' pass through the point P, that the points Q,R are on L, the points Q',R' on L' and the points Q'',R'' on L'', and that none of these points equals P. If finally $Q + Q' \neq R + R'$, $Q' + Q'' \neq R' + R''$ and $Q'' + Q \neq R'' + R$, then $(Q + Q') \cap (R + R')$, $(Q' + Q'') \cap (R' + R'')$ and $(Q'' + Q) \cap (R'' + R)$ are collinear points.*

The reduction of Proposition 2 to Proposition 1 may be effected in the usual fashion. [See e.g. Veblen-Young [1], p. 41.]

If we omit in Proposition 2 the hypothesis that $3 < r(A)$ or in Proposition 1 the hypothesis that L, L', L'' are not coplanar, then these Propositions cease to be true. At this point begins the investigation of the Theory of Projective Planes [which may or may not be Desarguesian]; and this theory is completely outside the scope of our investigation. It is, however, necessary for our purposes that we are assured of the validity of the Theorem of Desargues; and thus we impose the following further postulate.

IX. *If the three distinct lines L,L',L'' pass through the point P, if the points Q,R are on L, the points Q',R' on L' and the points Q'',R'' on L'', if none of these points equals P, if $Q + Q' \neq R + R'$, $Q' + Q'' \neq R' + R''$ and $Q'' + Q = R'' + R$, then $(Q + Q') \cap (R + R')$, $(Q' + Q'') \cap (R' + R'')$, $(Q'' + Q) \cap (R'' + R)$ are collinear points.*

VERIFICATION OF POSTULATE IX IN $S(F,A)$:

If the F-space A has rank not less than 4, then the validity of IX is a consequence of Proposition 2. If the F-space A has rank less than 4, then we form $A \oplus A$ which more comprehensive F-space has rank greater than 3, provided $r(A) = 3$ [if $r(A) < 3$, then IX is trivially satisfied]. Hence IX holds in $A \oplus A$ and consequently in the subspace A of $A \oplus A$. It would not have been difficult to verify the validity of IX by direct computation, a useful exercise.

Proposition 3: *Assume the validity of Postulates* I *to* IX. *If P,Q,R are three different, but collinear points, if P',P'',Q',Q'',R',R'' are points such that $P' + Q'$ and $P'' + Q''$ are different lines meeting in R, $Q' + R'$ and $Q'' + R''$ are different lines meeting in P, $R' + P'$ and $R'' + P''$ are different lines meeting in Q, and if $P' + P''$, $Q' + Q''$ and $R' + R''$ are three distinct lines, then these lines $P' + P''$, $Q' + Q''$ and $R' + R''$ are copunctual [have a point in common].*

This dualized theorem of Desargues may be verified as follows. If $3 < r(A)$, then the considerations used in proving Proposition 2 may be used [mutatis mutandis]; and if $3 = r(A)$, then the principle of duality may be applied to Postulate IX. We leave the details to the reader.

VII.4. The Imbedding Theorem

The objective of the present section is the proof of the following

Imbedding Theorem: *If the partially ordered set S satisfies Postulates* I *to* VIII, *and if the maximal element A in S has rank not less than* 3, *then the following condition is necessary and sufficient for the validity of the Theorem of Desargues* [*Postulate* IX] *in S*:

IX*. *S is contained in a partially ordered set T which is different from S and which satisfies Postulates* I *to* VIII.

If in particular the maximal element A has rank greater than 3, then the validity of IX is a consequence of VII.3, Proposition 2; and thus it follows from the Imbedding Theorem that S has also the Imbedding Property IX*. This is the principal reason why we have to prove the comparatively deep Imbedding Theorem instead of substituting for Postulate IX the Postulate IX*.

The Imbedding Theorem will be exploited in the next section. The following remark may indicate how this can be done. If S is properly contained in T, then there exists in T an element B such that A is a hyperplane with respect to B or B is a point over A. One may then construct the homothetic group of B over A; and this group contains in a way the additive group and the field of a linear manifold whose system of subspaces is projectively equivalent with S [see III, Appendix I].

That the Theorem of Desargues [Postulate IX] is a consequence of IX*, may be deduced from VII.3, Proposition 2, since the maximal element B of the containing system T certainly satisfies $A < B$ so that $3 < r(B)$.

Throughout the remainder of this section [which may be omitted in a first reading] we shall be concerned with the proof of the sufficiency of IX. Thus we assume throughout that Postulates I to IX are satisfied by S and that its maximal element A has rank not less than 3. We begin by constructing

THE BASIC QUADRANGLE:

It consists of four lines L_o, L_r, L_s, L_t in S which are coplanar, though no three of these lines are copunctual.

The existence of this basic quadrangle is easily deduced from the fact that there exist at least three independent points [$3 \leqslant r(A)$] and that every line carries at least three distinct points [VIII].

Then any two of these four lines are distinct lines in a plane; and thus they meet in a point. For this point we shall use the following systematic *notation*: If h,k,m,n is a permutation of the four symbols o,r,s,t, then

$$(h,k) = (k,h) = L_m \cap L_n.$$

One verifies easily that (h,k), (k,m), (m,h) are three different points on the line L_n and that

$$L_n = (h,k) + (k,m) = (k,m) + (m,h) = (m,h) + (h,k). \tag{1}$$

This basic quadrangle we shall retain throughout our discussion; and we

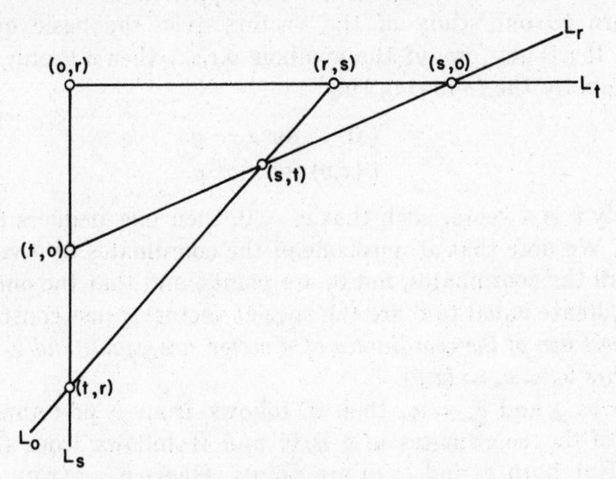

Fig. 14

shall use x,y,z for any three distinct symbols of the four symbols o,r,s,t.

Next we construct

THE VECTORS [*with respect to the basic quadrangle* L_o,L_r,L_s,L_t]:

$v = [v_o,v_r,v_s,v_t]$ is a vector with the x-coordinate v_x, if every v_x is an element in S such that

$$r(v_x) \leqslant 1, \tag{2.a}$$

$$v_x + (x,y) = (x,y) + v_y = v_y + v_x \textit{ for } x \neq y. \tag{2.b}$$

The construction of the vectors is very similar to a construction due to Hessenberg who drew his inspiration from constructions customary in descriptive geometry. In our present context it might be illuminating to reconstruct these concepts within the system $S(F,A)$ of subspaces of the F-space A. To do this we just select three independent elements r,s,t in the F-space A. Then

$$L_o = F(r-s) + F(s-t) + F(t-r),$$
$$L_r = Fs + Ft, \; L_s = Ft + Fr, \; L_t = Fr + Fs;$$

$(x,y) = F(x - y)$ for x,y any two different ones of the symbols o,r,s,t [where we let $o = 0$].

Clearly this is just such a basic quadrangle as we envisage here.

If a is an element in the F-space A, then

$$v(a) = [Fa, F(a - r), F(a - s), F(a - t)]$$

is a vector with x-coordinate $F(a - x)$; and the reader ought to verify that every vector arises from one and only one element in A in the fashion just indicated. This illustration should be kept in mind throughout our discussion. The reader might consult also III, Appendix I.

We return to our study of the vectors over the basic quadrangle L_o,L_r,L_s,L_t. If x is any one of the symbols o,r,s,t, then a vector, named x too, is defined by the following rule:

$$x_y = \begin{cases} 0 & \text{for } x = y \\ (x,y) & \text{for } x \neq y. \end{cases} \tag{3}$$

If conversely v is a vector such that $v_x = 0$, then one deduces from (2.b) that $v = x$. We note that at most one of the coordinates of a vector may be 0, that all the coordinates, not 0, are points, and that the only vectors with a coordinate equal to 0 are the special vectors x just constructed.

(4) *At most two of the coordinates of a vector are equal; and $v_x = v_y$ for $x \neq y$ implies $v_x = v_y = (x,y)$.*

PROOF: If $x \neq y$ and $v_x = v_y$, then it follows from a preceding remark that none of the coordinates of v is 0; and it follows from (2.b) that $(x,y) \leqslant v_x$. But both v_x and (x,y) are points. Hence $v_x = (x,y)$. If we had furthermore $v_x = v_z$, then it would follow likewise that $v_x = (x,z)$ so that $(x,y) = (x,z)$ which is incompatible with $y \neq z$. If finally h,k,m,n is a permutation of the four symbols o,r,s,t, and if $v_h = v_k$, $v_m = v_n$, then it would follow that $v_h = v_k = (h,k)$ and $v_m = v_n = (m,n)$. From (2.b) we would deduce now that $(h,m) \leqslant v_h + v_m = (h,k) + (m,n)$; and this would contradict the properties of the basic quadrangle. Thus at most two of the coordinates of the vector v can be equal.

Proposition 1: *If $P \neq Q$ are points in S such that*

$$P + Q = Q + (x,y) = (x,y) + Q,$$

then there exists one and only one vector v such that $v_x = P$, $v_y = Q$.

PROOF: Assume first that v and w are vectors which satisfy $P = v_x = w_x$, $Q = v_y = w_y$. Denote by h,k the remaining pair of indices. If z is different from x and y, then we deduce from (2.b) that

$$v_z + w_z \leqslant [P + (x,z)] \cap [Q + (y,z)].$$

Assume now that $v_z \neq w_z$. If v_z were 0, then $v = z$ so that $P = (x,z)$, and $Q = (y,z)$. But this would imply $w_z = 0$, contrary to our hypothesis that $v_z \neq w_z$. Likewise $w_z \neq 0$; and so v_z and w_z are distinct points. But then it follows from the above inequality and $P \neq Q$ that

$$v_z + w_z = P + (x,z) = Q + (y,z) = P + Q = (x,z) + (z,y).$$

Suppose now that both $v_h \neq w_h$ and $v_k \neq w_k$. Then it follows from what we have shown in the preceding paragraph that

$$L_k = (x,h) + (h,y) = v_h + w_h = P + Q = v_k + w_k = (x,k) + (y,k) = L_h;$$

and this is incompatible with $h \neq k$ and the properties of the basic quadrangle. Thus we may assume that $v_h = w_h$. It follows from (2.b) that

$$v_k + w_k \leqslant [v_h + (h,k)] \cap [P + (x,k)] \cap [Q + (y,k)].$$

If $v_k \neq w_k$, then it follows as before that

$$\begin{aligned} v_k + w_k &= v_h + (h,k) = P + (x,k) = Q + (y,k) = P + Q \\ &= (h,k) + (x,k) = (x,k) + (y,k); \end{aligned}$$

and this is again incompatible with the properties of the basic quadrangle. Hence $v_k = w_k$; and this completes the proof of $v = w$. Thus we have shown that there exists at most one vector with the required properties.

When constructing a vector with the desired properties, we distinguish two cases.

Case 1: At least one of the points P,Q is on at least one of the lines $(x,h) + (h,y)$ and $(x,k) + (k,y)$.

Without loss in generality we may assume that P is on $(x,h) + (h,y)$. Since this line carries the point (x,y), and since $P + (x,y) = P + Q$ is a line, it follows now that

$$P + Q = (x,h) + (h,y) = L_k.$$

If P or Q were on the line $(x,k) + (k,y)$, then it would follow likewise that $P + Q = (x,k) + (k,y)$; and this would imply the equality of L_k and L_h which is impossible. Hence neither P nor Q is on $(x,k) + (k,y)$. But then $P + (x,k)$ and $Q + (k,y)$ are two different lines which are both contained in the plane

$$P + Q + (x,k) + (k,y) = P + (x,y) + (x,k) + (k,y) = P + (x,k) + (k,y).$$

Different lines in a plane meet in a point; and so

$$K = [P + (x,k)] \cap [Q + (k,y)]$$

is a definite point. If K were equal to P, then $Q + (k,y)$ would be equal to $P + Q = (x,h) + (h,y)$ which is impossible; and if K were equal to (x,k), then $Q + (k,y)$ would be equal to

$$(x,k) + (k,y) = Q + (x,y) = P + Q = (x,h) + (h,y)$$

which is impossible too; and likewise we see that K cannot be equal to Q or (k,y). Consequently we find that

$$P + (x,k) = (x,k) + K = K + P,\ Q + (y,k) = (y,k) + K = K + Q,$$

and that $K + P$ and $K + Q$ are lines. If K were on $P + Q$, then these lines would equal $P + Q$ so that (x,k) and (y,k) would be on

$$P + Q = (x,h) + (h,y),$$

an impossibility. Hence $K \not\leq P + Q$.

If $K = (h,k)$, then we find that

$$P = [(x,h) + (h,x)] \cap [(x,k) + (k,h)] = (x,h),$$
$$Q = [(x,h) + (h,y)] \cap [(y,k) + (k,h)] = (y,h);$$

and h is the desired vector.

If $K \neq (h,k)$, then we see as before that $H = [P + Q] \cap [K + (h,k)]$ is a point. Since neither K nor (h,k) is on $P + Q = (x,h) + (h,y)$, the point H is different from K and (h,k); and consequently we have

$$K + (h,k) = (h,k) + H = H + K.$$

If $P = H$, then $H + K = P + K = (x,k) + (h,k)$ so that $P = (x,h)$. But a vector v is defined by the rule:

$$v_x = P = (x,h),\ v_y = Q,\ v_h = (x,h),\ v_k = K;$$

and this vector meets our requirements. Likewise we find a vector meeting our requirements, if $H = Q$. If finally H is different from P and Q, then we have

$$P + Q = Q + H = H + P = H + (x,h) = (y,h) + H;$$

since, for instance, $H = (x,h)$ would imply that $H + K$ contains the point (x,k) so that $H + K = H + P = P + Q$ which is impossible; and a vector v is defined by the rule:

$$v_x = P,\ v_y = Q,\ v_k = K,\ v_h = H;$$

and this vector meets our requirements.

Case 2: Neither P nor Q is on either of the lines $(x,h) + (h,y)$ and $(x,k) + (k,y)$.

Then $P + (x,z)$ and $Q + (y,z)$ are, for $z = h,k$, different lines which meet in a point $Z = [P + (z,x)] \cap [Q + (z,y)]$. One verifies that Z is different from P, (z,x), Q, (z,y); and thus it follows that

$$P + (z,x) = (z,x) + Z = Z + P,\ Q + (z,y) = (z,y) + Z = Z + Q$$

for $z = h,k$.

Naturally the line $P + Q$ is distinct from the lines $(x,h) + (h,y)$ and $(x,k) + (k,y)$. Thus the triangles P, (x,h), (x,k) and Q, (y,h), (y,k) are

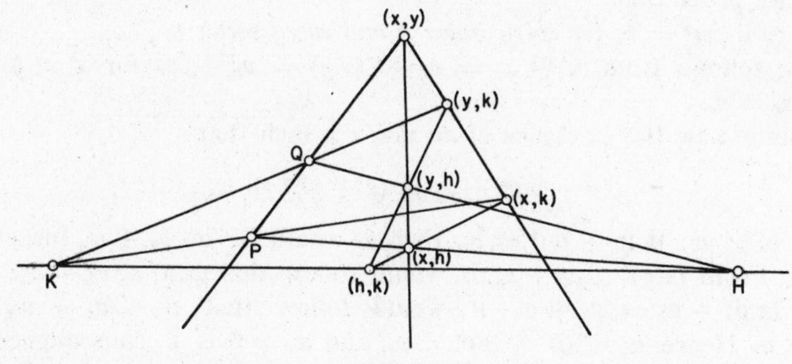

Fig. 15

perspective from the point (x,y)—here we use the collinearity of the three distinct points P, Q, (x,y) which is part of our hypothesis—and the Theorem of Desargues [Postulate IX] becomes applicable. But corresponding sides of our two triangles meet in the points

$$[P + (x,h)] \cap [Q + (y,h)] = H, \; [P + (x,k)] \cap [Q + (y,k)] = K,$$
$$[(x,h) + (x,k)] \cap [(y,h) + (y,k)] = (h,k);$$

and thus we find that $H + K = K + (h,k) = (h,k) + H$. Hence a vector v meeting our requirements is defined by the rule:

$$v_x = P, \; v_y = Q, \; v_h = H, \; v_k = K;$$

and this completes the proof.

Throughout we shall use the following *notation*:
If v and w are vectors, then

$$(v + w)^* = (v_o + w_o) \cap (v_r + w_r) \cap (v_s + w_s) \cap (v_t + w_t).$$

Thus $(v + w)^*$ is a well determined element in S.

Proposition 2: (*a*) $r[(v + w)^*] \leqslant 1$.

(*b*) $(v + w)^* = 0$ *if, and only if,* $v = w$.

(*c*) $(v + w)^* + v_x = (v + w)^* + w_x$ *for every* x.

PROOF: It is clear that $r[(v + w)^*] \leqslant 2$, since $(v + w)^* \leqslant v_x + w_x$. If this rank were 2, then we would have $(v + w)^* = v_x + w_x$ for every x so that in particular $(x,y) \leqslant v_x + v_y \leqslant (v + w)^*$ for $x \neq y$. But the sum of all the (x,y) has rank 3, a contradiction, proving (*a*).

If $v = w$, then $(v + w)^* = v_o \cap v_r \cap v_s \cap v_t$. Since none of the coordinates of a vector has rank exceeding 1, and since [by (4)] at most two of the coordinates of a vector are equal, it follows that $v = w$ implies $(v + w)^* = 0$.

Next we prove that

(5) $(v + x)^* = v_x$ *for every index x and every vector v.*

This follows from $v_y + x_y = v_y + (x,y) = v_y + v_x$ for $x \neq y$ and $v_x + x_x = v_x$.

Assume now the existence of an index x such that

$$(v + w)^* + v_x \neq (v + w)^* + w_x.$$

Then $v_x \neq w_x$. If $(v + w)^* = v_x$, then v_x would be on $v_y + w_y$ for every $y \neq x$. From $(x,y) \leqslant v_x + v_y$, it would follow that $(x,y) \leqslant v_y + w_y$; and from $(x,y) + w_y = w_y + w_x$ it would follow that $w_x \leqslant v_y + w_y$ for every y. Hence $w_x \leqslant (v + w)^* = v_x$; and $w_x = 0$ is a consequence of $v_x \neq w_x$, since v_x has rank at most 1. But then $w = x$ by (3) and $(v + w)^* = v_x$ by (5) so that

$$(v + w)^* + v_x = v_x = (v + w)^* + w_x,$$

a contradiction. Likewise we see that $(v + w)^* \neq w_x$. But

$$(v + w)^* \leqslant v_x + w_x\, ;$$

and so $(v + w)^* \neq 0$ would imply that

$$v_x + w_x = w_x + (v + w)^* = (v + w)^* + v_x$$

which is impossible. Hence

$$(v + w)^* = 0.$$

[Note that we have also shown the impossibility of $v_x = 0$ and of $w_x = 0$ so that v_x and w_x are distinct points and $v_x + w_x$ is a line.]

If $v_y = 0$, then $v = y$ and $(v + w)^* = w_y$ by (5). Hence $w_y = 0$ and $w = y$. But then $v = w$, contradicting $v_x \neq w_x$; and so it follows that none of the v_y and none of the w_y is 0.

Suppose now that $v_y = w_y$ and $v_z = w_z$ for $y \neq z$. Then it follows from $v \neq w$ and Proposition 1 that $v_y = v_z = w_y = w_z = (y,z)$. If u is a subscript different from y and z, then it follows from $v \neq w$ and Proposition 1 that $v_u \neq w_u$ and that therefore

$$v_u + (u,y) = v_u + v_y = v_u + (y,z) = (u,y) + (y,z) = w_u + (u,y) = v_u + w_u\, ;$$

and this would imply $(y,z) = (v + w)^* = 0$, an impossibility.

Assume now that $v_y = w_y$ for some y. Then it follows from the result of the preceding paragraph that v_z and w_z are different points for every $z \neq y$. Again we find that

$$v_z + (y,z) = (y,z) + v_y = (y,z) + w_y = w_z + (y,z) = v_z + w_z$$

so that $v_y = w_y \leqslant v_z + w_z$ for every z and hence $0 \neq v_y \leqslant (v + w)^* = 0$, an impossibility. Thus we have shown that

v_y *and* w_y *are different points for every index* y.

Consider now the lines $v_o + w_o$, $v_r + w_r$, $v_s + w_s$ and $v_t + w_t$. Any two of them are contained in a plane, since

$$v_x + w_x + v_y + w_y = v_x + (x,y) + w_x.$$

Thus two of these lines are either equal or they have just one point in common. If among these lines only two are different, then their intersection is a point equal to $(v + w)^* = 0$, an impossibility. Thus

at least three of these lines $v_x + v_y$ *are different.*

We show next:

(*d*) *If* $v_h + w_h \neq v_k + w_k$, *then* v_h, v_k, (h,k) *are three different points.*

It is a consequence of (2.*b*) that the equality of two of the points v_h, v_k,

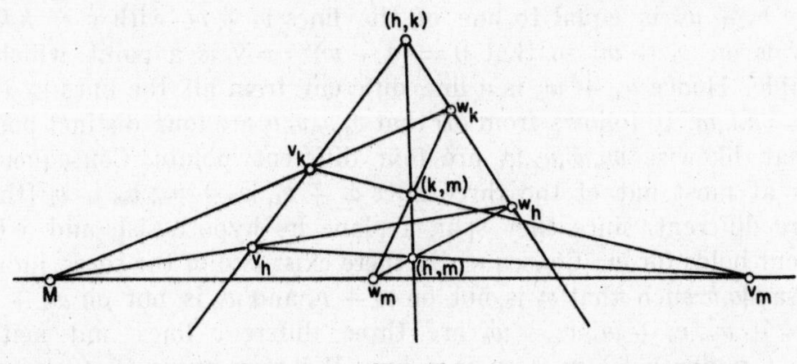

Fig. 16

(h,k) implies the equality of all three of them. If $v_h = v_k = (h,k)$, then $v_h + w_h = (h,k) + w_h = w_k + (h,k) = w_k + v_k$ by (2.*b*); and this proves (*d*).

(*e*) *If* $v_h + w_h$, $v_k + w_k$, $v_m + w_m$ *are three different lines, and if neither* $v_h + v_k + v_m$ *nor* $w_h + w_k + w_m$ *is a line, then*

$$(v_h + w_h) \cap (v_k + w_k) \cap (v_m + w_m) \textit{ is a point.}$$

It is clear from our hypotheses that $(v_h + w_h) \cap (v_k + w_k) = M$ is a point; and thus we need only show that M is on the line $v_m + w_m$. It follows from (*d*) that $v_h \neq v_k$ and so it follows from our hypothesis that $r(v_h + v_k + v_m) = 3$; and $r(w_h + w_k + w_m) = 3$ is seen likewise.

Next we note that the triangles v_h, w_h, (h,m) and v_k, w_k, (k,m) are perspective from the point (h,k) which is different from all the points in our triangles; and the lines $v_h + v_k$, $w_h + w_k$ are different, since otherwise

$v_h + w_h = v_k + w_k = v_h + v_k = w_h + w_k$. The lines $v_h + v_k$ and $(h,m) + (k,m)$ are different, since otherwise v_m would be on the line $v_h + v_k$; and $w_h + w_k \neq (h,m) + (k,m)$ is seen likewise.

The lines $v_h + w_h$ and $v_k + w_k$ meet in the point M. If the lines $v_h + (m,h) = v_h + v_m$ and $v_k + (m,k) = v_k + v_m$ were not different, then they would both be equal to the plane $v_h + v_k + v_m$ which is impossible. Hence

$$[v_h + (m,h)] \cap [v_k + (m,k)] = v_m.$$

$$[w_h + (m,h)] \cap [w_k + (m,k)] = w_m$$

is seen likewise. Now we have verified the applicability of the Theorem of Desargues [Postulate IX]. Hence M, v_m, w_m are collinear, as we desired to show.

(e') *The hypotheses of (e) are not satisfied for any triplet h,k,m of indices.*

Assume the validity of the hypotheses of (e) for the triplet h,k,m of indices. Then $N = (v_h + w_h) \cap (v_k + w_k) \cap (v_m + w_m)$ is a point. If the line $v_n + w_n$ is equal to one of the lines $v_z + w_z$ with $z = h,k,m$, then N is on $v_n + w_n$ so that $0 = (v + w)^* = N$ is a point which is impossible. Hence $v_n + w_n$ is a line different from all the lines $v_z + w_z$ with $z = h,k,m$. It follows from (d) that v_o,v_r,v_s,v_t are four distinct points and that likewise w_o,w_r,w_s,w_t are four different points. Consequently v_n is on at most one of the three lines $v_h + v_k$, $v_k + v_m$, $v_m + v_h$ [these lines are different, since they span a plane by hypothesis]; and a like statement holds for w_n. Consequently there exist two of the three indices h,k,m, say a,b such that v_n is not on $v_a + v_b$ and w_n is not on $w_a + w_b$. Then $v_n + w_n$, $v_a + w_a$, $v_b + w_b$ are three different lines and neither $v_n + v_a + v_b$ nor $w_n + w_a + w_b$ is a line. We may apply (e) again; and find that $N' = (v_n + w_n) \cap (v_a + w_a) \cap (v_b + w_b)$ is a point. Since $v_a + w_a$ and $v_b + w_b$ are different lines, and since a and b are two of the three indices h,k,m, it follows now that $N' = (v_a + w_a) \cap (v_b + w_b) = N$. Hence $0 = (v + w)^* = N = N'$ a contradiction which proves (e').

(f) *It is impossible that all four lines $v_z + w_z$ with $z = o,r,s,t$ are different.*

Assume that all the four lines are different. Then it follows from (e') that $v_h + v_k + v_m$ or $w_h + w_k + w_m$ is a line for each triplet h,k,m. If $v_h + v_k + v_m$ is a line, then this line equals the line $(h,k) + (k,m)$. If $w_h + w_k + w_m$ were a line too, then it too would equal $(h,k) + (k,m)$. But this would imply that $v_h + w_h = v_k + w_k = (h,k) + (k,m)$, an impossibility. Thus we have shown: if h,k,m is a triplet of indices, then one and only one of the sums $v_h + v_k + v_m$ and $w_h + w_k + w_m$ is a line and the other one is a plane.

We may assume without loss in generality that $v_o + v_r + v_s$ is a line and that $w_o + w_r + w_s$ is a plane. Since $v_o + v_r + v_s + v_t$ contains by (2.b) the plane spanned by all the (h,k), it is impossible that the point v_t is on $v_o + v_r + v_s$. Then $v_o + v_r + v_t$ is not a line; and it follows from

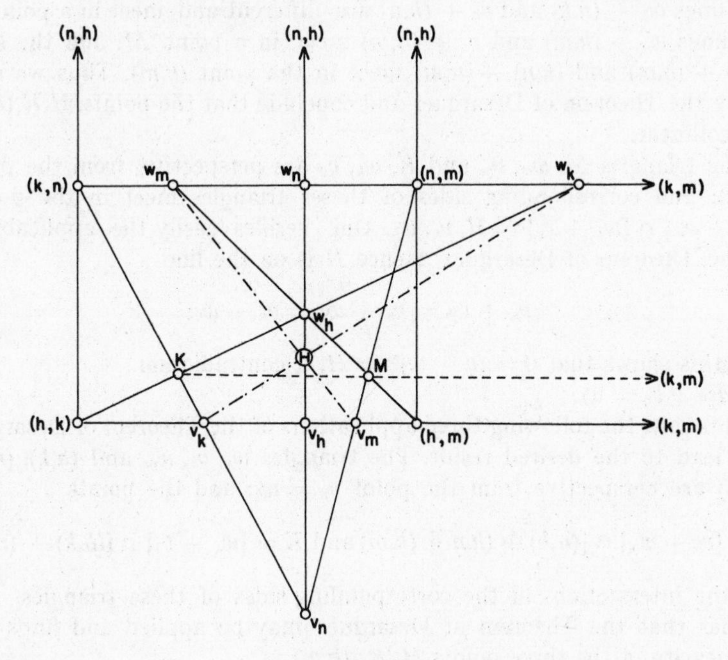

Fig. 17

the result of the preceding paragraph that $w_o + w_r + w_t$ is a line. That $w_r + w_s + w_t$ and $w_s + w_o + w_t$ are lines too, is shown likewise. It follows from (d) that all the points w_o,w_r,w_s,w_t are different. Hence w_o and w_s are on the line $w_r + w_t$; and thus it follows that $w_o + w_r + w_s + w_t$ is a line. This contradicts the fact that $w_o + w_r + w_s$ is a plane; and thus we have arrived at a contradiction which proves (f).

On the basis of (f) it follows that—choosing proper notation—the lines $w_h + v_h$, $w_k + v_k$, $w_m + v_m$ are all different whereas $w_n + v_n = w_h + v_h$. It follows from (e') that of the elements $v_h + v_k + v_m$ and $w_h + w_k + w_m$ one is a line and the other is a plane; and so we may assume without loss in generality that $v_h + v_k + v_m$ is a line and that $w_h + w_k + w_m$ is a plane.

Since the sum of all four v's contains the plane spanned by all the (i,j)'s, it follows that v_n is not on $v_h + v_k + v_m$. Hence $v_k + v_m + v_n$ is a plane; and now it follows from (e') that $w_k + w_m + w_n$ is a line; and this line cannot contain the point w_h. Now we distinguish two cases.

Case 1: $v_n \neq w_h$.

The triangles w_h, (h,k), (h,m) and v_n, (k,n), (n,m) are perspective from the point (n,h); and the three lines of perspectivity are different, since, for instance, $w_h + v_n = (h,k) + (k,n)$ would imply that w_h, w_k, w_n are collinear. The lines $w_h + (h,k)$ and $v_n + (k,n)$ are different and meet in a point K; the lines $w_h + (h,m)$ and $v_n + (n,m)$ meet in a point M; and the lines $(h,k) + (h,m)$ and $(k,n) + (n,m)$ meet in the point (k,m). Thus we may apply the Theorem of Desargues and conclude that the points $M,K,(k,m)$ are collinear.

The triangles M, w_m, v_m and K, w_k, v_k are perspective from the point (k,m); and corresponding sides of theses triangles meet in the points $[w_m + v_m] \cap [w_k + v_k] = H$, v_n, w_h. One verifies easily the applicability of the Theorem of Desargues. Hence H is on the line

$$v_n + w_h = v_n + w_n = v_h + w_h;$$

and this shows that $0 = (v + w)^* = H$, a contradiction.

Case 2: $v_n = w_h$.

This time the following three applications of the Theorem of Desargues will lead to the desired result. The triangles w_k, v_h, w_m and (h,k), (n,h), (h,m) are perspective from the point $v_n = w_h$; and the points

$$H = [v_h + w_m] \cap [(n,h) + (h,m)],\ (k,m) \text{ and } K = [w_k + v_h] \cap [(h,k) + (n,h)]$$

are the intersections of the corresponding sides of these triangles. One verifies that the Theorem of Desargues may be applied and finds the collinearity of the three points H, K, (k,m).

The triangles H, v_m, (m,k) and (n,h), v_n, (n,k) are perspective from the point (m,n); and corresponding sides of these triangles meet in the points v_k, K and $L = [v_m + H] \cap [v_n + (n,h)]$. We may apply the Theorem of Desargues and find that L is the intersection of the three lines $K + v_k$, $H + v_m$ and $v_n + (n,h) = v_n + v_h = w_h + v_h$.

The triangles H, v_m, w_m and K, v_k, w_k are perspective from the point (k,m), since we have shown in the first step of the discussion of Case 2 that H, K, (k,m) are collinear. From the result of the preceding paragraph of the proof we deduce that corresponding sides of these triangles meet in the points $[H + v_m] \cap [K + v_k] = L$, $[H + w_m] \cap [K + w_k] = v_h$ [as follows from the construction of H and K] and $[v_m + w_m] \cap [v_k + w_k] = M$. We have shown that L is on $w_h + v_h = L + v_h$, since $L \neq v_h$. Applying the Theorem of Desargues it follows that M is on $w_h + v_h$ too. Hence $0 = (v + w)^* = M$. This is the final contradiction which completes the proof of (c).

(b) is an immediate consequence of (c) and what we have shown in the beginning of our proof. Thus we have completed the proof of Proposition 2.

Corollary 1: *If v and w are vectors, then*

$$v_x + w_x = w_x + (v + w)^* = (v + w)^* + v_x \text{ for every } x.$$

PROOF: If $v = w$, then our statement is trivial, since $(v + v)^* = 0$ [Pro-

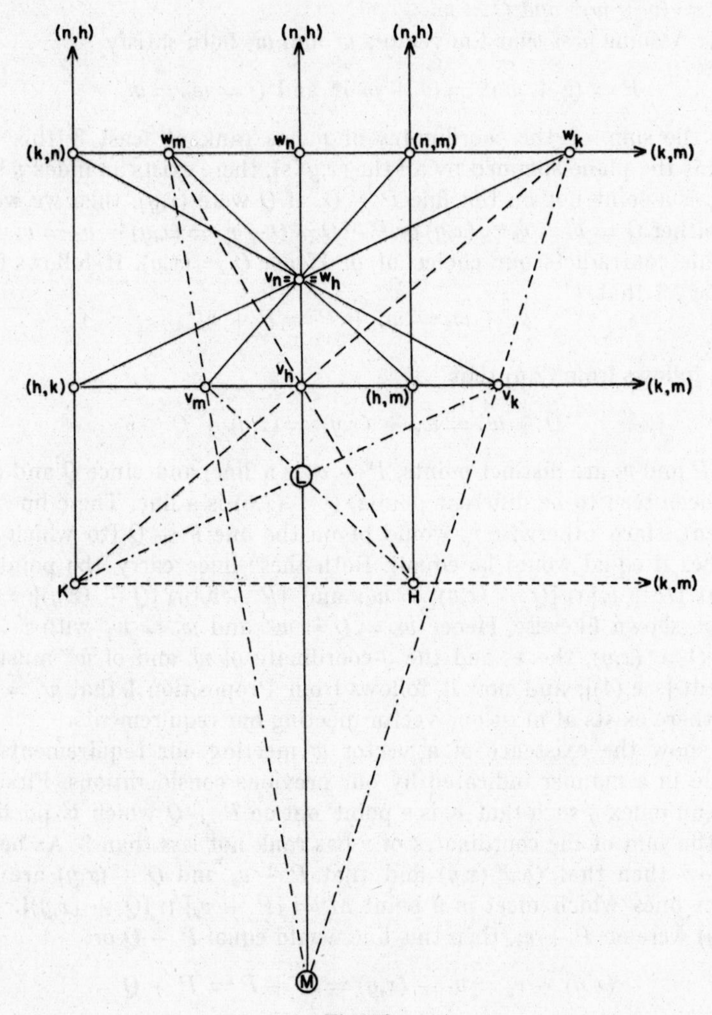

Fig. 18

position 2, (b)]. If $v \neq w$, then it follows from Proposition 2 that $(v + w)^*$ is a point. If $v_x = 0$, then $v = x$ and consequently $w_x \neq 0$ so that $(v + w)^* = w_x$ [see (5)]. If neither v_x nor w_x is 0, then our equations may be deduced from Proposition 2, (c), once one remarks that $v_x = w_x$ implies $v_x = w_x = (v + w)^*$ because of $(v + w)^* \leq v_x + w_x$.

Proposition 3: *To every vector v and to every pair of different points P,Q such that $P + Q = v_x + P$ there exists one and only one vector w such that $P = (v + w)^*$ and $Q = w_x$.*

PROOF: Assume first that the vectors w' and w'' both satisfy

$$P = (v + w')^* = (v + w'')^* \text{ and } Q = w'_x = w''_x.$$

Since the sum of the coordinates of v has rank at least 3 [this sum contains the plane spanned by all the (x,y)'s], there exists an index y such that v_y is a point not on the line $P + Q$. If Q were (x,y), then we would have either $Q = v_x = v_y = (x,y)$ or $P + Q = Q + v_x = (x,y) + v_x = v_x + v_y$; and this contradicts our choice of y. Hence $Q \neq (x,y)$. It follows from Corollary 1 that

$$v_y + w'_y = w'_y + P = P + v_y$$

and it follows from (2.b) that

$$Q + w'_y = w'_y + (x,y) = (x,y) + Q.$$

Since P and v_y are distinct points, $P + v_y$ is a line; and since Q and (x,y) have been seen to be different points, $Q + (x,y)$ is a line. These lines are different, since otherwise v_y would be on the line $P + Q$ [to which line the lines if equal would be equal]. Both these lines carry the point w'_y so that $[P + v_y] \cap [Q + (x,y)] = w'_y$; and $[P + v_y] \cap [Q + (x,y)] = w''_y$ may be shown likewise. Hence $w'_x = Q = w''_x$ and $w'_y = w''_y$ with $x \neq y$. Since $Q \neq (x,y)$, the x- and the y-coordinate of w' and of w'' must be different [see (4)]; and now it follows from Proposition 1 that $w' = w''$. Thus there exists at most one vector meeting our requirements.

To show the existence of a vector w meeting our requirements we procede in a manner indicated by our previous considerations. First we select an index y such that v_y is a point not on $P + Q$ which is possible, since the sum of the coordinates of v has rank not less than 3. As before we show then that $Q \neq (x,y)$ and that $P + v_y$ and $Q + (x,y)$ are two distinct lines which meet in a point $M = [P + v_y] \cap [Q + (x,y)]$. If Q or (x,y) were on $P + v_y$, then this line would equal $P + Q$ or

$$(x,y) + v_y = v_x + (x,y) = v_x + P = P + Q$$

which is impossible; and likewise we show that neither P nor v_y can be on $Q + (x,y)$. Hence M is a point different from P,Q,v_y and (x,y); and this implies that $Q + (x,y) = (x,y) + M = M + Q$. We infer from Proposition 1 the existence of one and only one vector w such that $w_x = Q$ and $w_y = M$. Since $w_y = M \neq v_y$, we have $w \neq v$; and it follows from Proposition 2 that $(v + w)^*$ is a point. This point is on the lines $v_x + w_x = v_x + Q = P + Q$ and $v_y + w_y = v_y + M = P + v_y$ which are different, since v_y is not on $P + Q$. Thus these two lines meet in the point P which is then necessarily the point $(v + w)^*$. Thus w meets all our requirements.

REMARK 1: If v is a vector and P is a point, then at least one of the coordinates of v, say v_x, is a point different from P [since the sum of the coordinates has rank not less than 3]. The line $P + v_x$ carries at least one third point Q [by Postulate VIII]. Then $P + v_x = v_x + Q = Q + P$; and it follows from Proposition 3 that there exists a vector w such that $P = (v + w)^*$ and $Q = w_x$. The vectors v and w are necessarily different, since $v_x \neq w_x$; and so it follows from Corollary 1 that

$$v_z + w_z = w_z + P = P + v_z$$

for every index z.

Proposition 4: *If u,v,w are three different vectors, then either*

$$(u + v)^* = (v + w)^* = (w + u)^*$$

or else

$(u + v)^*, (v + w)^*, (w + u)^*$ *are three distinct, but collinear points.*

PROOF: Assume first that $(v + u)^* = (v + w)^*$. Then we infer from Corollary 1 that

$$u_x + v_x = v_x + (u + v)^* = v_x + (v + w)^* = v_x + w_x \text{ for every } x;$$

and that

$$u_x + w_x = w_x + (u + w)^* = (u + w)^* + u_x \text{ for every } x.$$

If $u_x = 0$, then $u = x$ and $(u + w)^* = w_x$; and likewise we infer from $w_x = 0$ that $(u + w)^* = u_x$. If $u_x = w_x \neq 0$, then $(u + w)^* = u_x = w_x$, since it follows from Proposition 2 that $(u + w)^*$ is a point. If finally u_x and w_x are distinct points, then $u_x + v_x = v_x + w_x = w_x + u_x$ contains $(w + u)^*$. Thus we have shown that in every case

$$(u + w)^* \leqslant u_x + v_x = v_x + w_x \text{ for every } x;$$

and this implies that $(u + w)^* \leqslant (u + v)^* = (v + w)^*$. But $(u + w)^*$ and $(u + v)^*$ are points by Proposition 2; and hence they are equal.

It follows from Proposition 2 that $(u + v)^*$, $(v + w)^*$ and $(w + u)^*$ are points; and we have just shown that they are either all equal or pairwise different. Thus we assume now that $(u + v)^*$, $(v + w)^*$ and $(w + u)^*$ are three different points. If there exists an index x such that $r(u_x + v_x + w_x) < 3$, then we infer from

$$(u + v)^* + (v + w)^* + (w + u)^* \leqslant u_x + v_x + w_x$$

the collinearity of the three points $(u+v)^*$, $(v+w)^*$, $(w+u)^*$. Consequently we assume now that

(*) $$r(u_x + v_x + w_x) = 3 \text{ for every } x.$$

This implies in particular that u_x, v_x, w_x are points, and that, for every x, the lines $u_x + v_x, v_x + w_x$ and $w_x + u_x$ are pairwise different.

It is a consequence of (4) that at most two of the coordinates of a vector are equal. Thus there exists at most one pair of different indices x_o, y_o such that $w_{x_o} = w_{y_o}$. If $u_x + v_x$ were equal to $u_y + v_y$ for every pair $x \neq y$ such that $w_x \neq w_y$, then it would follow that

$$u_o + v_o = u_r + v_r = u_s + v_s = u_t + v_t;$$

and the sum of the coordinates of u [or of v] would have rank less than 3 which is impossible, since the sum of the coordinates of a vector contains the sum of all the (h,k) which has rank 3. Consequently there exists a pair of indices x,y such that

$$u_x + v_x \neq u_y + v_y \text{ and } w_x \neq w_y.$$

Since $u_x + v_x$ and $u_y + v_y$ are different lines, their sum has at least rank 3. If $v_x \leqslant u_x + u_y$, then

$$u_x + u_y = u_x + u_y + v_x = u_x + u_y + (x,y) + v_x = u_x + u_y + v_y + v_x$$

so that this sum would have rank 2. This is impossible; and thus it follows that v_x is not on $u_x + u_y$. We see likewise that v_y is not on $u_x + u_y$. If v_x were equal to v_y, then they would both equal (x,y) [by (4)] and thus would be on $u_x + u_y$; and thus we see that $v_x + v_y$ is a line. Likewise we see now that $u_x + u_y$ is a line too; and that $u_x + u_y$ and $v_x + v_y$ are two different lines.

Consider now the triangles $(u+w)^*$, u_x, u_y and $(v+w)^*$, v_x, v_y. It follows from Corollary 1 that

$$(u+w)^* + u_x = u_x + w_x = w_x + (u+w)^*,$$
$$(v+w)^* + v_x = v_x + w_x = w_x + (v+w)^*;$$

and it follows from (*) that these two lines are different. Hence

$$[(u+w)^* + u_x] \cap [(v+w)^* + v_x] = w_x;$$

and we see likewise that

$$[(u+w)^* + u_y] \cap [(v+w)^* + v_y] = w_y.$$

The lines $u_x + u_y$ and $v_x + v_y$ are different, as has been shown in the preceding paragraph of the proof; and thus it follows from (2.b) that

$$[u_x + u_y] \cap [v_x + v_y] = (x,y).$$

The points $w_x, w_y, (x,y)$ are certainly collinear; and it follows from (4) that the equality of two of these points would imply $w_x = w_y$ which is impossible by our choice of the indices x,y. Thus w_x, w_y and (x,y) are three collinear points. The lines $u_x + v_x$ and $u_y + v_y$ are different by our choice of x and y. If the lines $u_x + v_x$ and $(u + w)^* + (v + w)^*$ were equal, then we would deduce from $(u + v)^* \leqslant u_x + v_x$ the collinearity of the three points $(u + v)^*$, $(u + w)^*$, $(v + w)^*$; and this collinearity could also be deduced from the equality of the lines $u_y + v_y$ and $(u + w)^* + (v + w)^*$. Thus we may finally assume that

$$u_x + v_x \neq (u + w)^* + (v + w)^* \neq u_y + v_y.$$

But then we may apply the dualized Theorem of Desargues [VI.3, Proposition 3]; and it follows that the three lines $u_x + v_x$, $u_y + v_y$ and $(u + w)^* + (v + w)^*$ have a common point. But the first two lines meet in the point $(u + v)^*$ [by its definition]; and thus we have completed the proof of the collinearity of the three points $(u + v)^*, (v + w)^*$ and $(w + u)^*$.

Corollary 2: *Assume that $P \neq Q$ are points and $v \neq w$ vectors.*

(*a*) *There exists at most one vector u such that $(v + u)^* = P$ and $(w + u)^* = Q$.*

(*b*) *There exists a vector u such that $(v + u)^* = P$ and $(w + u)^* = Q$ if, and only if, P, Q and $(v + w)^*$ are three different, but collinear points.*

Proof: If there exists a vector u such that

$$(v + u)^* = P \text{ and } (w + u)^* = Q,$$

then we infer from $P \neq Q$ and Proposition 4 that P, Q and $(v + w)^*$ are three different, but collinear points. Consequently we assume now that P, Q and $(v + w)^* = R$ are three distinct collinear points. Since the sum of the coordinates of v has at least rank 3, there exists an index x such that v_x is not on the line $P + Q$ [and v_x is a point]. Assume now that $w_x \leqslant P + Q$. Then we deduce from Corollary 1 that

$$v_x + (v + w)^* = (v + w)^* + w_x \leqslant P + Q$$

which is impossible, since the point v_x is not on the line $P + Q$. Hence w_x is a point, not on the line $P + Q$. This implies that $P + v_x$, $Q + w_x$ and $P + Q$ are three different lines. Their sum is a plane, since, by Corollary 1,

$$P + v_x + Q + w_x = P + (v + w)^* + v_x + w_x = P + (v + w)^* + v_x.$$

Consequently $D = [P + v_x] \cap [Q + w_x]$ is a well determined point which is different from P and Q, since P is not on $Q + w_x$ nor Q on $P + v_x$.

Suppose now that u' and u'' are vectors such that

$$P = (v + u')^* = (v + u'')^* \text{ and } Q = (w + u')^* = (w + u'')^*.$$

Then it follows from Corollary 1 that

$$u'_x + (v + u')^* = (v + u')^* + v_x = P + v_x,$$
$$u'_x + (w + u')^* = (w + u')^* + w_x = Q + w_x.$$

This implies that $u'_x \neq 0$ is part of D. Since D and u'_x are both points, we have shown that $D = u'_x$; and $D = u''_x$ is shown likewise. Thus we have shown that the vectors u' and u'' satisfy $(v + u')^* = (v + u'')^*$ and $u'_x = u''_x$ and now it follows from Proposition 3 that $u' = u''$. This completes the proof of (a).

From $v_x + P = P + D$ and Proposition 3 we deduce the existence of a [uniquely determined] vector u such that $(u + v)^* = P$ and $u_x = D$. It follows from Proposition 4 and Corollary 1 that

$$(u + w)^* \leqslant (u + v)^* + (v + w)^* = P + Q,$$
$$(u + w)^* \leqslant (u + w)^* + w_x = u_x + w_x = D + w_x = D + Q,$$

since Q and D and w_x are collinear different points, and since $(v + w)^*$, P, Q are collinear different points. It is impossible that $u = w$, since otherwise $(v + w)^* = (v + u)^* + P$; and now we see that the point $(u + w)^*$ is the intersection Q of the two different lines $P + Q$ and $D + Q$. Thus the vector u meets all the requirements.

Remark 2: If $P = Q$, then (a) ceases to be true and instead of (b) one proves:

There exist vectors u such that $(v + u)^* = P = (w + w)^*$ if, and only if, $(v + w)^* = P$.

Construction of the Extended Projective Geometry

Within the partially ordered set S satisfying Postulates I to IX whose maximal element A has rank not less than 3 we have distinguished the basic quadrangle consisting of lines L_x and points (x,y). From this basic quadrangle we have derived the vectors. We are now ready to construct the extended projective geometry $E(S)$. The elements of $E(S)$ are pairs $X = [X_V, X_S]$ where X_V is a [possibly empty] set of vectors and where X_S is an element in S, subject to the following conditions:

(6.a) *If v and w are vectors in X_V, then $(v + w)^* \leqslant X_S$.*

(6.b) *If v and w are vectors such that $(v + w)^* \leqslant X_S$, and if v belongs to X_V, then w belongs to X_V.*

In the system $E(S)$ we introduce a partial ordering by the following rule:

(7) *$X' \leqslant X''$ if, and only if, X'_V is a subset of X''_V and $X'_S \leqslant X''_S$.*

By $H(S)$ we denote the subset of all those elements X in $E(S)$ where X_V is empty. It is clear that a projectivity of S upon $H(S)$ is obtained by mapping the element T in S upon the uniquely determined element X in $H(S)$ which satisfies $X_S = T$.

The elements in $E(S)$, but not in $H(S)$, have to be more closely investigated.

Proposition 5: *If X is an element in $E(S)$, and if v is a vector in X_V, then*

(*a*) *X_S is the sum of all the $(v + w)^*$ with w in X_V and*

(*b*) *X_V is the totality of vectors w with $(v + w)^*$ in X_S.*

PROOF: Denote by Y the sum of all the $(v + w)^*$ with w in X_V. Then we deduce $Y \leqslant X_S$ from (6.*a*). If P is a point and $P \leqslant X_S$, then we deduce from Proposition 3 and Remark 1 the existence of a vector p such that $(v + p)^* = P$. It follows from (6.*b*) that p belongs to X_V; and we deduce from the definition of Y that $P \leqslant Y$. It follows from VI.2, Proposition 1 that X_S is the sum of its points; and this implies $Y = X_S$. Hence (*a*) is true.

Denote by Z the totality of vectors z such that $(v + z)^* \leqslant X_S$. It follows from (6.*a*) that X_V is part of Z; and it follows from (6.*b*) that X_V contains Z; and this proves (*b*).

Corollary 2: *If X' and X'' are elements in $E(S)$ such that X'_V and X''_V are equal, but not empty, then $X' = X''$.*

This is an immediate consequence of Proposition 5, (*a*).

Proposition 6: *If K is an element in S and v a vector, and if Θ is the totality of vectors w such that $(v + w)^* \leqslant K$, then $[\Theta,K]$ is an element in $E(S)$.*

PROOF: Suppose that w' and w'' belong to Θ. Then it follows from Propositions 2 and 4 that

$$(w' + w'')^* \leqslant (v + w')^* + (w'' + v)^* \leqslant K;$$

and so (6.*a*) is satisfied by Θ,K. Suppose next that u',u'' are vectors such that $(u' + u'')^* \leqslant K$ and that u' belongs to Θ. Then it follows from the definition of Θ that $(v + u')^* \leqslant K$; and we find as before that

$$(v + u'')^* \leqslant (v + u')^* + (u' + u'')^* \leqslant K.$$

Hence u'' too belongs to Θ; and (6.*b*) is satisfied by Θ,K. Hence $[\Theta,K]$ is an element in $E(S)$.

Without any danger of confusion we may denote the empty set of vectors by 0. Then $[0,0]$ is the *null* [or minimal] *element* of the partially ordered set $E(S)$. If we denote by $V(S)$ the totality of all the vectors, then $[V(S),A]$ is the *maximal element* of $E(S)$. Furthermore $H(S) = [0,A]$.

If v is any vector, then we denote the set consisting of v alone also by v; and we find that $[v,0]$ belongs to $E(S)$. Clearly there are no elements between $[0,0]$ and $[v,0]$ so that the elements $[v,0]$ are points in $E(S)$. One verifies easily that all the other points in $E(S)$ have the form $[0,P]$ with P a point in S.

Proposition 7: *Postulates* I *to* IX *are satisfied by* $E(S)$; *and* $[0,A]$ *is a hyperplane in* $E(S)$.

PROOF: The system $E(S)$ is a partially ordered set under rule (7); and as such it satisfies Postulates I and II. Before verifying the other Postulates and Properties we introduce the following useful *notation*: If v is a vector and T an element in S, then $T + v$ is *the totality of vectors* w *such that* $(v + w)^* \leqslant T$. It is a consequence of Proposition 6 that $[T + v, T]$ is an element in $E(S)$; and it is a consequence of Proposition 5 that every element in $E(S)$ has either the form $[0, T]$ with T in $E(S)$ or the form $[T + v, T]$ with T in $E(S)$ and v a suitably selected vector.

Consider now a set Φ of elements in $E(S)$. Denote by Φ' the set of all the elements of the form $[0, T]$ in Φ and denote by Φ'' the totality of the remaining elements in Φ. If X_σ belongs to Φ'', then $X_{\sigma V}$ is not vacuous; and we may select in some way a vector v_σ in $X_{\sigma V}$. If Φ'' is vacuous, then the sum of Φ is clearly $[0,T]$ with T the sum of all the X with $[0,X]$ in Φ $[=\Phi']$. If Φ'' is not empty, then denote by v some of the vectors v_σ. Form the sum $R = \sum_{X \in \Phi} X_S + \sum_\sigma (v + v_\sigma)^*$. Then we claim that $[R + v, R]$ is the sum of Φ. From $(v + v_\sigma)^* + v = (v + v_\sigma)^* + v_\sigma$ and $(v + v_\sigma)^* \leqslant R$ we deduce that $R + v$ contains every v_σ. If X is in Φ, then $X_S \leqslant R$ and $X = [0, X_S]$ or $X = [X_S + v_\sigma, X_S]$ for some σ so that in either case $X \leqslant [R + v, R]$ for every X in Φ. Consider now an element U in $E(S)$ such that $X \leqslant U$ for every X in Φ. Then $X_S \leqslant U_S$ for every X in Φ. Furthermore v and v_σ are all contained in U_V and this implies $(v + v_\sigma)^* \leqslant U_S$ for every σ. Hence $R \leqslant U_S$; and $R + v \leqslant U_V$ is now easily deduced. Consequently $[R + v, R] \leqslant U$; and thus we have shown that $[R + v, R]$ is actually the sum of Φ. To construct the intersection of Φ denote by D the intersection of all the X_S with X in Φ. If there does not exist any vector v which belongs to all the sets X_V with X in Φ, then $[0, D]$ is the intersection of Φ; and if there exists a vector d which belongs to every X_V with X in Φ, then $[D + d, D]$ is the intersection of Φ. This shows the validity of Postulate III.

It will be convenient to give an explicite formula for the sum of two elements in $E(S)$. There are three cases to distinguish:

$$[0,X] + [0,Y] = [0,X + Y];$$

$$[0,X] + [Y + v, Y] = [(X + Y) + v, X + Y];$$

$$[X + x, X] + [Y + y, Y] = [X + Y + (x + y)^* + x, X + Y + (x + y)^*]$$

where x,y,v are vectors and X,Y elements in S.

We precede the verification of Dedekind's Law by the proof of the following

Lemma 1: *If M,N are elements in S and m,n are vectors in $V(S)$, then the following properties are equivalent.*

(I) *There exists a vector which belongs to both $M + m$ and $N + n$.*

(II) $(m + n)^* \leqslant M + N$.

(III) $M \cap N < N \cap [M + (m + n)^*]$ *or* $(m + n)^* \leqslant M$.

PROOF: Clearly there is nothing to prove, if $m = n$, since this implies $(m + n)^* = 0$ [Proposition 2]. Thus we shall assume that $m \neq n$.

Assume first the existence of a vector v which belongs to both $M + m$ and $N + n$. It follows from the definition of these vector sets that $(m + v)^* \leqslant M$ and $(n + v)^* \leqslant N$; and now we deduce from Proposition 4 that $(m + n)^* \leqslant (m + v)^* + (v + n)^* \leqslant M + N$. Hence (II) is a consequence of (I).

Assume next that $(m + n)^* \leqslant M + N$ and $(m + n)^* \not\leqslant M$. It follows from Dedekind's Law that

$$\begin{aligned} M < M + (m + n)^* &= M + [(m + n)^* \cap (M + N)] \\ &= [M + N] \cap [(m + n)^* + M] = M + [N \cap (M + (m + n)^*)]; \end{aligned}$$

and this implies clearly that $M \cap N < N \cap [M + (m + n)^*]$. Hence (III) is a consequence of (II).

Assume finally the validity of (III). If firstly $(m + n)^* \leqslant M$, then n belongs to $M + m$ [by definition]; and so n is the desired common vector. In the same way we see that m is the desired common vector, if $(m + n)^* \leqslant N$. Hence we assume next that $(m + n)^* \not\leqslant M$ and $(m + n)^* \not\leqslant N$. Then it follows from (III) that

$$M \cap N < N \cap [M + (m + n)^*];$$

and we may deduce from VI.2, Proposition 1 the existence of a point P in $N \cap [M + (m + n)^*]$ which does not belong to $M \cap N$. Since P is in N whereas the point $(m + n)^*$ does not belong to N, we have $P \neq (m + n)^*$. If $[P + (m + n)^*] \cap M$ were 0, then we could deduce from Dedekind's Law that

$$\begin{aligned} (m + n)^* &= (m + n)^* + ([P + (m + n)^*] \cap M) \\ &= [P + (m + n)^*] \cap [M + (m + n)^*] \\ &= (m + n)^* + ([M + (m + n)^*] \cap P) = (m + n)^* + P; \end{aligned}$$

and this would imply $P = (m + n)^*$ which is impossible. Since $(m + n)^*$ is not part of M, neither is the line $P + (m + n)^*$; and thus it follows that $Q = [P + (m + n)^*] \cap M$ is a point. Since Q is a point on M and since P, $(m + n)^*$ are not, it follows that $P, Q, (m + n)^*$ are three different collinear points. Now we deduce from Corollary 2 the existence of one and only one vector h such that $(h+m)^* = Q$ and $(h+n)^* = P$. Since $P \leqslant N$, h belongs to $N+n$; and since $Q \leqslant M$, h belongs to $M+m$; and thus h is the desired common vector. Hence (I) is a consequence of (III); and this completes the proof of the Lemma.

We turn now to the

VERIFICATION OF DEDEKIND'S LAW IN $E(S)$.

Assume that X, Y, Z are elements in $E(S)$ and that $X \leqslant Y$. Then it is immediately clear that

$$\text{(IV*)} \qquad X + (Y \cap Z) \leqslant Y \cap (X + Z).$$

We distinguish two cases.

Case 1: X_V or Z_V is vacuous.

Then at least one of the sets X_V and $(Y \cap Z)_V \leqslant Z_V$ is vacuous. Hence it follows from the formula for addition in $E(S)$ that

$$X_S + Z_S = (X+Z)_S,$$

$$[X + (Y \cap Z)]_S = X_S + (Y \cap Z)_S = X_S + (Y_S \cap Z_S) = Y_S \cap (X_S + Z_S)$$
$$= Y_S \cap (X+Z)_S = [Y \cap (X+Z)]_S.$$

If $[Y \cap (X+Z)]_V$ is empty, then we infer from (IV*) that $[X + (Y \cap Z)]_V$ is empty too; and this implies $X + (Y \cap Z) = Y \cap (X+Z)$. Assume therefore the existence of a vector w in $[Y \cap (X+Z)]_V$. Then w belongs to both Y_V and $(X+Z)_V$. If w is in Z_V, then w is in $(Y \cap Z)_V$ and then $[X + (Y \cap Z)]_V$ is not vacuous; and the same is true, if X_V is not vacuous. Thus we assume now that X_V is empty and that w is not in Z_V.Since $(X+Z)_V$ is not vacuous whereas X_V is vacuous, it is impossible that Z_V is vacuous. Hence $Z_V = Z_S + z$ for some vector z. Since w is in $(X+Z)_V$, we have $(X+Z)_V = (X+Z)_S + w = (X_S + Z_S) + w$. Since z is in Z_V, z is also in $(X+Z)_V$; and this implies

$$(z+w)^* \leqslant (X+Z)_S = X_S + Z_S.$$

Thus condition (II) of Lemma 1 is satisfied; and there exists consequently a vector v which belongs both to $X_S + w$ and to $Z_S + z$. Since w belongs to Y_V, we have $X_S + w \leqslant Y_S + w = Y_V$; and so the vector v belongs to $Y_V \cap Z_V$ and therefore to $[X + (Y \cap Z)]_V$. Thus we have shown that in every case $[X + (Y \cap Z)]_V$ contains a vector u which belongs to $[Y \cap (X+Z)]_V$ because of (IV*). Consequently we have

$$X + (Y \cap Z) = [(X + (Y \cap Z))_S + v, (X + (Y \cap Z))_S]$$
$$= [(Y \cap (X+Z))_S + v, (Y \cap (X+Z))_S] = Y \cap (X+Z);$$

and this completes the proof of Dedekind's Law in Case 1.

Case 2: Neither X_V nor Z_V is vacuous.

Denote by x and z vectors in X_V and Z_V respectively. From the addition formulas for $E(S)$ we deduce that

$$(X + Z)_S = X_S + Z_S + (x + z)^*;$$

and it follows from Dedekind's Law that

$$\begin{aligned}[Y \cap (X + Z)]_S &= Y_S \cap [X_S + Z_S + (x + z)^*] \\ &= X_S + (Y_S \cap [Z_S + (x + z)^*]).\end{aligned}$$

This is clearly equal to $X_S + (Y_S \cap Z_S)$, if

$$Y_S \cap Z_S = Y_S \cap [Z_S + (x + z)^*];$$

and from (IV*) we deduce then that

$$[Y \cap (X + Z)]_S = X_S + (Y_S \cap Z_S) \leqslant [X + (Y \cap Z)]_S \leqslant [Y \cap (X + Z)]_S$$

or

$$[Y \cap (X + Z)]_S = [X + (Y \cap Z)]_S.$$

If on the other hand

$$Y_S \cap Z_S < Y_S \cap [Z_S + (x + z)^*],$$

then we deduce from Lemma 1 the existence of a vector v which belongs both to $Y_S + x$ and $Z_S + z$. Thus v belongs to $(Y \cap Z)_V = Y_V \cap Z_V$; and it follows from the addition formulas for elements in $E(S)$ that

$$[X + (Y \cap Z)]_S = X_S + (Y \cap Z)_S + (x + v)^*.$$

Since x belongs to $X_V \leqslant Y_V$, and since v belongs to Y_V, we have $(x + v)^* \leqslant Y_S$; and since z and v belong to Z_V, we have $(z + v)^* \leqslant Z_S$. Now we infer from Proposition 4 that

$$(x + v)^* \leqslant (x + z)^* + (v + z)^* \leqslant (x + z)^* + Z_S;$$

If $(x + v)^*$ were part of Z_S, then we would infer from Proposition 4 that

$$(x + z)^* \leqslant (x + v)^* + (v + z)^* \leqslant Z_S$$

which contradicts our assumption that

$$Y_S \cap Z_S < Y_S \cap [Z_S + (x + z)^*].$$

Since $(x + z)^*$ and $(x + v)^*$ are points, this implies

$$Z_S + (x + z)^* = Z_S + (x + v)^*.$$

Since $(x + v)^* \leqslant Y_S$, we may deduce now from Dedekind's Law that

$$\begin{aligned}[Y \cap (X + Z)]_S &= X_S + (Y_S \cap [Z_S + (x + z)^*]) \\ &= X_S + (Y_S \cap [Z_S + (x + v)^*]) \\ &= X_S + (x + v)^* + (Y_S \cap Z_S) = [X + (Y \cap Z)]_S;\end{aligned}$$

and thus we have verified $[Y \cap (X + Z)]_S = [X + (Y \cap Z)]_S$ generally for our Case 2. Since the vector x belongs to $X_V \leqslant Y_V$; we see finally that

$$\begin{aligned} X + (Y \cap Z) &= [(X + (Y \cap Z))_S + x, (X + (Y \cap Z))_S] \\ &= [(Y \cap (X + Z))_S + x, (Y \cap (X + Z))_S] \\ &= Y \cap (X + Z); \end{aligned}$$

and this completes our verification of Dedekind's Law for $E(S)$.

VERIFICATION OF POSTULATE V [*The Complementation Postulate*]:

We note that [0,0] is the null-element and that $[V,A]$, for V the set of all vectors, is the maximal [or all-] element in $E(S)$. Consider now any element X in $E(S)$. There exists by Postulate V in S an element C such that $0 = X_S \cap C$, $A = X_S + C$. If $X_V = 0$, then we let $Y = [C + v, C]$ where v is any [random] vector; and if X_V is not vacuous, then we let $Y = [0,C]$. Thus one and only one of the sets X_V and Y_V is vacuous; and consequently we have $(X + Y)_S = X_S + C = A$. Since at least one of the sets X_V and Y_V contains a vector, say w, $(X + Y)_V$ contains w too. Hence $X + Y = [A + w, A] = [V,A]$. That $X \cap Y = [0, X_S \cap C] = [0,0]$ is a consequence of the fact that only one of the sets X_V and Y_V contains a vector. Thus Y is a complement of X.

VERIFICATION OF POSTULATE VI [*Existence of points*]:

If the element X in $E(S)$ is not the null-element [0,0], then either X_V contains a vector V and the point $[v,0]$ is part of X; or else $X_S \neq 0$ so that X_S contains [by VI] a point P in S and X contains the point $[0,P]$.

VERIFICATION OF POSTULATE VII [*Finite dependence*]:

Suppose that Θ is a set of elements in $E(S)$ and that Q is a point in $E(S)$ which is part of the sum of Θ. We distinguish two possibilities.

Case 1: $Q = [0,P]$.

We recall that the sum T of Θ has the following form: T_S is the sum of all the X_S with X in Θ and of all the $(v + w)^*$ with v and w in [different] X_V with X in Θ. Since P is a point, it follows from Postulate VII and from $P \leqslant T_S$ that there exists a finite number of elements X_i in Θ and a finite number of vectors v_j in sets X_V with X in Θ such that

$$P \leqslant \sum_i (X_i)_S + \sum_{i \neq j} (v_i + v_j')^*.$$

But now it is clear how to pick a finite number of elements Y_i in Θ such that $Q \leqslant \sum_i Y_i$.

Case 2: $Q = [v,0]$.

Since Q is part of the sum of Θ, there must exist at least one element X in Θ such that X_V is not vacuous. Then there exists a vector w in X_V.

If $v = w$, then $Q \leqslant X$. If $v \neq w$, then we infer from Proposition 2 that $(v + w)^*$ is a point. Since $[v,0]$ and $[w,0]$ are both contained in the sum T of Θ, so is their sum $[(v + w)^* + w, (v + w)^*]$; and the point $[0, (v + w)^*]$ is therefore also part of T. Now it follows from Case 1 that there exists a finite number of elements $X_1, \cdots, X_k$ in Θ such that $[0, (v + w)^*] \leqslant \sum_{i=1}^{k} X_i$; and from this we deduce that

$$Q \leqslant [(v + w)^* + w, (v + w)^*] = [w,0] + [0, (v + w)^*] \leqslant X + \sum_{i=1}^{k} X_i;$$

and this completes the proof of the validity of Postulate VII in $E(S)$.

Verification of Postulate VIII [*Existence of three points on a line*]:

Suppose that there are given two distinct points M and N in $E(S)$. Naturally we have to distinguish three possibilities.

Case 1: $M = [m,0]$ and $N = [n,0]$.

Then $m \neq n$ and it follows from Proposition 2 that $(m + n)^*$ is a point in S. Now it follows from the addition rule that $[0, (m + n)^*]$ is a third point on $M + N = [(m + n)^* + m, (m + n)^*]$.

Case 2: One and only one of the points M and N has the form $[0,P]$.

Without loss in generality we assume that $M = [0,P]$ with P a point in S. Then $N = [n,0]$. We infer from Remark 1 the existence of a vector v such that $(n + v)^* = P$. Then it follows from Proposition 2 that $n \neq v$; and consequently $[v,0]$ is a third point on $M + N = [P + n,P]$.

Case 3: $M = [0,P]$ and $N = [0,Q]$.

Then P and Q are different points; and we infer from Postulate VIII the existence of a third point R on the line $P + Q$ in S. Then $[0,R]$ is a third point on $M + N = [0,P + Q]$. This completes the verification of Postulate VIII in $E(S)$.

Verification of the Fact that $[0,A]$ is a Hyperplane:

Denote by v any one vector in $V(S)$. Then the maximal element in $E(S)$ has the form: $[V,A] = [A + v,A] = [v,0] + [0,A]$ where $[v,0]$ is a point in $E(S)$. Since $[0,A] = [V,A]$, this implies that $[V,A]$ is a point over $[0,A]$ so that $[0,A]$ is a hyperplane in $E(S)$.

Verification of Postulate IX [*Desargues' Theorem*]:

We recall that $r(A)$ is at least 3; and that the partially ordered set S [of elements contained in A] is projectively equivalent with the partially ordered set $H(S)$ of the elements of the form $[0,X]$ with X in S. Hence $3 \leqslant r([0,A])$. But $[0,A]$ is a hyperplane in $[V,A]$. Hence $3 < R([V,A])$; and Desargues' Theorem is a consequence of VI.3, Proposition 2. This completes the proof.

VII.5. The Group of a Hyperplane

Though we intend to apply the results of the present section onto the hyperplane $[0,A]$ in the extended partially ordered set $E(S)$, it will be more convenient to retain throughout this section the notations of the

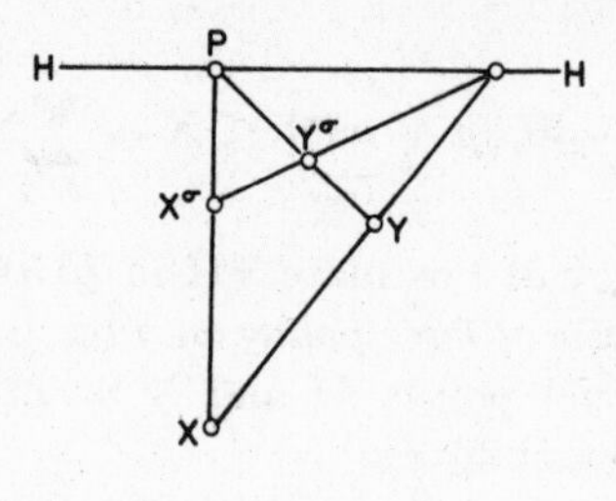

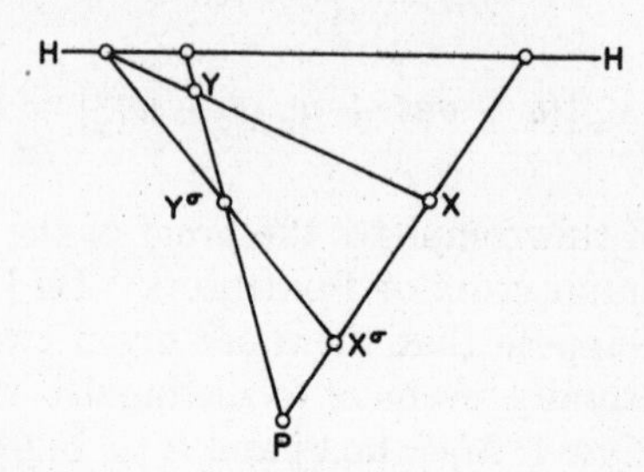

Fig. 19

first three sections of this chapter. Hence S will be a partially ordered set satisfying Postulates I to IX. The maximal element in S will be denoted by A; and the rank of A will be at least three. Finally we denote by H a hyperplane in A which will be kept fixed throughout our discussion.

The group $\Gamma = \Gamma(H)$ *of* H consists of all the permutations σ of S with the following two properties:

(1.*a*) $M \leqslant N$ *if, and only if*, $M\sigma \leqslant N\sigma$.

(1.*b*) $M = M\sigma$ *for every* $M \leqslant H$.

That the totality of permutations σ of S with the properties (1.*a*) and (1.*b*) actually forms a group, is quite obvious. In geometrical language we would term these permutations σ the perspectivities with axis H.

If σ is in Γ, and if the point P has the property

(2) $M + P = M\sigma + P$ *for every element* M *in* S,

then we term the point P *a center of* σ. Since σ maps points upon points, it is clear that centers of σ are fixed points of σ.

Lemma 1: *Every* $\sigma \neq 1$ *in* Γ *possesses one and only one center which we denote by* $C(\sigma)$.

PROOF: Assume that σ is in Γ and that the two different points P and Q are centers of σ. If the point X is on H, then $X = X\sigma$ by (1.*b*). If the point X is neither on H nor on the line $P + Q$, then it follows from (2) that

$$X + P = X\sigma + P \text{ and } X + Q = X\sigma + Q.$$

Since $X + P$ and $X + Q$ are different lines, it follows now that

$$X = [X + P] \cap [X + Q] = [X\sigma + P] \cap [X\sigma + Q] = X\sigma.$$

If the point X is not on H, but is on $P + Q$, then we consider some point R which is neither on H nor on $P + Q$ [the existence of such a point R is a consequence of $3 \leqslant r(A)$]. Then $R = R\sigma$ by what we have shown already. Since H is a hyperplane and $R + X$ is a line, not on H, the intersection $J = H \cap (R + X)$ is a point different from R and X; and it follows from (1.*b*) that $J = J\sigma$. Now we deduce from the fact that $R + X$ and $P + Q$ are distinct lines that

$$\begin{aligned} X &= (R + X) \cap (P + Q) = (R + J) \cap (P + Q) = (R\sigma + J\sigma) \cap (P\sigma + Q\sigma) \\ &= [(R + J) \cap (P + Q)]\sigma = X\sigma; \end{aligned}$$

and thus we have shown that every point in S is a fixed point of σ. But every element in S is a sum of points [VI.2, Proposition 1]; and $\sigma = 1$ is an immediate consequence of (1.*a*). Thus we have shown that elements, not 1, in Γ have at most one center.

Suppose now that $\sigma \neq 1$ is in Γ. We distinguish two cases.

Case 1: There exists a fixed point of σ, not on H.

If J is a fixed point of σ, J not on H, and if X is a point not on H and different from J, then the line $J + X$ meets the hyperplane H in a third point Q. Clearly $Q = Q\sigma$ so that

$$X + J = J + Q = J\sigma + Q\sigma = (J + Q)\sigma = (J + X)\sigma = J\sigma + X\sigma = J + X\sigma;$$

and now it is easy to see that J is the center of σ.

Case 2: Every fixed point of σ is on H.

Consider first any two distinct points X, Y neither of which is on H. Then $X + Y$ meets the hyperplane H in a fixed point J of σ. Furthermore $X + X\sigma$ and $Y + Y\sigma$ are well determined lines, since neither X nor Y is a fixed point. We have

$$\begin{aligned} X + X\sigma + Y + Y\sigma &= X\sigma + X + J + Y\sigma = X\sigma + X + J\sigma + Y\sigma \\ &= X\sigma + X + (J + Y)\sigma = X\sigma + X + (X + Y)\sigma = X + X\sigma + Y\sigma. \end{aligned}$$

Thus the lines $X + X\sigma$ and $Y + Y\sigma$ are either equal or at least coplanar. If $X + X\sigma \neq Y + Y\sigma$, then they meet in a well determined point K. The line $X + X\sigma$ meets the hyperplane H in a point K'; and we have

$$X + X\sigma = X\sigma + K' = X\sigma + K'\sigma = (X + K')\sigma = (X + X\sigma)\sigma;$$

and likewise we see that $(Y + Y\sigma)\sigma = Y + Y\sigma$. Consequently the intersection K of $X + X\sigma$ and $Y + Y\sigma$ is a fixed point of σ; and as such it is on H. Thus we have shown:

If X and Y are points not on H, then the lines $X + X\sigma$ and $Y + Y\sigma$ are invariant under σ; and $H \cap (X + X\sigma) = H \cap (Y + Y\sigma)$.

Thus all the lines $X + X\sigma$ with X not on H meet H in the same point which is the center of σ. This completes the proof.

Lemma 2: *If P,Q,R are three different, but collinear points, and if neither P nor Q is on H, then there exists one and only one σ in Γ such that $C(\sigma) = R$ and $P\sigma = Q$.*

PROOF: Assume first that σ' and σ'' are in Γ and that $C(\sigma') = C(\sigma'') = R$ and $P\sigma' = P\sigma'' = Q$. Then $\sigma'\sigma''^{-1} = \gamma$ belongs to Γ and satisfies $P = P\gamma$ and $M + R = R + M\gamma$ for every M in S. Since $A = P + H$ we find that $P + M = (P + M) \cap (P + H) = P + [(P + M) \cap H]$; and both P and $(P + M) \cap H$ are left invariant by γ. Hence

$$\begin{aligned} P + M &= P + [(P + M) \cap H] = P\gamma + [(P + M) \cap H]\gamma \\ &= (P + [(P + M) \cap H])\gamma = (P + M)\gamma = P\gamma + M\gamma = P + M\gamma; \end{aligned}$$

and thus we have shown that P and R are two different centers of the transformation γ in Γ. It follows from Lemma 1 that $\gamma = 1$ or $\sigma' = \sigma''$, showing the uniqueness of the desired transformation.

To prove the existence of at least one transformation meeting our requirements it will be convenient to exclude the case when one [and consequently every] line carries exactly three different points. In the case when lines carry just three points our method will not be workable, but a slightly simpler, though similar method may be used; and thus we leave the details in this case as an exercise to the reader.

Consider any line L through P which is different from $P + R$. On this line we can find points P',P'', not on H, such that P,P',P'' are three different points on L, since L carries one and only one point $L \cap H$ in H. Then $P + R$, $P' + R$ and $P'' + R$ are three different lines through the point R and in the plane $L + R$; and none of these three different lines belongs to H. It follows as usual that

$$Q' = [(L \cap H) + Q] \cap [P' + R],\ Q'' = [(L \cap H) + Q] \cap [P'' + R]$$

are well determined points which are not in H; and we note that the points $Q, Q', Q'', L \cap H$ are four different and collinear points.

Consider now a point X which is different from R and which is not on H. Then X can be on at most one of the three lines $P + R$, $P' + R$ and $P'' + R$. If X is neither on $P + R$ nor on $P' + R$, then we distinguish two cases:

Case 1: X,P,P' are collinear.

Then the line $L = X + P = P + P' = P' + X$ meets H in a point $H \cap L$; and the line $Z = (H \cap L) + Q = (H \cap L) + Q'$ meets the line $X + R$ in a well determined point $(X + R) \cap Z$.

Case 2: The points X,P,P' are not collinear.

Then the line $X + P$ meets H in a point $H \cap (X + P)$; and the line $X + P'$ meets H in another point $H \cap (X + P')$. The triangles $P, Q, (P + X) \cap H$ and $P', Q', (P' + X) \cap H$ are perspective from the point $(P + P') \cap H = (Q + Q') \cap H = L \cap H$, since the plane $L + X$ meets the hyperplane H in a line carrying the points $L \cap H$, $(P + X) \cap H$, $(P + X') \cap H$; and corresponding sides of these triangles meet in the points

$$(P + Q) \cap (P' + Q') = R,$$
$$(P + [(P + X) \cap H]) \cap (P' + [(P' + X) \cap H]) = (P + X) \cap (P' + X) = X,$$
$$(Q + [(P + X) \cap H]) \cap (Q' + [(P' + X) \cap H]).$$

Application of the Theorem of Desargues [Postulate IX] shows then that the lines $X + R$, $Q + [(P + X) \cap H]$ and $Q' + [(P' + X) \cap H]$ have a point in common so that

$$(X + R) \cap (Q + [(P + X) \cap H]) = (X + R) \cap (Q' + [(P' + X) \cap H]).$$

We restate the result of this discussion in the following form:

(*) If the point X is neither on $P + R$ nor on $P' + R$ [nor on H], then

$$(X + R) \cap (Q + [(P + X) \cap H]) = (X + R) \cap (Q' + [(P' + X) \cap H])$$

is a well determined point.

Analoguous statements may be made, if X is not on the lines $P' + R$, $P'' + R$ etc.

Now we are ready to define a point to point transformation mapping points X upon points X^* according to the following rules:

$$X = X^*, \text{ if } X = R \text{ and if } X \leqslant H;$$
$$X^* = (X + R) \cap (Q + [(P + X) \cap H]), \text{ if } X \text{ is not on } P + R;$$
$$X^* = (X + R) \cap (Q' + [(P' + X) \cap H]), \text{ if } X \text{ is not on } P' + R;$$
$$X^* = (X + R) \cap (Q'' + [(P'' + X) \cap H]), \text{ if } X \text{ is not on } P'' + R.$$

It is a consequence of the lemma (*) that this mapping X^* is a single-valued mapping from points to points. Interchanging throughout P and Q we obtain another single-valued point to point mapping X^+; and it is not difficult to verify that these mappings are reciprocals of each other [Dedekind's Law]. Consequently the mapping X^* effects a permutation of the points whose only fixed points are R and the points on H.

Suppose now that X and Y are two different points neither of which is R and neither of which is on H. Then it is impossible that X and Y are on all three lines $P + R$, $P' + R$, $P'' + R$; and thus we may assume without loss in generality that neither X nor Y is on $P + R$. If the lines $X + Y$ and $X^* + Y^*$ happen to be equal, then they meet H in the same point. Thus we assume now that $X + Y$ and $X^* + Y^*$ are different lines. This implies in particular that the lines $X + X^*$ and $Y + Y^*$ are different. Consider now the triangles P, X, Y and Q, X^*, Y^*. They are perspective from the point R [by definition of the *-mapping and by choice of

P,Q] and the three lines of perspectivity are different. Corresponding sides of these triangles meet in the points

$$(P + X) \cap (Q + X^*) = (P + X) \cap H,$$
$$(P + Y) \cap (Q + Y^*) = (P + Y) \cap H,$$
$$(X + Y) \cap (X^* + Y^*).$$

We may apply the Theorem of Desargues [Postulate IX] and find that the points

$$(P + X) \cap H, (P + Y) \cap H, (X + Y) \cap (X^* + Y^*)$$

are collinear. Since the first two points are on H, so is the last one; and since neither of the lines $X + Y$ and $X^* + Y^*$ is on H, this signifies that the lines $X + Y$ and $X^* + Y^*$ meet H in the same point. Now one checks easily the validity of the following statement:

(**) If X and Y are two different points neither of which is on H, then $(X + Y) \cap H = (X^* + Y^*) \cap H$.

We note finally the obvious fact that $P^* = Q$.

Consider now any element M in S. Then we define:

$M\sigma =$ sum of all the points X^* with X a point on M.

It is clear that $0\sigma = 0, A\sigma = A$ and $X\sigma = X^*$ for X a point. Consider now a point $Y \leqslant M\sigma$. There exists one and only one point X such that $X^* = Y$. We want to show that $X \leqslant M$. It is an immediate consequence of the definition of $M\sigma$ and of Postulate VII that there exists a [minimal] finite number of points $X_1, \cdots, X_k$ on M such that $X^* \leqslant \sum_{i=1}^{k} X_i^*$. If $k = 1$, then $X^* = X_1^*$; and $X = X_1$ is a consequence of the fact that the *-operation is a permutation. Thus we may make the inductive hypothesis that a point Z is on $\sum_{i=1}^{k-1} X_i$ whenever $Z^* \leqslant \sum_{i=1}^{k-1} X_i^*$. If it happens that $X^* = X_k^*$, then we conclude as before that $X = X_k$. Thus we may assume that $X^* \neq X_k^*$. Then $X^* + X_k^*$ is a line; and it follows from Dedekind's Law that

$$X^* + X_k^* = [X^* + X_k^*] \cap \left[\sum_{i=1}^{k} X_i^*\right] = X_k^* + [(X^* + X_k^*) \cap \sum_{i=1}^{k-1} X_i^*].$$

Since $X^* + X_k^*$ is a line and X_k^* is a point, it follows that

$$(X^* + X_k^*) \cap \sum_{i=1}^{k-1} X_i^*$$

is a point Z^* since X^* is not on $\sum_{i=1}^{k-1} X_i^*$ [minimality of k]. We apply the inductive hypothesis to see that the point Z is on $\sum_{i=1}^{k-1} X_i$, since $Z^* \leqslant \sum_{i=1}^{k-1} X_i^*$. If Z^* were equal to X_k^*, then X_k^* would be on $\sum_{i=1}^{k-1} X_i^*$ which would contradict the minimal choice of k; and likewise we see that $Z^* \neq X^*$. Consequently $L^* = X^* + Z^* = Z^* + X_k^* = X_k^* + X^*$ is a line. If one of the three points X,Z,X_k is on H, then the collinearity of the three points X,Z,X_k is a consequence of (**) and the collinearity of Z^*,X^*,X_k^*. If none of the three points X,Z,X_k is on H, then $L^* \cap H$ is a point which is different from X,Z,X_k [since H is a hyperplane, and since points on H are fixed points of the *-operation]. But then we infer from (**) that this point $L^* \cap H$ is on the three lines $X + Z$, $Z + X_k$ and $X_k + X$ so that X,X_k,Z are collinear. It follows thus in every case that

$$X \leqslant Z + X_k \leqslant \sum_{i=1}^{k-1} X_i + X_k \leqslant M.$$

This completes the induction and the proof of the fact that the point X is on M whenever X^* is on $M\sigma$. Now it is an almost immediate consequence of VI.2, Proposition 1 that σ is a permutation of the elements in S which meets requirement (1.*a*). Since $X^* = X$ for X a point on H, it follows also that $M = M\sigma$ follows from $M \leqslant H$. We have finally $P\sigma = P^* = Q$; and $X + R = X^* + R = X\sigma + R$ for points X. Hence R is a center of σ; and this completes the proof of Lemma 2.

Lemma 3: *If the transformations σ',σ'' and $\sigma'\sigma''$ in Γ are all different from 1, then $C(\sigma'\sigma'') \leqslant C(\sigma') + C(\sigma'')$.*

Proof: If $C(\sigma') = C(\sigma'') = Z$, then it follows from (2) that

$$M + Z = M\sigma' + Z = (M\sigma')\sigma'' + Z = M(\sigma'\sigma'') + Z;$$

and it follows from Lemma 1 that $C(\sigma'\sigma'') = Z = C(\sigma') = C(\sigma'')$. We assume now that $C(\sigma')$ and $C(\sigma'')$ are different points. We select a line L, not on H, and such that neither $C(\sigma')$ nor $C(\sigma'')\,\sigma'^{-1}$ nor $C(\sigma'\sigma'')$ is on L. Then L carries apart from the point $L \cap H$ on H two different points X and Y neither of which is on H. From

$$L \cap H = (L \cap H)\sigma' = (L \cap H)\sigma'' = (L \cap H)\sigma'\sigma''$$

we deduce that the lines

$$L = X + Y,\ L\sigma' = X\sigma' + Y\sigma' \text{ and } L\sigma'\sigma'' = X\sigma'\sigma'' + Y\sigma'\sigma''$$

all pass through $L \cap H$. We see now that the triangles $X, X\sigma', X\sigma'\sigma''$ and $Y, Y\sigma', Y\sigma'\sigma''$ are perspective from the point $L \cap H$; and that the three lines $X + Y = L$, $L\sigma'$ and $L\sigma'\sigma''$ are all different, since otherwise $C(\sigma'')$ would be on $L\sigma'$ or $C(\sigma'\sigma'')$ would be on L etc. One verifies that the lines $X + X\sigma'$ and $Y + Y\sigma'$ are different and meet in $C(\sigma')$, that the lines $X\sigma' + X\sigma'\sigma''$ and $Y\sigma' + Y\sigma'\sigma''$ meet in $C(\sigma'')$ and that the lines $X + X\sigma'\sigma''$ and $Y + Y\sigma'\sigma''$ meet in $C(\sigma'\sigma'')$. We apply the Theorem of Desargues and find that the points $C(\sigma')$, $C(\sigma'')$ and $C(\sigma'\sigma'')$ are collinear; and this completes the proof of Lemma 3.

If σ is in Γ, then either $\sigma = 1$ and every point in S is a center; or else $\sigma \neq 1$ and σ possesses [by Lemma 1] one and only one center, namely $C(\sigma)$. If $M \neq 0$ is an element in S, then we infer from Lemma 3 that *the totality* $\Gamma(M)$ *of transformations* σ *in* Γ *with center on* M *is a subgroup of* Γ.

We let $\Gamma(0) = 1$. It is obvious that $\Gamma = \Gamma(A)$. We are especially interested in $\Gamma(H) = B$.

Theorem 1: (*a*) *B is a normal subgroup of* Γ.

(*b*) *B is a commutative [abelian] group.*

(*c*) *Mapping the element* M *in* S *with* $M \leqslant H$ *upon* $\Gamma(M) \leqslant B$ *is a projectivity.*

(*d*) *The endomorphisms* η *of the group B which satisfy* $\Gamma(M)^{\eta} \leqslant \Gamma(M)$ *for* $M \leqslant H$ *form a field F.*

(*e*) *The subgroup T of B satisfies* $T^F \leqslant T$ *if, and only if,* $T = \Gamma(N)$ *for some* $N \leqslant H$.

PROOF: If σ' and σ'' are in Γ, $\sigma' \neq 1$, then $C(\sigma''^{-1}\sigma'\sigma'') = C(\sigma')\sigma''$ is easily deduced from (2), since

$$M + C(\sigma')\sigma'' = [M\sigma''^{-1} + C(\sigma')]\sigma'' = [M\sigma''^{-1}\sigma' + C(\sigma')]\sigma''$$
$$= M\sigma''^{-1}\sigma'\sigma'' + C(\sigma')\sigma''.$$

If in particular $C(\sigma') \leqslant H$, then $C(\sigma''^{-1}\sigma'\sigma'') = C(\sigma')\sigma'' = C(\sigma')$; and consequently we have $\sigma^{-1}B\sigma = B$ for every σ in Γ. This implies (*a*).

Suppose next that σ', σ'' are elements, not 1, in B such that $C(\sigma') \neq C(\sigma'')$. Then $C(\sigma')$ and $C(\sigma'')$ are two different points on H. Consider any point P, not on H. Then $P, P\sigma', C(\sigma')$ are collinear points and $P, P\sigma'', C(\sigma'')$ are collinear points. Since the line $L' = P + P\sigma'$ meets H in $C(\sigma')$ and $L'' = P + P\sigma''$ meets H in $C(\sigma'')$, the lines L' and L'' are different and meet in P. [Note that $P \neq P\sigma'$ and $P \neq P\sigma''$ may be deduced from Lemma 1.] Now one verifies easily that

$$P\sigma'\sigma'' = [P\sigma' + C(\sigma'')] \cap [P\sigma'' + C(\sigma')] = P\sigma''\sigma';$$

and $\sigma'\sigma'' = \sigma''\sigma'$ may be deduced from Lemma 2.

If σ' and σ'' belong to B, but have the property that $C(\sigma') = C(\sigma'')$ then we deduce from Lemma 2 the existence of a transformation $\sigma \neq 1$

in B such that $C(\sigma) \neq C(\sigma') = C(\sigma'')$. Then it follows easily from (2) that $C(\sigma^{-1}) = C(\sigma)$; and we deduce from the result of the first paragraph of this proof that $\sigma\sigma' = \sigma'\sigma$ and $\sigma^{-1}\sigma'' = \sigma''\sigma^{-1}$. If $C(\sigma'\sigma)$ were equal to $C(\sigma^{-1}\sigma'')$, then it would follow from Lemma 3 that this center would also equal $C(\sigma'\sigma\sigma^{-1}\sigma'') = C(\sigma'\sigma'') = C(\sigma') = C(\sigma'')$. But $C(\sigma'\sigma) = C(\sigma')$ implies also $C(\sigma') = C(\sigma'^{-1}\sigma'\sigma) = C(\sigma)$, contrary to our choice of σ. Hence $C(\sigma'\sigma) \neq C(\sigma^{-1}\sigma'')$; and now it follows from the results already obtained that

$$\sigma'\sigma'' = \sigma'\sigma\sigma^{-1}\sigma'' = \sigma^{-1}\sigma''\sigma'\sigma = \sigma''\sigma^{-1}\sigma\sigma' = \sigma''\sigma';$$

and this completes the proof of the commutativity of the group B [we note that in this proof we used the fact $1 < r(H)$].

Suppose that T' and T'' are elements in S such that $T' < T'' \leqslant H$. It follows immediately from the definition of the groups $\Gamma(T)$ that $\Gamma(T') \leqslant \Gamma(T'') \leqslant \Gamma(H) = B$. From $T' < T''$ and VI.2, Proposition 1 we deduce the existence of a point P on T'' such that P is not on T'. There exists by Lemma 2 a transformation $\sigma \neq 1$ in Γ such that $C(\sigma) = P$. Then σ belongs to $\Gamma(T'')$, but not to $\Gamma(T')$; and hence $\Gamma(T') < \Gamma(T'')$; and from this fact our proposition (c) is an immediate consequence.

Denote by F the system of all the endomorphisms η of the group B which satisfy

$$\Gamma(M)^{\eta} \leqslant \Gamma(M) \text{ for every } M \leqslant H.$$

Since B is a commutative group, sum, difference and product of endomorphisms of B are again endomorphisms of B [see V.1]; and now one verifies easily that F is a ring.

Consider an element η in F. Suppose that σ' and σ'' are elements, not 1, in B such that $C(\sigma') \neq C(\sigma'')$. Then $C(\sigma'\sigma'')$ is a point on the line $C(\sigma') + C(\sigma'')$ which is different from both $C(\sigma')$ and $C(\sigma'')$. Assume that $\sigma'^{\eta} = 1$. Then $(\sigma'\sigma'')^{\eta}$ belongs to $\Gamma[C(\sigma'\sigma'')]$; but since $(\sigma'\sigma'')^{\eta} = \sigma''^{\eta}$, it belongs also to $\Gamma[C(\sigma'')]$. Since $C(\sigma'')$ and $C(\sigma'\sigma'')$ are different points, it follows that $\Gamma[C(\sigma'\sigma'')] \cap \Gamma[C(\sigma'')] = 1$ or $\sigma''^{\eta} = 1$. Assume now that $\eta \neq 0$. Then there exists a transformation σ in B such that $\sigma^{\eta} \neq 1$. If σ' has the property that $C(\sigma') \neq C(\sigma)$, then it follows from what we have shown just now that $\sigma'^{\eta} \neq 1$. If $\sigma' \neq 1$, but $C(\sigma) = C(\sigma')$, then we deduce from Lemma 2 the existence of a transformation σ'' in B such that

$$C(\sigma'') \neq C(\sigma) = C(\sigma').$$

It follows that $\sigma''^{\eta} \neq 1$; and this implies that $\sigma'^{\eta} \neq 1$. Thus we have shown:

(d.1) *If $\eta \neq 0$ is in F, then η is one to one.*

If σ is some element in Γ, then an automorphism of B is obtained by mapping the element β in B upon the element $\beta^{\sigma} = \sigma^{-1}\beta\sigma$, since—as we have shown already—B is a normal subgroup of Γ. Since $C(\beta) \leqslant H$ for β

in B, we deduce from a former remark that $C(\beta) = C(\sigma^{-1}\beta\sigma) = C(\beta^{\sigma})$. Now one verifies easily that $\Gamma(M)^{\sigma} = \Gamma(M)$ for every $M \leqslant H$; and thus we have shown that the automorphism, induced by σ in B, belongs to the ring F. Next we show that the converse of this fact is true [apart from a trivial exception], namely

(*d*.2) *If* $\eta \neq 0$ *is an element in* F, *then there exists an element* σ *in* Γ *such that* $\beta^{\eta} = \beta^{\sigma}$ *for every* β *in* B.

There is nothing to prove, if η happens to be the identity transformation. Consequently we may assume without loss in generality that there exists an element τ in B such that $\tau \neq \tau^{\eta}$. It is a consequence of (*d*.1) that neither τ nor τ^{η} is the identity. Since η is in F, we have $C(\tau) = C(\tau^{\eta})$. Since $\tau \neq \tau^{\eta}$, there exists a point P such that $P\tau \neq P\tau^{\eta}$. Since τ and τ^{η} are in B, their only fixed points are on H. Since every point on H is a fixed point, it follows now that P is not on H and that $P, P\tau, P\tau^{\eta}$ are three different points. From (2) we infer that the points $P, P\tau, C(\tau)$ and the points $P, P\tau^{\eta}$, $C(\tau^{\eta}) = C(\tau) = Z$ are collinear. Hence $P, P\tau, P\tau^{\eta}$, Z are four collinear points. We deduce now from Lemma 2 the existence of one and only one transformation σ in Γ such that $P = C(\sigma)$ and $P\tau^{\eta} = P\tau^{\sigma}$. We note that τ^{σ} belongs to B and $C(\tau) = C(\tau^{\sigma})$. Since P is a fixed point of σ and hence of σ^{-1}, we find that

$$P\tau^{\sigma} = P\sigma^{-1}\tau^{\sigma} = P\tau^{\sigma} = P\tau^{\eta}.$$

Thus τ^{η} and τ^{σ} are transformations in B with the same center Z; and they have the same effect on the point P which is neither their center nor on H. It follows from Lemma 1 that $\tau^{\sigma} = \tau^{\eta}$. Consider now the endomorphism $\varkappa$ in F which is defined by the rule $\beta^{\varkappa} = \beta^{\eta - \sigma}$ for β in B [so that $\varkappa$ is the difference of η and the automorphism induced by σ]. Then $\tau^{\varkappa} = \tau^{\eta - \sigma} = 1$, though $\tau \neq 1$. It follows from (*d*.1) that $\varkappa = 0$; and this implies $\beta^{\eta} = \beta^{\sigma}$ for every β in B. This completes the proof of (*d*.2).

From (*d*.2) it follows immediately that every element $\eta \neq 0$ in F is an automorphism of B and possesses an inverse in F. Consequently F is a field; and this completes the proof of (*d*).

It follows from the definition of F that $\Gamma(M)^{F} \leqslant \Gamma(M)$ for $M \leqslant H$ [actually equality holds, since F contains an identity element]. Suppose now conversely that T is a subgroup of B satisfying $T^{F} \leqslant T$. We show first:

(*e*.1) *If* $\sigma \neq 1$ *is in* T, *and if* σ' *is a transformation such that* $C(\sigma) = C(\sigma')$, *then* σ' *is in* T.

Let $Z = C(\sigma) = C(\sigma')$. Then Z is a point on H, since σ is in $T \leqslant B$. Since σ, σ' are transformations, not 1, in B, their only fixed points are on H [and every point on H is a fixed point]. Let P be any point not on H. Then $P \neq P\sigma$ and $P \neq P\sigma'$; and the points $P, P\sigma, P\sigma', Z$ are collinear. If

$P\sigma = P\sigma'$, then $\sigma = \sigma'$ by Lemma 2; and we assume therefore that $P\sigma \neq P\sigma'$. Now we deduce from Lemma 2 the existence of a transformation σ'' in Γ such that $P = C(\sigma'')$ and $P\sigma = P\sigma'\sigma''$. As before [in the proof of $(d.2)$] we verify that $\sigma = \sigma''^{-1}\sigma'\sigma''$ or $\sigma' = \sigma''\sigma\sigma''^{-1} = \sigma^{\sigma''^{-1}}$. But σ'' and σ''^{-1} induce elements in F; and it follows that σ' belongs to $\sigma^F \leqslant T^F \leqslant T$, as we desired to show.

(*e*.2) *If σ' and σ'' are transformations in T neither of which is* 1, *and if $C(\sigma) \leqslant C(\sigma') + C(\sigma'')$, then σ is in T.*

This is an immediate consequence of $(e.1)$, in case $C(\sigma') = C(\sigma'')$. Thus we assume now that $C(\sigma') \neq C(\sigma'')$ and we note that the line $C(\sigma') + C(\sigma'')$ is on H. Our proposition may likewise be deduced from $(e.1)$ in case $C(\sigma) = C(\sigma'')$ or $C(\sigma')$; and thus we assume now that $C(\sigma)$, $C(\sigma')$ and $C(\sigma'')$ are three different collinear points. Consider now any point P, not on H. Then P, $P\sigma'$, $C(\sigma)'$ are three different collinear points; and the lines

$$P + C(\sigma) \text{ and } P\sigma' + C(\sigma'')$$

are two different lines in the plane $P + C(\sigma') + C(\sigma'')$. Thus they meet in a well determined point $Q = [P + C(\sigma)] \cap [P\sigma' + C(\sigma'')]$ which is not on H. We infer from Lemma 2 the existence of one and only one transformation τ in Γ such that $C(\tau) = C(\sigma'')$ and $P\sigma'\tau = Q$. It is clear that τ is in B; and it follows from $(e.1)$ that τ is in T, since σ'' is in T. Hence $\sigma'\tau$ is in T, since σ' is in the subgroup T of B. We note that $P\sigma'\tau = Q$. Hence $C(\sigma'\tau) = H \cap [P + Q] = C(\sigma)$; and now we deduce from $(e.1)$ that σ itself is in T; as we desired to prove.

(*e*.3) *If $\sigma_1, \cdots, \sigma_n$ are a finite number of transformations in T, and if $C(\sigma) \leqslant \sum_{i=1}^{n} C(\sigma_i)$, then σ is in T.*

This proposition is certainly true for $n = 1$ and $n = 2$, as follows from $(e.1)$ and $(e.2)$. Thus we may make the inductive hypothesis that our statement holds true for n-1. It is a consquence of $(e.1)$ that σ is in T whenever $C(\sigma) = C(\sigma_n)$; and it follows from the inductive hypothesis that σ is in T whenever $C(\sigma) \leqslant \sum_{i=1}^{n-1} C(\sigma_i)$. Thus we assume now that $C(\sigma) \neq C(\sigma_n)$ and that $C(\sigma)$ is not on $K = \sum_{i=1}^{n-1} C(\sigma_i)$. This implies in particular that $C(\sigma_n)$ is not on K. We form the line $C(\sigma) + C(\sigma_n) = L$; and verify that $L \cap K$ is a point different from $C(\sigma)$ and $C(\sigma_n)$. We infer from Lemma 2 the existence of transformations $\sigma' \neq 1$ in Γ such that $C(\sigma') = L \cap K$; and it follows from the inductive hypothesis that σ' is in T. Since $C(\sigma)$ is on the line $L = C(\sigma_n) + C(\sigma')$, and since σ_n and σ' are in

T, it follows from (e.2) that σ is in T. This completes the inductive proof of (e.3).

Denote now by N the sum of all the points $C(\sigma)$ with $\sigma \neq 1$ in T [so that $N = 0$ in case $T = 1$]. It is clear then that $T \leqslant \Gamma(N)$. If σ is in $\Gamma(N)$, then $\sigma = 1$ or else $C(\sigma) \leqslant N$. In the latter case we deduce from the definition of N and from Postulate VII the existence of a finite number of elements $\sigma_1, \cdots, \sigma_n$ in T such that $C(\sigma) \leqslant \sum_{i=1}^{n} C(\sigma)_i$; and now we deduce from (e.3) that σ is in T. Thus we have shown that $T = \Gamma(N)$; and this completes the proof of (e).

REMARK: Using the facts derived in the course of the preceding proof the reader will find it easy to verify the following interesting fact:

If P is a point, not on H, then every coset of Γ modulo B contains one and only one element in $\Gamma(P)$; and the multiplicative group of the elements, not 0, in F is isomorphic to $\Gamma(P)$.

VII.6. The Representation Theorem

We are now ready to prove the principal result of the present chapter.

Theorem: *If the partially ordered set S satisfies Postulates* I *to* IX, *if the maximal element M in S has rank not less than* 3, *then there exists a linear manifold A such that S and the partially ordered set $S(A)$ of the subspaces of A are projectively equivalent.*

PROOF: It is a consequence of VII.4, Proposition 7 [Imbedding Theorem] that there exists an extension T of the partially ordered set S such that Postulates I to IX are satisfied by T and such that M is a hyperplane in T. We denote by M' the maximal element in T.

Now we form the group $\Gamma(M)$ of the hyperplane M in T according to VII.5. We denote by A the subgroup of $\Gamma(M)$ consisting of all those transformations σ in $\Gamma(M)$ with center on M. It follows from VII.5, Theorem 1 that A is an abelian group. We denote by F the set of all the endomorphisms η of A such that $\Gamma(X)^{\eta} \leqslant \Gamma(X)$ for $X \leqslant M$ [or equivalent: for X in S] where we denote by $\Gamma(X)$ the totality of transformations in A with center on X. It follows from VII.5, Theorem 1 that F is a field, that a subset V of A is a subspace of the F-space A if, and only if, $V = \Gamma(X)$ for some X in S, and that mapping X in S upon $\Gamma(X)$ in $S(A)$ constitutes a projectivity of S upon $S(A)$. Hence the F-space A is the desired linear manifold representing S.

VII.7. The Principles of Affine Geometry

It is our objective in this section to sketch the relations between the principles of affine geometry and those of projective geometry. We shall show in particular that the considerations of Section VII.5 will suffice for building up affine geometry and that the comparatively deep Imbedding Theorem of Section VII.4 is not needed for this purpose. In this respect the foundations of affine geometry constitute a much simpler problem than those of projective geometry.

We recall first that every linear manifold (F,A) defines both a projective geometry and an affine geometry. The projective geometry defined by (F,A) is the partially ordered set $S(F,A)$ of all the F-subspaces of A; and *the affine geometry defined by* (F,A) *is the partially ordered set* $V(F,A)$ *of all the F-flocks of* A [see I.1]. Here we define an F-flock as a non-empty subset W of A with the following property:

If a,b,c are elements in W and if f is an element in F, then $f(a-b)+c$ belongs to W.

Using the terminology customary in group theory we may say that the flocks are just the cosets modulo the F-subspaces of A. The following simple property will prove useful.

Lemma: *The non-empty subset W of A is an F-flock if, and only if,* $\sum_{i=1}^{n} f_i w_i$ *belongs to W whenever the* w_i *belong to W and the numbers* f_i *in F satisfy* $\sum_{i=1}^{n} f_i = 1$.

Proof: The sufficiency of these conditions is almost obvious. Assume now that W is a flock. If the elements $w_1, \cdots, w_n$ belong to W, and if the numbers f_i in F satisfy $\sum_{i=1}^{n} f_i = 1$, then we have clearly

$$w = \sum_{i=1}^{n} f_i w_i = \sum_{i=1}^{n-1} f_i(w_i - w_n) + w_n.$$

It follows from the flock property that $f_{n-1}(w_{n-1} - w_n) + w_n$ belongs to W; and consequently we may deduce from the flock property that $f_{n-2}(w_{n-2} - w_n) + f_{n-1}(w_{n-1} - w_n) + w_n$ belongs to W. Proceding like this we verify by complete induction that w itself belongs to W, as we wanted to show.

We shall call $V(F,A)$ *an algebraical model of affine geometry.* We turn next to

THE CONSTRUCTION OF A PROJECTIVE MODEL OF AFFINE GEOMETRY:

To do this we consider a projective geometry S which may be supposed to be either given in the form $S(F,A)$ or just as some partially ordered set S satisfying the postulates I to IX. In S we distinguish a hyperplane H. Then *the affine geometry* $[S,H]$ *is the partially ordered set of all those elements* [= *subspaces*] X *in* S *which are not part of* H.

In customary geometrical terminology we may say that the affine geometry $[S,H]$ arises from the projective geometry S by deletion of the hyperplane H [the hyperplane "at infinity"].

We recall first that the partial order in $[S,H]$ is the natural partial order induced by that in S. If Θ is a set of elements in $[S,H]$, then none of the elements in Θ is part of H; and the same is clearly true of their sum. Thus sums of elements in $[S,H]$ are well determined elements in $[S,H]$. But the situation changes completely once we consider intersections. For if X and Y belong to $[S,H]$, then their intersection $X \cap Y$ is a well determined element in S; but $X \cap Y$ may be part of H in spite of the fact that neither X nor Y is part of H. Consequently it may very well happen that the intersection of elements in $[S,H]$ does not exist in $[S,H]$. This means in the customary geometrical language that the elements X and Y in $[S,H]$ are parallel.

ALGEBRAICAL PROOF OF EQUIVALENCE OF ALGEBRAICAL AND PROJECTIVE MODELS OF AFFINE GEOMETRY:

We precede the equivalence proof proper by the following considerations. Suppose that (F,A) is a linear manifold and that H is a hyperplane in A. Then we may select [in many ways] an element t in A which does not belong to H; and it is clear that

$$A = H \oplus Ft.$$

If the F-subspace X of A is not part of H, then we denote by X^* the totality of elements h in H such that $h + t$ belongs to X. We show that

(1) X^* *is an F-flock in H for every X in* $[S(F,A),H]$.

PROOF: It is almost obvious that X^* is not vacuous, since X is not part of H. If a,b,c are elements in X^*, and if f belongs to F, then a,b,c are elements in H and $a + t$, $b + t$, $c + t$ belong to X. But X is an F-subspace of A. Hence $(a + t) - (b + t) = a - b$ belongs to $[X \cap H]$; and this implies that $f(a - b) + (c + t) = [f(a - b) + c] + t$ is an element in X. Consequently $f(a - b) + c$ belongs to X^*; and we have shown that X^* is a flock.

It follows from (1) that mapping X in $[S(F,A),H]$ upon X^* constitutes a single-valued and order preserving mapping of $[S(F,A),H]$ into $V(F,H)$.

If Y is an F-flock in H, then we denote by Y^* the F-subspace of A which is spanned by the elements $y + t$ for y in Y. It is clear that mapping Y onto Y^* constitutes a single-valued and order preserving mapping of $V(F,H)$ into $[S(F,A),H]$. We shall refer to the two mappings just constructed as to the *star-mappings*; and we show that

(2) *the two star-mappings are reciprocals of each other.*

PROOF: It is clear that

$$Y \leqslant Y^{**} \text{ for every } Y \text{ in } V(F, H).$$

Consider now an element z in Y^{**}. Then $z + t$ belongs to Y^* and z belongs to H. It follows from the definition of Y^* that there exist elements $y_1, \cdots, y_n$ in Y and numbers $f_1, \cdots, f_n$ in F such that

$$z + t = \sum_{i=1}^{n} f_i(y_i + t).$$

Since z and the y_i belong to H, and since $A = H \oplus Ft$, it follows that

$$z = \sum_{i=1}^{n} f_i y_i, \qquad 1 = \sum_{i=1}^{n} f_i.$$

Since Y is an F-flock, z belongs to Y [Lemma]. Hence $Y = Y^{**}$.

Consider next some subspace X of A which is not part of H. One verifies without difficulty that X is spanned by the elements of the form $t + x$ with x in X. In other words X is spanned by $X^* + t$; and this shows that $X = X^{**}$. Our contention (2) is essentially a restatement of these two equations $X = X^{**}$ and $Y = Y^{**}$.

We have already shown that the two star-mappings are single-valued and preserve the partial order. Using (2) one verifies now easily that the two star-mappings are one to one and exhaustive too. Thus they are what we usually call projectivities and what here might better be called affine equivalences. Consequently we may restate our result in the following form:

(3) *The two star-mappings constitute reciprocal affine equivalences between* $[S(F,A),H]$ *and* $V(F,H)$.

It is now very simple to prove the equivalence of algebraical and projective models of affine geometry. Suppose firstly that there is given some algebraical model $V(F,A)$ of an affine geometry. There exists clearly a linear manifold (F,B) such that A is a hyperplane in (F,B). There exists an element b such that $B = Fb \oplus A$. Now we may construct the star-mapping of $V(F,A)$ into $[S(F,B),A]$; and it follows from (3) that this is an affine equivalence. Thus $V(F,A)$ has been represented as a projective model of an affine geometry. Suppose next that there is given some projective model $[S,H]$ of an affine geometry. We represent the projective geometry S by its algebraical model: $S = S(F,A)$ where (F,A) is a suitably selected linear manifold. Then H is a hyperplane in A and there exists an element t such that $A = Ft \oplus H$. Now we may construct the star-mapping of $[S(F,A),H]$ into $V(F,H)$; and it follows from (3) that it is an affine equivalence. Thus $[S,H] = [S(F,A),H]$ has been represented as an algebraical model of affine geometry; and this completes our equivalence proof.

SYNTHETIC CONSTRUCTION OF ALGEBRAICAL MODELS OF AFFINE GEOMETRY:

When we constructed an algebraical model $V(F,H)$ equivalent to a given projective model $[S,H]$ of affine geometry, we assumed that the projective geometry S was given by an algebraical model $S(F,A)$. In other words we made use of the Representation Theorem of VII.6. We want to show now that it suffices to use the considerations of VII.5 and that the use of the Imbedding Theorem of VII.4 may be avoided.

To do this we consider the totality A of all the auto-projectivities σ of S with the following properties:

(*a*) σ leaves invariant every element in S which is part of H.

(*b*) σ possesses a center on H.

It follows from the results of VII.5 that A is an abelian group. If X is a subspace of H, then we denote by $\Gamma(X)$ the totality of elements in A whose center is on X; and we denote by F the totality of endomorphism η of the abelian group A which satisfy $\Gamma(X)^{\eta} \leqslant \Gamma(X)$ for every subspace X of H. It follows from VII.5, Theorem 1 that F is a field and that a projectivity of the partially ordered set of subspaces of H upon $S(F,A)$ is obtained by mapping the subspace X of H upon $\Gamma(X)$.

Next we select in some way a point P in S which does not belong to the hyperplane H. If X belongs to $[S,H]$, then we denote by X^* the totality of elements σ in A which map P upon a point in X. One verifies now without great difficulty that mapping X upon X^* constitutes an affine equivalence between $[S,H]$ and $V(F,A)$; and thus we have constructed an algebraical model of the affine geometry $[S,H]$. We leave the details to the reader.

INTRINSIC CHARACTERIZATION OF PROJECTIVE MODELS OF AFFINE GEOMETRY:

The problem is to characterize those partially ordered sets L which are affinely equivalent to a projective model $[S,H]$ of an affine geometry. As $L^* = [S,H]$ consists just of those elements in S which are not part of H, the main problem will be the construction within L^* of those elements in S which are part of H. These elements are just the soc. ideal elements of affine geometry.

The proper framework for such an undertaking seems to be provided by the concept of exchange lattice due to S. MacLane: A Lattice Formulation for Transcendence Degrees and p-Bases, *Duke Math. Journal*, vol. 4 (1938), pp. 455-468. In order to construct the ideal elements one has to develop a theory of parallelism. Such a theory has been developed, for instance, in the important paper by K. Menger which we quoted in the Bibliography at the beginning of this chapter. [Note that part of the energy saved in avoiding the Imbedding Theorem has to be invested in this theory of parallelism; but it seems to the author that in spite of this the principles of affine geometry are still considerably simpler than those of projective geometry.]

It should be noted, however, that the synthetic construction of the algebraical model of affine geometry which we outlined just now can be modified quite easily in such a way that no explicite use is made of the ideal elements [as we did]; the reader is advised to consult in this context Artin's beautiful paper which we quoted in the Bibliography at the beginning of this chapter.

APPENDIX S

A Survey of the Basic Concepts and Principles of the Theory of Sets

In this survey we are going to procede in a completely unsophisticated fashion. No proofs will be given and no notice will be taken of the axioms and the problems connected with them. Concepts and principles will be stated in a form convenient for their application; and this whole survey shall only serve as a ready collection of the needed tools from the theory of sets. Any one wanting to go beyond this completely superficial presentation is advised to consult one of the books enumerated below.

A Selection of Suitable Introductions into the Theory of Sets

G. Birkhoff : Lattice Theory. *Amer. Math. Soc. Coll. Publ.*, 25. New York, 1948.

N. Bourbaki : Éléments de mathématiques. *Actualités scient. et ind.*, 840, 858, 916, 934, 1029. Paris, 1939-1947.

Jean Cavaillès : Transfini et continu. Paris, 1947.

A. Denjoy : L'énumération transfini. Paris, 1946.

A. Fränkel : Einleitung in die Mengenlehre. 3. Aufl. Berlin, 1928.

K. Gödel : The Consistency of the Continuum Hypothesis. *Ann. of Math. Studies*, 3, Princeton, N. J.

F. Hausdorff : Mengenlehre. Berlin, 1914.

H. Hermes and G. Köthe : Die Theorie der Verbände. *Enzyklopädie der math. Wiss.*, I, 1, 13.

E. Kamke : Allgemeine Mengenlehre. *Enzyklopädie der math. Wiss.*, I, 1, 5.

E. Kamke : Theory of Sets. Translated from 2nd German ed. by F. Bagemihl. New York, 1950.

W. Sierpinski : Leçons sur les nombres transfinis. Paris, 1928.

W. Sierpinski : Hypothèse du continu. Warszawa, 1934.

A. Tarski : Cardinal Algebras. New York, 1949.

Sets and Subsets

A set is any aggregate of elements which may be given either in a completely abstract fashion or which may arise from some other structure like sets of points, sets of numbers, sets of functions, sets of sets etc.

If S is some definite set, then we have to consider the subsets of S. These are sets all of whose elements belong to S. The totality of subsets of S has a certain mathematical structure. If U and V are subsets of S, then U may or may not be a subset of V; in the first case we write $U \leqslant V$

and in the second case we write $U \lessdot\leqslant V$. This partial ordering of the subsets of S is often referred to as the partial ordering defined by inclusion.

If Φ is a set of subsets of S, then the intersection $\bigcap_{X \in \Phi} X$ of the subsets in Φ is the totality of elements which belong to all the subsets X in Φ; and the join of Φ is the totality of elements which belong to at least one subset in Φ.

Mappings

If S and T are sets, then a single-valued mapping σ of S into T assigns to every element s in S its image s' in T in a unique fashion. The image of the element s under σ will be denoted in various ways [according to circumstances] like $s\sigma$, s^σ, σs, $\sigma(s)$, ${}^\sigma s$, etc, and the totality of single-valued mappings of S into T is denoted by T^S.

The single-valued mapping σ of S into T is called exhaustive or a mapping of S upon T, if every element in T is the σ-image of at least one element in S.

The single-valued mapping σ of S into T is called a one-to-one mapping if different elements in S have different images in T.

The one-to-one and exhaustive mappings of S upon T are characterized by the fact that they possess inverses [reciprocals]; and the one-to-one and exhaustive mappings of S upon itself are called permutations of S [independent of finiteness or infinity of S etc].

Partially Ordered Sets

The set S is partially ordered by a relation "$\leqslant$", if this relation meets the following requirements:

(*a*) If a and b are elements in S, then $a \leqslant b$ or $b \leqslant a$ or neither of these relations is satisfied;

(*b*) $a \leqslant b$ and $b \leqslant a$, if, and only if, $a = b$;

(*c*) $a \leqslant b$ and $b \leqslant c$ imply $a \leqslant c$.

A typical example of a partially ordered set is the set of subsets of a set S, ordered by inclusion.

If $a \neq b$, but $a \leqslant b$, then we may say $a < b$ [because of (*b*)].

Every subset of a partially ordered set S is itself a partially ordered set with respect to the basic partial ordering of S. The subset T of S is called *an ordered subset of* S, if it meets the following requirement:

(*d*) If a and b are in T, then $a \leqslant b$ or $b \leqslant a$.

If T is some subset of S, and if u is some element in T such that $t \leqslant u$ for every t in T, then u is *an upper bound of* T; and lower bounds are defined accordingly. The element m in T is termed *a maximal element of*

T, if $m \leqslant t$ and t in T imply $m = t$. [Note that maximal elements need not be upper bounds.] Minimal elements are defined accordingly.

We are now ready to state the fundamental

MAXIMUM PRINCIPLE OF SET THEORY:

The [non-vacuous] subset T of the partially ordered set S possesses a maximal element, if every ordered subset of T possesses an upper bound in T.

This principle has first been discovered [in an equivalent form] by Hausdorff [1, p. 140]; see also A. D. Wallace: A Substitute for the Axiom of Choice. *Bull. Amer. Math. Soc.*, vol. 50 (1944), p. 278.

For a discussion of the application of this principle the reader might consult:

W. Teichmüller: Braucht der Algebraiker das Auswahlaxiom ? *Deutsche Mathematik*, vol. 4 (1939), pp. 567-577;

M. Zorn: A Remark on Method in Transfinite Algebra. *Bull. Amer. Math. Soc.*, vol. 51 (1935), pp. 667-670.

In most applications of this principle the partially ordered set under discussion will be a set S of subsets [of some set] which is ordered by inclusion. The subsets T of S on which to apply the maximum principle will have the following property: If U is an ordered subset of T, then the join of the sets in U belongs to T.

[By interchanging upper bound and lower bound, maximal element and minimal element a minimum principle is obtained.]

We note finally that one-to-one and order preserving mappings of partially ordered sets are called projectivities whereas one-to-one and order inverting mappings are termed dualities.

Well Ordering

An ordered set [which satisfies all the rules (*a*) to (*d*) of ordering] is called well ordered, if every non-vacuous subset contains a first element. The celebrated Well Ordering Theorem of Zermelo asserts that every set possesses at least one well ordering. In its power it is equivalent to the maximum principle.

Ordinal Numbers

Two well ordered sets S and T are said to have the same ordinal number if, and only if, there exists a one-to-one and order preserving mapping of S upon T. The ordinal number of the well ordered set S is said not to exceed the ordinal number of the well ordered set T, if there exists a one-to-one and order preserving mapping of S into T. If there is given any set of ordinal numbers, then the partial ordering just defined actually produces a well ordering of them.

If S is some well ordered set, then every element s in S may be "numbered" by the ordinal number of the set of elements in S preceding it. This we might indicate as follows:

$$S = \{ s_0, s_1, \cdots, s_\nu, \cdots \}.$$

Every ordinal number ν possesses an immediate successor which we denote by $\nu + 1$. But there exist ordinal numbers apart from 0 which are not of this form; they are referred to as limit ordinals. The first ordinal of this type is usually denoted by ω; it is the ordinal number of the positive integers in their natural order.

PROOF AND DEFINITION BY TRANSFINITE INDUCTION.

They are very similar to proof and definition by complete induction. For instance a function $f(\nu)$ is defined for every ordinal ν not exceeding σ, if

(a) $f(0)$ is defined,

(b) a rule is given for finding $f(\nu + 1)$ once $f(\nu)$ is known,

(c) a rule is given for finding $f(\nu)$ in case ν is a limit ordinal and $f(\tau)$ is known for every $\tau < \nu$.

Likewise a proposition $P(\nu)$ is true for every ordinal ν not exceeding σ, if

(a) $P(0)$ is true,

(b) $P(\nu + 1)$ is a consequence of $P(\nu)$,

(c) $P(\nu)$ is true in case ν is a limit ordinal and $P(\tau)$ has been verified for every $\tau < \nu$.

In very many cases it is possible to substitute for proofs by transfinite induction an application of the Maximum Principle; but which of these methods will be more convenient to apply has to be decided in each case.

Cardinal Numbers

The sets S and T are said to have the same cardinal number, if there exists a one-to-one correspondence mapping S upon T; and we shall denote the cardinal number of S by $|S|$. We say that $|S| \leqslant |T|$ whenever there exists a one-to-one mapping of S into T. This partial ordering of the cardinal numbers proves again to be a well ordering.

The infinite cardinal numbers are usually denoted by $\aleph$. The first of them is $\aleph_0$, the cardinal number of the set of integers. If S is any set, then the cardinal number of the set of all subsets of S is denoted by $2^{|S|}$. Cantor's famous theorem asserts that

$$a < 2^a \text{ for every cardinal number } a \neq 0.$$

Whether or not we have for every infinite cardinal number the equation

$$2^{\aleph_\nu} = \aleph_{\nu+1}$$

is an open question. This equation is known under the name "general continuum hypothesis."

Whereas the set of all subsets has greater cardinal number than the basic set, the situation changes when considering special subsets only. In this respect the following theorem is important and useful: If S is an infinite set, and if T is a set of finite subsets of S such that every element in S belongs to at least one subset in T, then $|T| = |S|$.

The algebraic operations with cardinal numbers are defined as follows. *Addition*: If A and B are disjoint sets of cardinal numbers a and b respectively, then $a + b$ is the cardinal number of the join of the sets A and B. *Multiplication*: If A and B are sets of cardinal numbers a and b respectively, then ab is the cardinal number of the set of all pairs (x,y) with x in A and y in B [of the cartesian product of the sets A and B]. *Exponentiation*: If A and B are sets of cardinal numbers a and b respectively, then a^b is the cardinal number of the set A^B of all single-valued mappings of B into A. [Note that this is in agreement with our previous definition of 2^a.]

ABSORPTION LAWS:

If a and b are cardinal numbers such that $a \leqslant b$ and b is infinite, then

$$a + b = ab = b.$$

For the further information concerning the algebra of cardinal numbers the reader is referred to the literature.

BIBLIOGRAPHY

A Selected Bibliography of Useful Works on Algebra and Geometry with Special Reference to the Needs of the Present Study

A. A. Albert:
[1] Modern Higher Algebra. Chicago, 1947.
[2] Structures of Algebras. *Amer. Math. Soc. Coll. Pub.*, 24. New York, 1939.
E. Artin:
[1] Galois Theory. *Notre Dame Math. Lectures*, 2, 1944.
E. Artin, C. J. Nesbitt and R. M. Thrall:
[1] Rings with Minimum Condition. Ann Arbor, Mich., 1944.
G. Birkhoff:
[1] Lattice Theory, 2nd ed. *Amer. Math. Soc. Coll. Pub.*, 25. New York, 1948.
G. Birkhoff and S. MacLane:
[1] A Survey of Modern Algebra. New York, 1948.
N. Bourbaki:
[1] Eléments de mathématiques. Paris, 1937-47.
H. S. M. Coxeter:
[1] The Real Projective Plane. New York, 1949.
J. Dieudonné:
[1] Sur les groupes classiques. Paris, 1948.
[2] On Automorphisms of the Classical Groups. *Memoirs of the Amer. Math. Soc.*, No. 2, 1950.
P. Dubreil:
[1] Algèbre, I. Paris, 1946.
P. Halmos:
[1] Finite Dimensional Vector Spaces. *Annals of Math. Studies*, 7. Princeton, N. J., 1940.
H. Hasse:
[1] Höhere Algebra, 2 Bde; 2. Aufl. Berlin and Leipzig, 1933.
O. Haupt:
[1] Einführung in die höhere Algebra, 2 Bde. Leipzig, 1929.
H. Hermes and G. Köthe:
[1] Die Theorie der Verbände. *Enzyklopädie der math. Wiss.*, I, 1, 13.
W. W. D. Hodge and D. Pedoe:
[1] Methods of Algebraic Geometry. Cambridge, 1947.
L. Hua :
[1] Supplement to the Paper of Dieudonné on the Automorphisms of Classical Groups. *Memoirs of the Amer. Math. Soc.*, No. 2, 1950.
N. Jacobson:
[1] The Theory of Rings. *Math. Surveys*, II. New York, 1943.
[2] Lectures in Abstract Algebra. Vol. I: Basic Concepts. New York, 1951.

W. Krull:

[1] Elementare Algebra vom höheren Standpunkt. Berlin, 1939.

F. Levi:

[1] Geometrische Konfigurationen. Leipzig, 1929.

[2] Algebra. Calcutta, 1942.

[3] Finite Geometrical Systems. Calcutta, 1942.

C. C. MacDuffee:

[1] An Introduction to Abstract Algebra. New York, 1940.

[2] Vectors and Matrices. *Carus Math. Monographs*, 7. Ithaca, N. Y., 1943

W. Magnus:

[1] Allgemeine Gruppentheorie. *Enzyklopädie der math. Wiss.*, I, 1, 9.

J. von Neumann:

[1] Continuous Geometry. Princeton, N. J., 1935-37.

G. Pickert:

[1] Einführung in die höhere Algebra. *Studia Mathematica /Mathematische Lehrbücher*, Band 7. Göttingen, 1951.

C. E. Rickart:

[1] Isomorphisms of Infinite Dimensional Analogues of the Classical Groups. *Bull. Amer. Math. Soc.*, vol. 51 (1951), pp. 435-448.

O. Schreier and E. Sperner:

[1] Einführung in die analytische Geometrie und Algebra, 2 Bde. Leipzig and Berlin, 1931-1935.

[2] Introduction to Modern Algebra and Matrix Theory. Translated [from [1]] by M. Davis and M. Hausner. New York, 1951.

H. Schwerdtfeger:

[1] Introduction to Linear Algebra and the Theory of Matrices. Groningen, 1951.

B. Segre:

[1] Lezioni di geometria moderna, vol. 1. Fondamenti di geometria sopra un corpo qualsiasi. Bologna, 1948.

E. Sperner:

[1] Einführung in die analytische Geometrie und Algebra, Teil 1 and 2. *Studia Mathematica /Mathematische Lehrbücher*, Band 1 and 6. Göttingen, 1950.

O. Veblen and J. W. Young:

[1] Projective Geometry, 2 vols., 2nd ed. Boston, 1918-1938.

B. L. van der Waerden:

[1] Moderne Algebra, 2 Bde, 2. Aufl. Berlin, 1937.

[2] Gruppen von linearen Transformationen. *Ergebnisse der Mathematik*, Bd. 4, 2. Berlin, 1935.

H. Weyl:

[1] The Classical Groups. Princeton, N. J., 1946.

H. Zassenhaus:

[1] The Theory of Groups. New York, N. Y., 1949.

INDEX

A

Absorption law, 312.
Adjoint space, 25-26.
Adjunct space, 36.
Admissible
  involutions of first and second kind, 147.
  projectivities, 144.
Annulet, 172.
Anti-
  automorphism, 78.
  isomorphism, 96.
  homomorphism, 103.
  semi-linear transformation, 99, 190.
Auto-projectivities, group of, 52.

B

Basis, 13, 265.
Betweenness, 223.
Bilinear form, 101.

C

Canonical dual space, 98.
Center of a group, 201.
Center of a perspectivity, 292.
Centralizer
  first, 63, 203.
  second, 203.
Characteristic two, criteria for
  Fano's postulate, 38.
  IV. 4. Proposition 1, 114.
  IV. 4. Corollary 3, 116.
Class of a linear transformation, 207.
Collineation, 62.
Commutativity, criteria for — of the field of scalars
  III. 3. Proposition 2, 67.
  Second fundamental theorem of projective geometry, 68.
Commutativity, criteria for — of the field of scalars
  Postulate of Pappus, 71.
  III. 4. Remark 4, 75-76.
  III. 4. Theorem 2, 87.
  IV. 2. Lemma 1 (*c*), 106.
  IV. 2. Theorem, 108.
  IV. 4. Proposition 1, 114.
  IV. App. II. Proposition 1 (*a*), 132.
  IV. App. III. Theorem 1, 141.
  IV. App. III. Remark 2, 143.
Complement, 12.
Complementation
  principle, 260.
  theorem, 12.
Conjugate numbers in fields, 72.
Containedness, 8.
Cross ratio, 71-72.
Crossed homomorphisms, 248.

D

Decomposition endomorphisms, 169.
Dedekind's (or modular) law, 9, 259.
Dependence of
  elements (vectors), 13.
  points, 41, 264.
Desargues' theorem, 266-267.
Dilatation, 53, 57.
Dimension, 16, 265.
Direct sum, 12.
Dualities, 32, 95, 310.
Duality
  principle of projective geometry, 100, 263.
  theorem, 97.

E

Euclidean fields, 144.
  anti-isomorphisms of endomorphism rings, 192-193.

Existence theorem for
basis, 14.
dualities, 96.
polarities, 159.

F

Fano's postulate, 37.
Field, 7.
Finiteness, criteria for — of rank
II. 3. Corollary 1, 26.
II. 3. Theorem 2, 28.
II. 3. Proposition 3 (*c*), 30.
IV. 1. Existence theorem (for dualities), 96.
IV. 1. Duality theorem, 97.
IV. 1. Self-duality theorem, 97.
IV. 1. Proposition 2, 103.
IV. 2. Theorem, 108.
IV. 3. Theorem 2, 111.
V. 3. Theorem 2, 182.
V. 5. Proposition 2, 189.
V. 5. Existence theorem, 192.
V. 5. General existence theorem, 193.
VI. App. I. Proposition 3, 243.
VI. 7. Proposition 2, 250.
VI. 7. Corollary 1, 252.
Flock, 3, 303.
Formally real fields, 144.
Forms,
bilinear, 101.
linear, 25.
semi-bilinear, 101.
Function space [*F*, *C*], 19.
Fundamental theorem of projective geometry
first, 44.
second, 68.
complement to second — ; III. 4. Theorem 2, 87.

G

Galois correspondence
between a space and its adjoint space, 28.
between a space and its ring, 172-173.

Galois correspondence
between a space and its group; VI. 4. Theorem 4, 227.
between direct decomposition into points and maximal groups of involutions ; VI. App. I. Remark 1, 244.
Geometry defined by a bilinear form; IV. 5. Remark 3, 152.
Group
automorphism —
of endomorphism ring; V. 4. Theorem 2, 187.
of full linear group, 237.
of full semi-linear group, 256.
of collineations, 62.
of a duality, 144.
extended — of automorphisms of endomorphism ring, 196.
homothetic, 52.
of a hyperplane, 292.
of involutions, 237.
of a linear manifold (survey), 63.
of multiplications, 52, 63, 246.

H

Harmonic set; III. 4. Remark 5, 76.
Homothetic transformation, 53, 57.
Hyperplane, 19, 264.

I

Ideals, 172.
Idempotents, 169.
Imbedding theorem, 268.
Inclusion, 8.
Independent
elements (vectors), 13.
points, 41, 264.
Index
of inertia (Sylvester), 130.
of polarity, 121.
Induced isomorphisms of first kind
of full linear group, 229.
of full semi-linear group, 249.
Induced isomorphisms of second kind
of full linear group, 229.
of full semi-linear group, 251.
Intersection of subspaces, 9.
Involutions, 144.

Involutorial mappings, 111.
Isomorphism law, 11.
Isomorphism theorem
 for endomorphism ring; V. 4. Theorem 1, 183.
 for full linear group, 231.
 for full semi-linear group, 252.
Isotropic subspaces, 113.
 strictly —, 109, 113.

K

Kernel, 168.

L

Line, 18, 71, 265.
Linear
 anti-form, 36.
 dependence and independence, 13.
 form, 25.
 manifold, 7.
 mapping, 91.
 transformation, 42.

M

Maximal elements in partially ordered sets, 310.
Maximum principle of set theory, 310.
Multiplications
 characterization of — ; III. 1. Proposition 3, 43.
 group of —, 52, 63, 246.

N

Non-isotropic subspaces, 101, 113.
Normalizer, 219.
Nullity
 of an endomorphism, 168.
 of a polarity, 116.
Null systems and N-subspaces, 106.

O

Ordering
 algebraic — of a field, 94, 127.
 projective — of a space, 94.
Orthogonal
 group, 158.
 idempotents, 169.

P

Paired spaces, 35.
Pappus
 property of, 69.
 postulate of, 71.
Partially ordered set, 258, 309.
Pascal, theorem of, 143.
Pascalean polarity, 139.
Permutations, 309.
Perspectivity, 64, 292.
Plane, 18, 265.
Point, 18, 260.
Polarity; pole-polar relation, 109.
Positivity, domain of, 93, 127.
Preservation of forms by transformations, 151.
Projection of a line from a point, 91.
Projectivity, 40, 310.
Pythagoras, theorem of, 152.

Q

Quotient spaces, 10.

R

Rank
 of an endomorphism, 168.
 of a subspace, 16, 265.
Rank formulas
 for a pair of subspaces, 17-18.
 for a subspace and its dual, 33.
Regular rings, 179.
Ring of endomorphisms of a linear manifold, 169.

S

Self-duality theorem, 97.
Semi-
 automorphism, 83.
 bilinear form, 101.
 linear transformation, 42
Simplex, 66.
Singular automorphisms
 of full linear group, 230.

Singular automorphisms
of full semi-linear group, 247.
Strictly isotropic, 113.
Structure theorem
for endomorphism ring of linear manifolds, 183.
for full linear groups, 229.
for full semi-linear groups, 247.
for linear manifolds, 16.
for projective spaces, 51.
Subspace, 8.
Sum of subspaces, 9, 259.
Survey of groups of a linear manifold, 63.
Sylvester's theorem of inertia, 129.
Symmetrical forms, 111.
Symplectic group, 155.

T

Transitive groups, 153.
Translation, 53, 57.

U

Uniqueness theorem
for anti-semi-linear transformations, 194.
for polarities, 148, 159.
for rank, 14.
Unitary group, 158.